AF356147

Yeast Biotechnology 5.0

Yeast Biotechnology 5.0

Editor

Ronnie Willaert

Basel • Beijing • Wuhan • Barcelona • Belgrade • Novi Sad • Cluj • Manchester

Editor
Ronnie Willaert
Vrije Universiteit Brussel
Brussels
Belgium

Editorial Office
MDPI AG
Grosspeteranlage 5
4052 Basel, Switzerland

This is a reprint of articles from the Special Issue published online in the open access journal *Fermentation* (ISSN 2311-5637) (available at: https://www.mdpi.com/journal/fermentation/special_issues/yeast_5).

For citation purposes, cite each article independently as indicated on the article page online and as indicated below:

Lastname, A.A.; Lastname, B.B. Article Title. *Journal Name* **Year**, *Volume Number*, Page Range.

ISBN 978-3-7258-2399-4 (Hbk)
ISBN 978-3-7258-2400-7 (PDF)
doi.org/10.3390/books978-3-7258-2400-7

© 2024 by the authors. Articles in this book are Open Access and distributed under the Creative Commons Attribution (CC BY) license. The book as a whole is distributed by MDPI under the terms and conditions of the Creative Commons Attribution-NonCommercial-NoDerivs (CC BY-NC-ND) license.

Contents

About the Editor

Ronnie Willaert

Prof. Ronnie Willaert (Vrije Universiteit Brussel, Brussels, Belgium) has extensive expertise in yeast research (*Saccharomyces cerevisiae* and *Candida* species), yeast space biology research, cell (yeast) immobilisation biotechnology, fermentation technology, brewing science and technology, single-molecule biophysics (atomic force microscopy), micro(bio)fabrication and bone cell mechanobiology.

Article

Utilization of Yeast Waste Fermented Citric Waste as a Protein Source to Replace Soybean Meal and Various Roughage to Concentrate Ratios on In Vitro Rumen Fermentation, Gas Kinetic, and Feed Digestion

Chaichana Suriyapha [1], Anusorn Cherdthong [1,*], Chanon Suntara [1] and Sineenart Polyorach [2]

[1] Tropical Feed Resources Research and Development Center (TROFREC), Department of Animal Science, Faculty of Agriculture, Khon Kaen University, Khon Kaen 40002, Thailand; chaichana_s@kkumail.com (C.S.); chanon_su@kkumail.com (C.S.)

[2] Department of Animal Production Technology and Fisheries, Faculty of Agricultural Technology, King Mongkut's Institute of Technology, Bangkok 10520, Thailand; sineenart.po@kmitl.ac.th

* Correspondence: anusornc@kku.ac.th; Tel./Fax: +66-4320-2362

Citation: Suriyapha, C.; Cherdthong, A.; Suntara, C.; Polyorach, S. Utilization of Yeast Waste Fermented Citric Waste as a Protein Source to Replace Soybean Meal and Various Roughage to Concentrate Ratios on In Vitro Rumen Fermentation, Gas Kinetic, and Feed Digestion. *Fermentation* **2021**, *7*, 120. https://doi.org/10.3390/fermentation7030120

Academic Editor: Ronnie G. Willaert

Received: 23 June 2021
Accepted: 16 July 2021
Published: 17 July 2021

Publisher's Note: MDPI stays neutral with regard to jurisdictional claims in published maps and institutional affiliations.

Abstract: The objective of this study was to determine the application of citric waste fermented yeast waste (CWYW) obtained from an agro-industrial by-product as a protein source to replace soybean meal (SBM) in a concentrate diet. We also determined the effect of various roughage to concentrate ratios (R:C) on the gas production kinetics, ruminal characteristics, and in vitro digestibility using an in vitro gas production technique. The experiment design was a 3×5 factorial design arranged in a completely randomized design (CRD), with three replicates. There were three R:C ratios (60:40, 50:50, and 40:60) and five replacing SBM with CWYW (SBM:CWYW) ratios (100:0, 75:25, 50:50, 25:75, and 0:100). The CWYW product's crude protein (CP) content was 535 g/kg dry matter (DM). There was no interaction effect between R:C ratios and SBM:CWYW ratios for all parameters observed ($p > 0.05$). The SBM:CWYW ratio did not affect the kinetics and the cumulative amount of gas. However, the gas potential extent and cumulative production of gas were increased with the R:C ratio of 40:60, and the values were about 74.9 and 75.0 mL/0.5 g, respectively ($p < 0.01$). The replacement of SBM by CWYW at up to 75% did not alter in vitro dry matter digestibility (IVDMD), but 100% CWYW replacement significantly reduced ($p < 0.05$) IVDMD at 24 h of incubation and the mean value. In addition, IVDMD at 12 h and 24 h of incubation and the mean value were significantly increased with the R:C ratio of 40:60 ($p < 0.01$). The SBM:CWYW ratio did not change the ruminal pH and population of protozoa ($p > 0.05$). The ruminal pH was reduced at the R:C ratio of 40:60 ($p < 0.01$), whereas the protozoal population at 4 h was increased ($p < 0.05$). The SBM:CWYW ratio did not impact the in vitro volatile fatty acid (VFA) profile ($p > 0.05$). However, the total VFA, and propionate (C3) concentration were significantly increased ($p < 0.01$) by the R:C ratio of 40:60. In conclusion, the replacement of SBM by 75% CWYW did not show any negative impact on parameters observed, and the R:C ratio of 40:60 enhanced the gas kinetics, digestibility, VFA, and C3 concentration.

Keywords: citric waste; yeast waste; industrial by-product; protein source

Copyright: © 2021 by the authors. Licensee MDPI, Basel, Switzerland. This article is an open access article distributed under the terms and conditions of the Creative Commons Attribution (CC BY) license (https://creativecommons.org/licenses/by/4.0/).

1. Introduction

The security of livestock feed is a fairly frequent topic of discussion in terms of quality and quantity, especially in terms of the lack of protein sources, which results in low performance. High-quality protein feed sources such as soybean meal are expensive and lead to an increase in the cost of livestock production [1]. Many researchers have attempted to look for alternative sources of protein that could help improve the production and productivity of livestock [2,3]. The utilization of agro-industrial by-products as animal feed is an interesting consideration, since it could reduce feed costs and help reduce environmental pollution [4].

Citric waste is a by-product of the citric acid industry and is generated from rice, corn, cassava, or cassava pulp that is fermented with *Aspergillus niger* [5]. The citric waste still has some nutritive value of 30–70 g/kg DM of crude protein and contain high fiber content (861.3 g/kg DM of neutral detergent fiber (NDF) and 197.4 g/kg DM of acid detergent fiber (ADF), respectively) [6]. Uriyapongson et al. [6] reported that the inclusion of 10% citric waste in the diet of buffalos did not negatively affect feed intake, average daily gain (ADG), and the feed conversion ratio (FCR), whereas the digestibility was decreased with more than 10% citric waste. This might be due to the high content of fiber limiting digestion by animals, resulting in them being able to utilize only a low amount of nutrients [7]. Therefore, for the use of citric waste as animal feed, we should improve the quality by reducing the fiber composition and enhancing another nutrients, particularly the protein content.

Yeast (*Saccharomyces cerevisiae*) is a source of probiotics that enable a positive effect on the rumen fermentation of ruminants. It has been used as a biological method to improve the protein quality of feedstuff [8]. In ethanol production processes, the initial substrates are molasses and inoculants of the yeast *S. cerevisiae*. Yeast waste is a residue generated from ethanol production. Díaz et al. [9] reported that yeast waste contains a high content of live yeast cells (about 60–70%). Furthermore, Cherdthong et al. [10] revealed that yeast waste contains around 264.0 g/kg of crude protein and is rich in vitamins and minerals. Cherdthong et al. [11] found that dried yeast waste can replace up to 100% of soybean meal in animal feed with no negative effect to feed intake, feed utilization, and ruminal ecology in beef cattle. Therefore, yeast waste obtained from industrial by-products containing many live yeast cells might be beneficial for enhancing feed quality and feed utilization in animals.

The ratio of roughage to concentrate (R:C) in feeds is important for good nutrient utilization for production. Feeds have strong ties with the ecology of the rumen, ruminal bacteria, and trends of rumen fermentation [12]. Suitable amounts of concentrate can provide fiber utilization by enhancing fermentable organic matter, nitrogen, and energy sources for ruminal microbes [3]. Generally, feeding with a concentrate can provide more fermentation end-products than feeding with only roughage. Rice straw is easily obtainable from rice cultivation areas and is usually collected by farmers for feeding cattle [13]. Hence, the feeding of ruminants with rice straw and a concentrated diet that contains a high amount of protein and energy would be beneficial for ruminant productivity [14].

It was hypothesized that inoculated yeast waste obtained from industrial by-products could improve the quality of citric waste and could be used as a potential alternative protein source. Therefore, we investigated the utilization of citric waste fermented yeast waste (CWYW) obtained from agro-industrial by-products as a protein source to replace soybean meal in a concentrated diet, as well as the effect of various roughage to concentrate ratios (R:C) on gas production kinetics, rumen characteristics, and in vitro feed digestibility using an in vitro gas measuring technique.

2. Materials and Methods

The experimental cattle involved in this research were approved by Khon Kaen University's Animal Ethics Committee (record no. IACUC-KKU-27/64).

2.1. Preparation of Citric Waste Fermented Yeast Waste (CWYW)

Yeast waste was received as a by-product of ethanol production from KSL Green Innovation Public Company Limited (KGI), Nam Phong District, Khon Kaen Province, Thailand. Citric waste was obtained as a by-product of the citric acid industry from Sam Mor Farm Limited Partnership, Muang District, Udon Thani Province. Commercial grade urea and molasses were purchased from a local shop.

CWYW was prepared with the procedure that follows. First, 100 mL of yeast waste was added to a flask (A). Next, 20 g of brown sugar was weighed and dissolved in 100 mL of distilled water, and then 50 g of urea was mixed in (B). Yeast waste media solution was made by mixing A and B at a ratio of 1:1 and then flushing them for 16 h with air using an air pump at room temperature. The pH was adjusted to a range of 3.9 to 4.5. After 16 h, we transferred the yeast waste media solution and mixed it with citric waste at a ratio of 1 mL to 1 g. After that, anaerobic fermentation was performed in container bottles for 14 days, followed by 48 h of sun-drying to obtain less than 10% moisture. The CWYW was packed in a plastic bag for subsequent use as an experimental ingredient for dietary treatment.

2.2. Experimental Design and Dietary Treatments

The present experiment was performed at different incubation intervals using a gas production technique. A 3 × 5 factorial experiment design was conducted and arranged according to a completely randomized design (CRD) with three replication runs. The experimental diets had three roughage to concentrate (R:C) ratios of 60:40, 50:50, and 40:60 with five replacing soybean meal ratios of replacement in diets (SBM:CWYW) of 100:0, 75:25, 50:50, 25:75, and 0:100. All of the experimental dietary samples were oven-dried at 72 °C and ground to pass a 1 mm sieve (Cyclotech Mill, Tecator, Sweden) for the analysis of the chemical composition and the gas production test. Experimental diets including concentrate, rice straw, and CWYW were analyzed for dry matter (DM; ID 967.03), ash (ID 492.05), and crude protein (CP; ID 984.13) content using the standard analysis of the AOAC [15]. NDF and ADF contents were determined using the procedures of Van Soest et al. [16]. Table 1 shows the diet compositions and ingredients of the concentrate, rice straw, and CWYW used in this experiment. The concentrate diets were prepared with 141.0–142.0 g/kg of CP, which is recommended for beef cattle.

Table 1. Ingredient and chemical composition of concentrates, rice straw and citric waste fermented yeast waste used in the experiment.

Item	SBM:CWYW [1]					RS [2]	YW [3]	CW [4]	CWYW [5]
	100:0	75:25	50:50	25:75	0:100				
Ingredients (kg of dry matter)									
Cassava chips	59.3	58.5	59.5	58.8	59.7	-	-	-	-
Rice bran	9.0	9.0	9.0	9.0	9.0	-	-	-	-
Soybean meal	15.0	12.3	7.5	3.8	0.0	-	-	-	-
Palm kernel meal	13.0	13.0	13.0	13.0	13.0	-	-	-	-
CWYW [1]	0.0	3.8	7.5	12.3	15.0	-	-	-	-
Urea	1.3	1.0	1.0	0.7	0.8	-	-	-	-
Mineral premix	1.0	1.0	1.0	1.0	1.0	-	-	-	-
Molasses, liquid	1.0	1.0	1.0	1.0	1.0	-	-	-	-
Pure sulfur	0.5	0.5	0.5	0.5	0.5	-	-	-	-
Salt	1.0	1.0	1.0	1.0	1.0	-	-	-	-
Chemical composition									
Dry matter (g/kg)	926	923	922	922	918	944	360	919	882
g/kg of dry matter									
Organic matter	888	882	877	861	860	827	899	889	894
Ash	112	118	123	139	140	173	101	111	106
Crude protein	141	142	141	141	142	26	315	110	535
Neutral detergent fiber	159	167	169	171	174	779	217	709	402
Acid detergent fiber	72	84	91	94	99	536	44	426	294

[1] SBM:CWYW = replacing soybean meal with citric waste fermented yeast waste ratio; [2] RS = rice straw, [3] YW = yeast waste, [4] CW = citric waste; [5] CWYW = citric waste fermented yeast waste.

2.3. Animals and Ruminal Inoculums Preparation

Two male 3-year-old ruminally fistulated crossbreed (Thai × Holstein) cattle with body weights (BW) of 280 ± 15.0 kg were used as ruminal liquor donors. Ruminal liquor was obtained while the animals were fed ad libitum with roughage (rice straw) and concentrate (140 g/kg CP and 805 g/kg TDN) at 0.5% of BW daily (6:30 a.m. and 4:30 p.m.). The cattle were housed in individual pens, and clean water and mineral blocks were freely available. The cattle were fed with the diet for 21 d before collecting the rumen liquor.

From each of the cattle, 1000 mL of ruminal liquor was collected before feeding time in the morning. Ruminal liquor was filtered through five layers of cheesecloth and then moved to the laboratory in pre-warmed thermos bottles.

The medium was prepared using the procedures reported by Makkar et al. [17] which consisted of combining 3000 mL of reduced medium with 1500 mL of ruminal liquor from cattle (2:1; reduced medium: ruminal liquor). The medium mixture was then kept under stirring at 39 °C under CO_2 with a hot plate. Each experimental bottle was filled with 40 mL of ruminal fluid mixture and incubated in a water bath at 39 °C.

2.4. In Vitro Gas Production and Ruminal Fermentation Characteristics

The gas production was measured instantly after incubation for 0, 0.5, 1, 2, 4, 6, 8, 12, 18, 24, 48, 72, and 96 h according to the modified procedures of Cherdthong et al. [10]. Ørskov and McDonald [18] models were used for curve fitting and analysis of the kinetics of gas as follows:

$$y = a + b\,(1 - e^{(-ct)})$$

where a = soluble fraction from gas production, b = insoluble fraction from gas production, c = rate of gas production constant for the insoluble fraction (b), t = incubate time, $(|a| + b)$ = the potential extent of gas production, and y = gas produced at time 't'.

The ruminal pH was recorded using a digital pH meter (HANNA Instrument (HI) 8424 microcomputer, Singapore) at incubation times of 0 and 4 h. The incubated ruminal liquor was divided into two parts. The first part (20 mL) was kept in 5 mL of 1 M H_2SO_4 and stored at -20 °C for ammonia nitrogen (NH_3-N) analysis according to the micro-Kjeldahl methods, and in vitro volatile fatty acid (VFA) concentration was performed according to the procedures of Samuel et al. [19] using high performance liquid chromatography (HPLC machine; Shimadzu LC-20A, Kyoto, Japan) equipped with an Inertsil ODS-3 C18 (250 mm $\times$ 4.6 mm i.d., 5 µm,) column and mobile phase: phosphoric acid 25 mM, flow rate: 1 mL/minute, detection (UV): 210 nm: injection: 20 microliters (Shimadzu LC-20A, Kyoto, Japan). The second part (1 mL) was collected in 9 mL of 10% formalin for the direct counting of protozoa [20].

In vitro dry matter digestibility (IVDMD) and in vitro organic matter digestibility (IVOMD) were analyzed after incubation for 12 and 24 h using the procedures of Tilley and Terry [21].

2.5. Statistical Analysis

The data of the experiment were statistically evaluated with a 3 $\times$ 5 factorial arrangement according to CRD using the Proc. GLM procedure of SAS software version 9.4 (SAS Inst. Inc., Cary, NC, USA) [22]. All data were analyzed by the following equation:

$$Yij = \mu + Ai + Bj + ABij + \varepsilon ij$$

where: Y = observations; μ = overall mean; Ai = effect of factor A (R:C ratio at 60:40, 50:50, and 40:60; i = 1 to 3); Bj = effect of factor B (SBM:CWYW ratio at 100:0, 75:25, 50:50, 25:75, and 0:100; j = 1 to 5), ABij = the interaction effect of R:C ratio and SBM:CWYW, and εij = the residual effect. Duncan's New Multiple Range Test (DMRT) performed multiple comparisons between treatment methods [23]. Differences between mean values of $p < 0.05$ were considered to represent statistically significant differences.

3. Results

3.1. Dietary Chemical Composition

Experimental dietary ingredients and chemical compositions of rice straw, concentrate, yeast waste, citric waste and CWYW are presented in Table 1. The concentrate diets were provided with almost the same protein content in each group, ranging from 141 to 142 g/kg DM, and urea was administered to adjust the CP content. The yeast waste's CP content was 315 g/kg DM. The citric waste had low CP and high fiber contents (NDF and ADF) of 110,

709, and 426 g/kg DM, respectively. After quality improvement, the CWYW product's CP was increased, and the fiber (NDF and ADF) value was reduced to 535, 402, and 294 g/kg DM, respectively.

3.2. Kinetics and Cumulative Production of Gas

The data obtained for the substrates analyzed from the kinetics and cumulative production of gas are shown in Table 2. The data demonstrated that the soluble fractions of gas production (a), insoluble fraction of gas production (b), rate of constants for the insoluble fraction (c), potential extent of gas production ($|a| + b$) and the cumulative production of gas were not affected by the interaction between the R:C ratio and SBM:CWYW ratio. In addition, the SBM:CWYW ratio did not affect the kinetics and cumulative amount of gas. However, soluble fractions of gas production (a) ranged from -3.7 to -5.3 mL/0.5 g and were decreased at the R:C ratio of 40:60 ($p < 0.01$). The value of the insoluble fraction of gas production (b) was also decreased at the R:C ratio of 40:60 ($p < 0.01$). The value of the rate of gas production (c) ranged from 0.06 to 0.08 mL/h and was increased at the R:C ratio of 40:60 ($p < 0.01$). The potential extent of gas ($|a| + b$) value and the cumulative production of gas (at 96 h of incubation) were increased ($p < 0.01$) at the R:C ratio of 40:60, and the values were about 74.5 and 77.0 mL/0.5 g, respectively.

Table 2. Effect of R:C ratio level combined with SBM:CWYW ratio level on gas kinetics and cumulative gas at 96 h after incubation.

R:C [1]	SBM:CWYW [2]	Gas Kinetics [3]				Cumulative Gas (mL/0.5g)
		a	b	c	$|a| + b$	
	100:0	−3.9	70.4	0.06	74.4	74.7
	75:25	−3.8	70.2	0.06	74.0	73.9
60:40	50:50	−3.7	70.2	0.06	73.9	74.0
	25:75	−3.8	70.4	0.06	74.2	73.6
	0:100	−4.0	70.4	0.06	74.4	74.3
	100:0	−4.9	69.8	0.07	74.8	74.8
	75:25	−5.0	69.8	0.07	74.9	75.7
50:50	50:50	−5.0	69.7	0.07	74.8	75.6
	25:75	−5.0	69.9	0.07	74.9	76.0
	0:100	−5.1	69.7	0.07	74.9	75.7
	100:0	−5.3	69.9	0.07	75.3	76.4
	75:25	−5.4	69.2	0.08	74.6	77.1
40:60	50:50	−5.4	69.5	0.07	74.6	76.0
	25:75	−5.2	69.5	0.07	74.8	75.8
	0:100	−5.2	69.7	0.07	75.1	76.3
SEM		0.21	0.26	0.002	0.30	1.31
Comparison						
R:C ratio		<0.01	<0.01	<0.01	<0.01	<0.01
60:40		−3.8 [a]	70.3 [a]	0.063 [c]	74.2 [b]	74.5 [c]
50:50		−5.0 [b]	69.8 [b]	0.068 [b]	74.8 [a]	75.8 [b]
40:60		−5.3 [c]	69.6 [b]	0.071 [a]	74.9 [a]	77.0 [a]
SBM:CWYW ratio		0.31	0.58	0.96	0.40	0.98
100:0		−4.6	70.1	0.07	74.8	75.9
75:25		−4.6	69.9	0.07	74.5	75.8
50:50		−4.7	69.9	0.07	74.4	75.8
25:75		−4.7	69.8	0.07	74.6	75.7
0:100		−4.8	69.7	0.07	74.8	75.6
Interaction		0.94	0.96	0.39	0.95	0.99

[a–c] Value on the same row with different superscripts differ ($p < 0.05$); [1] R:C = roughage to concentrate ratio. [2] SBM:CWYW = replacing soybean meal with citric waste fermented yeast waste ratio; [3] a = the gas production from the immediately soluble fraction, b = the gas production from the insoluble fraction, c = the gas production rate constant for the insoluble fraction (b), $|a| + b$ = the gas potential extent of gas production.

3.3. In Vitro Digestibility

Table 3 shows the influence of substituting SBM for CWYW in combination with the R:C ratio on IVDMD and IVOMD. It was found that IVDMD and IVOMD did not show interaction with each other. There were no changes in IVOMD when the SBM:CWYW ratio was included. The replacement of SBM by CWYW by up to 75% did not alter IVDMD, but 100% CWYW replacement significantly reduced ($p < 0.05$) IVDMD at 24 h of incubation and the mean value ($p < 0.05$). In addition, IVDMD at 12 h and 24 h of incubation and the mean value were significantly increased ($p < 0.01$) with the R:C ratio of 40:60 (about 559, 701, and 631 g/kg, respectively). Moreover, the R:C ratio of 40:60 significantly increased ($p < 0.01$) IVOMD at 12 h, 24 h, and the mean value (706, 793, and 750 g/kg, respectively).

Table 3. Effect of R:C ratio level combined with SBM:CWYW ratio level on in vitro dry matter digestibility (IVDMD) and in vitro organic matter digestibility (IVOMD).

R:C [1]	SBM:CWYW [2]	IVDMD (g/kg)			IVOMD (g/kg)		
		12 h	24 h	Mean	12 h	24 h	Mean
	100:0	541	682	622	683	770	726
	75:25	542	676	618	675	773	724
60:40	50:50	540	675	612	675	774	725
	25:75	540	674	619	674	774	724
	0:100	539	678	611	679	768	723
	100:0	553	691	630	692	781	736
	75:25	553	690	623	693	782	737
50:50	50:50	551	688	621	692	781	736
	25:75	551	689	622	691	775	733
	0:100	551	688	622	691	775	733
	100:0	560	704	634	709	792	751
	75:25	560	702	632	707	795	751
40:60	50:50	559	702	630	706	793	749
	25:75	559	701	626	705	793	749
	0:100	559	700	621	705	793	748
SEM		0.21	0.13	0.10	0.33	0.31	0.19
	Comparison						
	R:C ratio	<0.01	<0.01	<0.01	<0.01	<0.01	<0.01
	60:40	540 [c]	677 [c]	609 [c]	677 [c]	772 [c]	724 [c]
	50:50	552 [b]	689 [b]	621 [b]	692 [b]	778 [b]	735 [b]
	40:60	559 [a]	701 [a]	631 [a]	706 [a]	793 [a]	750 [a]
	SBM:CWYW ratio	0.86	<0.01	0.02	0.50	0.32	0.23
	100:0	551	692 [a]	622 [a]	693	781	738
	75:25	551	689 [ab]	620 [ab]	692	783	737
	50:50	551	688 [ab]	620 [ab]	690	782	737
	25:75	551	688 [ab]	620 [ab]	691	780	735
	0:100	549	686 [c]	617 [c]	692	779	735
	Interaction	0.95	0.28	0.78	0.92	0.73	0.94

[a–c] Value on the same row with different superscripts differ ($p < 0.05$); [1] R:C = roughage to concentrate ratio. [2] SBM:CWYW = replacing soybean meal with citric waste fermented yeast waste ratio.

3.4. Ruminal NH₃-N, pH and Protozoal Population

Table 4 shows the influence of substituting SBM for CWYW in combination with the R:C ratio on the ruminal NH_3-N, pH, and population of protozoa. There was no interaction effect between factors on the ruminal NH_3-N, pH, and protozoal population ($p > 0.05$). The SBM:CWYW ratio of 0:100 significantly increased the impact ($p < 0.05$) on ruminal NH_3-N at 2 h, 4 h, and the mean value with the highest values of 17.6, 19.4 and 18.2 mg/dL, respectively. However, the pH and protozoal population were not affected by the SBM:CWYW ratio ($p > 0.05$).

Table 4. Effect of R:C ratio level combined with SBM:CWYW ratio level on ruminal NH_3-N, pH and protozoal population.

R:C [1]	SBM:CWYW [2]	NH_3-N (mg/dL)			pH			Protozoal Count ($\times 10^5$ cell/mL)		
		2 h	4 h	Mean	2 h	4 h	Mean	2 h	4 h	Mean
60:40	100:0	15.4	16.5	16.4	7.04	6.99	7.01	3.8	3.9	3.8
	75:25	14.8	15.8	16.0	7.10	7.04	7.07	3.8	3.9	3.8
	50:50	15.6	16.6	16.6	7.11	7.05	7.08	3.7	4.0	3.9
	25:75	15.7	16.7	17.2	7.11	7.04	7.07	3.8	3.9	3.8
	0:100	16.0	17.1	17.2	7.10	7.04	7.08	3.7	4.0	3.9
50:50	100:0	16.7	17.0	16.8	6.97	6.93	6.94	3.7	4.1	3.9
	75:25	16.7	17.8	17.6	6.97	6.93	6.95	3.7	4.1	3.9
	50:50	17.3	17.7	17.4	7.00	6.93	6.95	3.7	4.1	3.9
	25:75	17.8	18.4	17.9	6.97	6.94	6.95	3.6	4.2	3.8
	0:100	17.8	18.9	18.3	6.97	6.92	6.94	3.7	4.2	4.0
40:60	100:0	17.9	18.8	18.8	6.97	6.73	6.84	3.8	4.3	4.0
	75:25	18.6	19.6	19.3	6.97	6.73	6.84	3.8	4.3	4.0
	50:50	19.2	20.2	19.7	6.97	6.74	6.85	3.7	4.3	4.0
	25:75	19.3	20.4	19.8	6.97	6.73	6.84	3.7	4.3	4.0
	0:100	19.3	20.4	19.8	6.99	6.74	6.87	3.9	4.3	4.1
SEM		0.42	0.45	0.44	0.04	0.04	0.06	0.48	0.07	0.26
Comparison										
R:C ratio		<0.01	<0.01	<0.01	0.04	<0.01	<0.01	0.43	<0.01	0.92
60:40		15.4 [b]	16.6 [c]	16.0 [c]	7.09 [a]	7.03 [a]	7.06 [a]	3.8	3.9 [b]	3.9
50:50		16.9 [b]	17.9 [b]	17.4 [b]	6.97 [b]	6.90 [b]	6.94 [b]	3.7	4.1 [b]	3.9
40:60		18.9 [a]	19.8 [a]	19.4 [a]	6.96 [b]	6.83 [c]	6.85 [c]	3.6	4.2 [a]	3.8
SBM:CWYW ratio		0.02	0.01	0.02	0.98	0.93	0.96	0.98	0.58	0.98
100:0		16.4 [c]	17.5 [c]	16.9 [c]	7.02	6.90	6.93	3.7	4.1	3.9
75:25		16.9 [bc]	17.6 [bc]	17.2 [bc]	7.01	6.90	6.95	3.7	4.1	3.9
50:50		17.2 [abc]	18.3 [abc]	17.6 [abc]	7.01	6.90	6.95	3.7	4.1	3.9
25:75		17.4 [ab]	18.5 [ab]	18.0 [ab]	7.00	6.90	6.96	3.7	4.1	3.9
0:100		17.6 [a]	19.4 [a]	18.2 [a]	6.98	6.88	6.96	3.8	4.2	4.0
Interaction		0.54	0.70	0.63	0.99	0.98	0.99	0.99	0.77	0.99

[a–c] Value on the same row with different superscripts differ ($p < 0.05$); [1] R:C = roughage to concentrate ratio. [2] SBM:CWYW = replacing soybean meal with citric waste fermented yeast waste ratio.

Additionally, the R:C ratio of 40:60 significantly increased ($p < 0.01$) the ruminal NH_3-N at 2 h, 4 h, and the mean value and the values, which were 18.9, 19.8, and 19.4 mg/dL, respectively. The ruminal pH at 2 h, 4 h, and the mean values ranged from 6.73 to 7.10, which were reduced ($p < 0.01$) by the R:C ratio of 40:60 group. The protozoal population at 2 h and the mean values remained unchanged ($p > 0.05$), while the protozoal population at 4 h was increased ($p < 0.05$) by the R:C ratio of 40:60.

3.5. In Vitro VFAs Concentration

Table 5 shows the influence of substituting SBM for CWYW in combination with the R:C ratio level on in vitro VFAs. An interaction effect was not detected between factors on the in vitro VFA concentration ($p > 0.05$). In addition, the SBM:CWYW ratio did not impact the in vitro VFA profile ($p > 0.05$). However, the total VFA at 2 h, 4 h, and the mean value were significantly increased ($p < 0.01$) by decreasing the R:C ratio to 40:60. Propionate (C3) at 2 h, 4 h, and the mean value were significantly increased ($p < 0.01$) by decreasing the R:C ratio to 40:60. Furthermore, the decrease in the R:C ratio to 40:60 decreased ($p < 0.01$) acetate (C2) at 2 h, 4 h, and the mean value. The C2 to C3 ratio (C2:C3) at 2 h, 4 h, and the mean value were decreased ($p < 0.01$) by decreasing the R:C ratio, while butyrate (C4) remained similar ($p > 0.05$).

Table 5. Effect of R:C ratio level combined with SBM:CWYW ratio level on in vitro volatile fatty acids (VFAs).

R:C [1]	SBM:CWYW [2]	Total VFA (mmol/L)			C2 (mol/100 mol)			C3 (mol/100 mol)			C4 (mol/100 mol)			C2:C3 Ratio		
		2 h	4 h	Mean	2 h	4 h	Mean	2 h	4 h	Mean	2 h	4 h	Mean	2 h	4 h	Mean
60:40	100:0	85.7	86.9	81.3	65.0	63.0	63.9	25.1	27.0	26.1	10.0	10.1	10.0	2.6	2.3	2.5
	75:25	84.7	86.9	80.8	65.3	63.3	64.3	26.1	28.0	27.0	8.7	8.7	8.6	2.5	2.3	2.4
	50:50	84.6	86.8	80.7	65.8	63.8	63.7	25.9	27.8	26.8	8.4	8.5	8.4	2.5	2.3	2.4
	25:75	84.3	86.5	80.4	65.6	64.0	64.7	25.7	27.8	26.8	8.7	8.3	8.4	2.6	2.3	2.5
	0:100	84.1	86.6	80.3	65.8	63.8	64.8	25.6	27.5	26.6	8.6	8.7	8.6	2.6	2.3	2.5
50:50	100:0	93.2	88.7	85.9	62.9	61.3	62.2	26.7	28.6	27.6	10.4	10.1	10.2	2.4	2.1	2.5
	75:25	93.5	88.5	85.9	63.2	61.5	62.3	26.9	28.9	27.9	10.0	9.7	9.8	2.4	2.1	2.3
	50:50	93.1	88.6	85.8	63.5	61.0	62.3	26.6	28.6	27.6	10.0	10.4	10.1	2.4	2.1	2.3
	25:75	94.6	88.6	86.6	63.5	61.5	62.5	26.0	28.0	27.0	10.5	10.5	10.5	2.4	2.2	2.2
	0:100	93.2	87.7	85.4	63.6	61.6	62.6	26.1	27.9	26.9	10.3	10.6	10.4	2.4	2.2	2.3
40:60	100:0	98.3	91.2	89.8	61.4	59.4	60.4	29.2	30.3	29.7	9.4	10.3	9.8	2.1	2.0	2.2
	75:25	97.0	91.1	89.1	61.9	60.4	61.1	29.0	30.5	29.7	9.1	9.1	9.1	2.1	2.0	2.0
	50:50	97.7	90.6	88.8	61.5	59.5	60.5	28.8	30.1	29.4	9.7	10.5	10.0	2.1	2.0	2.1
	25:75	97.1	90.1	88.9	61.8	59.8	60.8	28.6	30.5	29.5	9.6	9.8	9.7	2.2	2.0	2.0
	0:100	97.2	90.5	88.8	62.0	60.2	61.2	28.6	29.6	29.3	9.4	9.9	9.6	2.2	2.0	2.1
SEM		0.52	0.93	0.60	0.73	0.61	0.65	1.14	1.13	1.14	1.39	1.26	1.31	0.11	0.10	0.10
Comparison																
R:C ratio		<0.01	<0.01	<0.01	<0.01	<0.01	<0.01	<0.01	<0.01	<0.01	0.32	0.22	0.27	<0.01	<0.01	<0.01
60:40		74.7 [c]	86.7 [c]	80.7 [c]	65.4 [a]	63.5 [a]	64.5 [a]	25.7 [b]	27.7 [b]	26.7 [b]	8.9	8.8	8.9	2.6 [a]	2.3 [a]	2.4 [a]
50:50		83.5 [b]	88.4 [b]	86.0 [b]	63.4 [b]	61.3 [b]	62.4 [b]	26.4 [b]	28.2 [b]	27.4 [b]	9.4	10.2	10.2	2.3 [b]	2.1 [b]	2.2 [b]
40:60		87.3 [a]	90.7 [a]	89.1 [a]	61.8 [c]	59.8 [c]	60.8 [c]	28.8 [a]	30.2 [a]	29.5 [a]	10.2	9.9	9.4	2.2 [b]	1.9 [c]	2.0 [c]
SBM:CWYW ratio		0.18	0.88	0.60	0.76	0.72	0.75	0.97	0.95	0.97	0.97	0.90	0.95	0.92	0.97	0.93
100:0		82.3	88.9	85.7	63.8	61.9	62.2	27.0	28.6	27.8	8.9	8.8	8.9	2.4	2.1	2.3
75:25		82.2	88.8	85.3	63.6	61.8	62.6	27.3	29.1	28.2	9.4	10.2	10.2	2.3	2.1	2.3
50:50		81.7	88.6	85.1	61.6	61.7	62.5	27.1	28.8	27.9	10.2	9.9	9.4	2.4	2.1	2.2
25:75		81.6	88.3	85.3	65.4	61.4	62.7	26.8	28.7	27.8	8.9	8.8	8.9	2.4	2.1	2.3
0:100		81.5	88.2	84.9	63.1	61.3	62.8	26.8	28.5	27.6	9.4	10.2	10.2	2.4	2.1	2.3
Interaction		0.42	0.98	0.95	0.98	0.96	0.998	0.99	0.99	0.99	0.99	0.99	0.99	0.99	0.99	0.99

[a–c] Value on the same row with different superscripts differ ($p < 0.05$);[1] R:C = roughage to concentrate ratio. [2] SBM:CWYW = replacing soybean meal with citric waste fermented yeast waste ratio.

4. Discussion

4.1. Dietary Chemical Composition

In this study, the CP content of non-fermented citric waste was 110 g/kg DM, which was comparable to the result reported by Silva et al. [5] (79.3–110.8 g/kg DM of CP). However, a lower content of CP of 61.1 g/kg DM in citric waste was found in a study of Tanpong et al. [7]. The NDF and ADF contents in non-fermented citric waste were lower than in the report by Uriyapongson et al. [6] (861.3 and 197.4 g/kg DM, respectively). This was probably due to the differences between raw materials used for citric production, such as the variety, age of harvest, and soil fertilizer [24,25]. In addition, the processing of citric production might influence the different nutrient compositions [5].

The CP content in the yeast waste used in this study was 315 g/kg DM, which is close to the result from Bátori et al. [26], who reported 320 g/kg DM. However, the CP content in yeast waste could vary from about 182.5 to 296 g/kg DM [10,27–29]. This might be caused by substrate used in ethanol production, heat from the fermentation process, yeast strain, and freshness of the yeast stillage, which is a complex medium [26,30,31].

After improving the citric acid waste quality by yeast waste fermentation, CWYW was found to have increased CP to 535 g/kg DM when compared with non-fermented citric waste. This could be due yeast waste containing high protein content and being rich in essential amino acids [9,10]. In addition, the inclusion of media solution containing urea in the CWYW fermented product during the fermentation process might enhance the CP content [3]. Fiber contents were reduced when citric acid waste was fermented with yeast waste. Similarly, Suntara et al. [32] indicated that some strains of yeast could produce cellulolytic enzymes to break down fiber in plant materials. Furthermore, the fermentation process with a media containing urea as an alkaline agent might degrade the structure fiber of CWYW and lead to the fiber structure decreasing [33].

4.2. Kinetics and Cumulative Production of Gas

Yeast waste rich in *S. cerevisiae* could promote the microorganism in the rumen and improve the incubated substrate's digestibility, thus improving the kinetics of gas production and increasing the total production of gas [34]. However, in this study, the kinetics of gas was not changed by the SBM:CWYW ratio. This is probably due to the fermentation of protein not leading to the production of gas [35]. This agreed with the results of Cherdthong et al. [10], who stated that the gas and fermentation kinetics was not changed by the substitution of soybean meal with yeast waste.

Moreover, soluble fraction (a), rate of gas production (c), potential extent of gas production ($|a| + b$), and the cumulative production of gas were significantly affected by the R:C ratio. The addition of a high concentration ratio contributes to an improvement in the rate of fermentation and soluble fraction of the rumen [34]. Starch degradation is an important factor in regulating energy utilization for the growth of rumen microorganisms, increasing the rumen population, and increasing digestion [36]. The potential extent of gas production ($|a| + b$) is known to be essentially the result of the carbohydrates fermented into acetate, propionate, and butyrate [37].

The present results demonstrated that the intercept value of (a) was negative in this study. This was a result of the delay in ruminal microbial growth of the substrates during the early stage of incubation. The data show that there is a lag period after the soluble part of the substrate is ingested but before the cell walls are fermented [38,39]. Several researchers [35,40] have also stated that, when using mathematical models to match the kinetics of gas output, there were negative values for different substrates. It is understood that it is possible to use the absolute value of a, ($|a|$), to define the ideal fermentation of the soluble fraction. In this experiment, the absolute gas production was the highest for the R:C ratio of 40:60. The soluble fraction makes it easy for rumen microbes to bind and contribute to the greater production of gas [41]. The results revealed that the insoluble fraction (b) at the R:C ratio of 60:40 was significantly the highest value. The high fiber in feed had an effect of the increase in (b), which increases the polysaccharides and activities of glycoside hydrolase against lignified plant tissues [3]. Particularly, the NDF degradability was substantially associated with the NDF fraction. Similarly, Phesatcha et al. [34] revealed that (b) of gas production increased as the ratio of concentrates in the diet decreased.

4.3. In Vitro Digestibility

The yeast *S. cerevisiae* can scavenge the accessible oxygen to support metabolic activity, thus reducing the ruminal redox potential and stimulating the ruminal microbes to have a higher rate of feed digestion. This improves the digestibility of nutrients [34]. In addition, the findings of Cherdthong et al. [11] showed that 100% of yeast waste could be used to replace SBM as a source of protein in concentrated diets without detrimental effects on digestibility. The present results indicated that CWYW can replace SBM at up to 75% without a negative impact on IVDMD and IVOMD.

During its metabolic activities, *S. cerevisiae* may be responsible for secreting extracellular enzymes into the citric waste mash, such as lignocellulose peroxidase, lignin peroxidase, cellulase, and hemicellulose [32,42]. This results in yeast proliferation. Additionally, this could happen because the alkaline agents (ammonium hydroxide; NH_4OH) produced from urea during the fermentation process of CWYW cause the hemicellulose–lignin complex in citric waste to swell [33]. The concentrated alkaline agents can physically swell structural fibers by chemically degrading their ester bonds [43]. This could help enable the extracellular enzymes from *S. cerevisiae* to attack the structural carbohydrates more easily and increase the degradability of CWYW.

However, replacement with SBM:CWYW at up to 100% decreased IVDMD at 24 h and the mean value. This could be due to the structural carbohydrates content in the CWYW negatively affecting digestibility in vitro. Uriyapongson et al. [6] reported that the use of citric waste at more than 10% in the diet results in the digestibility decreasing because of the high fiber content. It was concluded that changes in cell-wall composition involving structural carbohydrate contents in CWYW restricted the possible degree of digestion, while chemical factors other than the crystalline or physical nature of the fiber limited the rate of digestion [44].

The R:C ratio of 40:60 improved the in vitro digestibility. This may have been due to increased levels of concentrate, which would supply energy that is more readily available, thereby improving the subsequent degradability by ruminal microbes. The concentrate diet has a pronounced stimulatory effect on the ruminal microflora that is achieved more readily from carbohydrates than from forages in the rumen [45]. These studies agree with Cherdthong et al. [12], who demonstrated that when the fiber value was reduced, particularly with a higher concentrate level, ruminal microbe activity could be encouraged [46]. However, in buffalo, the in vitro organic matter digestibility (IVOMD) increased with the increase in concentrate in the diet, while the cumulative gas production showed an irregular trend and was not closely correlated to digested OM [47].

4.4. In Vitro Ruminal NH_3-N Concentration and Ruminal pH

In the present study, the ruminal NH_3-N concentration was increased with higher levels of concentrate diet and levels of CWYW used to replace soybean meal. This is probably due to CWYW containing yeast waste, which has a high protein content of 315 g/kg DM. Thus, substantial increases in NH_3-N concentrations occur in response to the microbial degradation of yeast cells [48,49]. Additionally, it could be due to the ability to provide stimulatory factors and even protein [50,51] to ruminal bacteria, or by changing in the abundance of microbes with proteolytic activity [52].

Another reason is likely the NPN-urea level in CWYW, which was higher than in previous studies by Polyorach et al. [3], where increased levels of urea-N in feed resulted in an increase in ruminal NH_3-N concentration from the dissolution of urea. The rapid hydrolysis of NPN-urea to rumen NH_3-N by microbial enzymes is another possible cause in the present study [10]. The amount of N actually digested in the rumen increased as the proportion of concentrate in the diet increased, which is likely to be a key explanation for enhancing in the concentration of NH_3-N in the rumen [14]. Additionally, decreasing the R:C ratio from 60:40 to 40:60 in the diet increased the ruminal NH_3-N concentration. Similarly, Suriyapha et al. [46] and Matra et al. [53] revealed that the concentrations of rumen NH_3-N increased significantly with a decreasing R:C ratio.

The ruminal pH is an important parameter that reflects the internal homeostasis of the rumen environment. Normally, ruminants have a highly balanced ecology for preserving a ruminal pH range of 6.0–7.0 [14]. The yeast motivates lactate users and enhances their population, but it also serves as a contender with the producers of lactate [54]. However, the data of this study revealed that the ruminal pH was not changed by the influence of the SBM:CWYW ratio. Similarly, Cherdthong et al. [10] found that 100% of yeast waste used to replace soybean meal did not change the ruminal pH in vitro, and saw no negative impact on the ruminal pH in Thai native bulls [11]. Additionally, ruminal pH at 4 h and the mean value were decreased by the higher R:C ratio. This agrees with Cherdthong et al. [12], who reported that a high ratio of concentrate diet usually results in a significant drop in ruminal pH, which decreases the activity of cellulolytic bacteria and slows digestion.

4.5. In Vitro Protozoal Population

This study revealed that the number of protozoa did not change when changing the SBM:CWYW ratio. However, the protozoal counts at 4 h was increased by the highest concentrate ratio, which agrees with Cherdthong et al. [12]. This might have happened because of the role of protozoal in starch utilization, which progressively increase when a carbohydrate with fast fermentation is added [34]. In contrast, Van Soest [55] demonstrated that feeding over a certain level of concentrate diet could reduce the population of protozoa. Suriyapha et al. [46] and Matra et al. [53] revealed that an experimental diet with an R:C ratio higher than 30:70 decreased the protozoal population. This is probably due to the increased concentrate diet leading to a high fermentation rate, which results in a lower pH that is unsuitable for the rumen ecology and decreases protozoal populations [3].

4.6. In Vitro VFAs

When replacing SBM with CWYW at up to 100%, the concentration of VFA and VFA profiles could be maintained. Similar results on VFA production between CWYW and SMB indicate that CWYW has similar nutritional quality and that it could be comparable to SBM when used to enhance ruminal end-products. Increased concentrate levels enhanced in vitro VFA, which could be supported by the fact that a concentrate diet contains a fraction of highly degradable carbohydrates, particularly starch. The high level of starch in concentrate diet appeared to increase the total VFA and C3, while C2 and the C2:C3 ratio were decreased with an expanding concentrate level [45,56].

In particular, C3 is obtained by the fermentation of soluble carbohydrates with more concentrate diet by ruminal bacteria activity [57]. This agrees with Cherdthong et al. [11], who also reported that the fermentation of a high concentrate level resulted in a greater molar concentration of ruminal C3. In addition, Phesatcha et al. [34] reported that increasing the ratio of a concentrate diet to 80% could increase VFA and C3, whereas C2 and the C2:C3 ratios decreased.

5. Conclusions

Citric waste can improve the nutritional values by being fermented with yeast waste and appropriate media solutions. No interaction effect was found between the R:C ratio and SBM:CWYW for all parameters. CWYW could be substituted for SBM in concentrate diets at up to 75% by without negative impact on gas kinetics, ruminal parameters, and in vitro digestibility. In addition, the R:C ratio of 40:60 could be beneficial for gas kinetics, ruminal ecology, digestibility, volatile fatty acids, and propionic acid concentration. However, more in vivo trials should be conducted in order to determine the success of animal production.

Author Contributions: Planning and design of the study, C.S. (Chaichana Suriyapha) and A.C.; conducting and sampling, C.S. (Chaichana Suriyapha), A.C. and C.S. (Chanon Suntara); sample analysis, C.S. (Chaichana Suriyapha), A.C. and C.S.(Chanon Suntara); statistical analysis, C.S. (Chaichana Suriyapha) and A.C.; manuscript drafting, C.S. (Chaichana Suriyapha), A.C. and C.S. (Chanon Suntara); manuscript editing and finalizing, C.S. (Chaichana Suriyapha), A.C. and C.S. (Chanon Suntara); and S.P.; All authors have read and agreed to the published version of the manuscript.

Funding: The authors express their most sincere gratitude to the Research Program on the Research and Development of Winged Bean Root Utilization as Ruminant Feed, Increase Production Efficiency and Meat Quality of Native Beef and Buffalo Research Group and Research and Graduate Studies, Khon Kaen University (KKU), and Office of National Higher Education Science Research and Innovation Policy Council through the Program Management Unit for Competitiveness (PMUC) (contract grant C10F640078). Mr. Chaichana Suriyapha was granted by Animal Feed Inter Trade Co., Ltd. and Thailand Science Research and Innovation (TSRI) through the Researcher for Industry (RRi) program (contract grant PHD62I0021).

Institutional Review Board Statement: The study was conducted under approval procedure no. IACUC-KKU-27/64 of Animal Ethics and Care issued by Khon Kaen University.

Informed Consent Statement: Not applicable.

Data Availability Statement: Not applicable.

Acknowledgments: The authors would like to express our sincere thanks to the KSL Green Innovation Public Co. Limited, Thailand and the Tropical Feed Resources Research and Development Center (TROFREC), Department of Animal Science, Faculty of Agriculture, Khon Kaen University (KKU) for the use of the research facilities.

Conflicts of Interest: The authors declare no conflict of interest.

References

1. Tsinas, A.; Tzora, A.; Peng, J. Alternative protein sources to soybean meal in pig diets. *J. Food. Agric. Environ.* **2014**, *12*, 655–660.
2. Wanapat, M.; Rowlinson, P. Nutrition and feeding of swamp buffalo: Feed resources and rumen approach. *Ital. J. Anim. Sci.* **2007**, *6*, 67–73. [CrossRef]
3. Polyorach, S.; Wanapat, M.; Cherdthong, A. Influence of yeast fermented cassava chip protein (YEFECAP) and roughage to concentrate ratio on ruminal fermentation and microorganisms using in vitro gas production technique. *Asian-Australas. J. Anim. Sci.* **2014**, *27*, 36–45. [CrossRef] [PubMed]
4. Ajila, C.M.; Brar, S.K.; Verma, M.; Tyagi, R.D.; Godbout, S.; Valéro, J.R. Bio-processing of agro-byproducts to animal feed. *Crit. Rev. Biotechnol.* **2012**, *32*, 382–400. [CrossRef] [PubMed]
5. Silva, C.E.; Gois, G.N.; Silva, L.M.; Almeida, R.M.; Abud, A.K. Citric waste saccharification under different chemical treatments. *Acta Sci. Technol.* **2015**, *37*, 387–395. [CrossRef]
6. Uriyapongson, S.; Wachirapakorn, C.; Nananukraw, C.; Phoemchalard, C.; Panatuk, J.; Tonuran, W. Digestibility and performance of buffalo fed total mixed ration with different levels of citric waste. *Buffalo Bull.* **2013**, *32*, 829–833.
7. Tanpong, S.; Cherdthong, A.; Tengjaroenkul, B.; Tengjaroenkul, U.; Wongtangtintharn, S. Evaluation of physical and chemical properties of citric acid industrial waste. *Trop. Anim. Health Prod.* **2019**, *51*, 2167–2174. [CrossRef] [PubMed]
8. Polyorach, S.; Wanapat, M.; Wanapat, S. Enrichment of protein content in cassava (*Manihot esculenta* Crantz) by supplementing with yeast for use as animal feed. *Emir. J. Food Agric.* **2013**, *25*, 142–149. [CrossRef]
9. Díaz, A.; Ranilla, M.J.; Saro, C.; Tejido, M.L.; Pérez-Quintana, M.; Carro, M.D. Influence of increasing doses of a yeast hydrolyzate obtained from sugarcane processing on in vitro rumen fermentation of two different diets and bacterial diversity in batch cultures and rusitec fermenters. *Anim. Feed Sci. Technol.* **2017**, *232*, 129–138. [CrossRef]
10. Cherdthong, A.; Prachumchai, R.; Supapong, C.; Khonkhaeng, B.; Wanapat, M.; Foiklang, S.; Milintawisamai, N.; Gunun, N.; Gunun, P.; Chanjula, P.; et al. Inclusion of yeast waste as a protein source to replace soybean meal in concentrate mixture on ruminal fermentation and gas kinetics using in vitro gas production technique. *Anim. Prod. Sci.* **2018**, *59*, 1682–1688. [CrossRef]
11. Cherdthong, A.; Sumadong, P.; Foiklang, S.; Milintawisamai, N.; Wanapat, M.; Chanjula, P.; Gunun, N.; Gunun, P. Effect of post-fermentative yeast biomass as a substitute for soybean meal on feed utilization and rumen ecology in Thai native beef cattle. *J. Anim. Feed Sci.* **2019**, *28*, 238–243. [CrossRef]
12. Cherdthong, A.; Wanapat, M.; Kongmun, P.; Pilajun, R.; Khejornsart, P. Rumen fermentation, microbial protein synthesis and cellulolytic bacterial population of swamp buffaloes as affected by roughage to concentrate ratio. *J. Anim. Vet. Adv.* **2010**, *9*, 1667–1675. [CrossRef]
13. Cherdthong, A.; Wanapat, M.; Wachirapakorn, C. Influence of urea calcium mixture supplementation on ruminal fermentation characteristics of beef cattle fed on concentrates containing high levels of cassava chips and rice straw. *Anim. Feed Sci. Technol.* **2011**, *163*, 43–51. [CrossRef]

14. Wanapat, M.; Gunun, P.; Anantasook, N.; Kang, S. Changes of rumen pH, fermentation and microbial population as influenced by different ratios of roughage (rice straw) to concentrate in dairy steers. *J. Agric. Sci.* **2014**, *152*, 675–685. [CrossRef]
15. Association of Official Analytical Chemists. *Official Methods of Analysis*, 16th ed.; Association of Official Analytical Chemists: Arlington, VA, USA, 1995.
16. Van Soest, P.J.; Robertson, J.B.; Lewis, B.A. Methods for dietary fiber neutral detergent fiber, and non-starch polysaccharides in relation to animal nutrition. *J. Dairy Sci.* **1991**, *74*, 3583–3597. [CrossRef]
17. Makkar, H.P.S.; Blumme, M.; Becker, K. Formation of complexes between polyvinyl pyrrolidones or polyethylene glycols and tannins, and their implication in gas production and true digestibility in in vitro techniques. *Br. J. Nutr.* **1995**, *73*, 897–913. [CrossRef]
18. Ørskov, E.R.; McDonald, I. The estimation of protein degradability in the rumen from incubation measurements weighted according to rate of passage. *J. Agric. Sci.* **1979**, *92*, 499–503. [CrossRef]
19. Samuel, M.; Sagathewan, S.; Thomus, J.; Mathen, G. An HPLC method for estimation of volatile fatty acids of rumen fluid. *Indian J. Anim. Sci.* **1997**, *67*, 805–807.
20. Galyean, M. *Laboratory Procedure in Animal Nutrition Research*; Department of Animal and Range Sciences, New Mexico State University: Las Cruces, NM, USA, 1989.
21. Tilley, J.M.A.; Terry, R.A. A two-stage technique for the digestion of forage crops. *J. Br. Grassl. Soc.* **1963**, *18*, 104–111. [CrossRef]
22. SAS (Statistical Analysis System). *SAS/STAT User's Guide*, 4th ed.; Statistical Analysis Systems Institute, Version 9; SAS Institute Inc.: Cary, NC, USA, 2013.
23. Steel, R.G.D.; Torrie, J.H. *Principles and Procedures of Statistics: A Biometerial Approach*, 2nd ed.; McGraw-Hill: New York, NY, USA, 1980.
24. Shewfelt, R.L. Sources of variation in the nutrient content of agricultural commodities from the farm to the consumer. *J. Food Qual.* **1990**, *13*, 37–54. [CrossRef]
25. Haile, E.; Njonge, F.K.; Asgedom, G.; Gicheha, M. Chemical composition and nutritive value of agro-industrial by-products in ruminant nutrition. *Open J. Anim. Sci.* **2017**, *7*, 8–18. [CrossRef]
26. Bátori, V.; Ferreira, J.A.; Taherzadeh, M.J.; Lennartsson, P.R. Ethanol and protein from ethanol plant by-products using edible fungi *Neurospora intermedia* and *Aspergillus oryzae*. *Biomed. Res. Int.* **2015**, *2015*, 1–10. [CrossRef]
27. Mumtaz, S.; Sheikh, M.A.; Iqbal, T.; Rehman, K.; Rashid, S. Bioconversion of distillery sludge (treated) to lysine and its biological evaluation. *Int. J. Agri. Biol.* **2000**, *2*, 274–277.
28. Sharif, M.; Shahzad, M.A.; Rehman, S.; Khan, S.; Ali, R.; Khan, M.L.; Khan, K. Nutritional evaluation of distillery sludge and its effect as a substitute of canola meal on performance of broiler chickens. *Asian-Australas. J. Anim. Sci.* **2012**, *25*, 401–409. [CrossRef]
29. Sharif, M.; Shoaib, M.; Saif Ur Rehman, M.; Fawwad, A.; Asif, J. Use of distillery yeast sludge in poultry: A review. *J. Agric. Sci.* **2016**, *6*, 242–256.
30. Azhara, S.H.M.; Abdullaab, R.; Jamboa, S.A.; Marbawia, H.; Gansaua, J.A.; Faika, A.A.M.; Rodriguesc, K.F. Yeasts in sustainable bioethanol production: A review. *Biochem. Biophys. Rep.* **2017**, *10*, 52–61.
31. Khan, M.U.; Shariati, M.A.; Kadmi, Y.; Elmsellem, H.; Majeed, M.; Khan, M.R.; Fazel, M.; Rashidzadeh, S. Design, development and performance evaluation of distillery yeast sludge dryer. *Process. Saf. Environ. Prot.* **2017**, *111*, 733–739. [CrossRef]
32. Suntara, C.; Cherdthong, A.; Uriyapongson, S.; Wanapat, M.; Chanjula, P. Comparison effects of ruminal crabtree-negative yeasts and crabtree-positive yeasts for improving ensiled rice straw quality and ruminal digestion using in vitro gas production. *J. Fungi.* **2020**, *6*, 109. [CrossRef] [PubMed]
33. Mapato, C.; Wanapat, M.; Cherdthong, A. Effect of urea treatment of straw and dietary level of vegetable oil on lactating dairy cows. *Trop. Anim. Health Prod.* **2010**, *42*, 1635–1642. [CrossRef] [PubMed]
34. Phesatcha, K.; Phesatcha, B.; Wanapat, M.; Cherdthong, A. Roughage to concentrate ratio and *Saccharomyces cerevisiae* inclusion could modulate feed digestion and in vitro ruminal fermentation. *Vet. Sci.* **2020**, *7*, 151. [CrossRef] [PubMed]
35. Khazaal, K.; Dentinho, M.T.; Riberiro, J.M.; Ørskov, E.R. A comparison of gas production during incubation with rumen contents in vitro and nylon bag degradability as predictors of the apparent digestibility in vivo and the voluntary intake of hays. *Anim. Sci.* **1993**, *57*, 105–112. [CrossRef]
36. Nocek, J.E.; Tamminga, S. Site of digestion of starch in the gastrointestinal tract of dairy cows and its effect on milk and composition. *J. Dairy Sci.* **1991**, *74*, 3598–3629. [CrossRef]
37. Getachew, G.; Blummel, M.; Makkar, H.P.S.; Becker, K. In vitro gas measuring techniques for assessment of nutritional quality of feeds: A review. *Anim. Feed Sci. Technol.* **1998**, *72*, 261–281. [CrossRef]
38. Chanthakhoun, V.; Wanapat, M. The in vitro gas production and ruminal fermentation of various feeds using rumen liquor from swamp buffalo and cattle. *Asian J. Anim. Vet. Adv.* **2012**, *7*, 54–60. [CrossRef]
39. Soriano, A.P.; Mamuad, L.L.; Kim, S.H.; Choi, Y.J.; Jeong, C.D.; Bae, C.D.; Chang, M.B.; Lee, S.S. Effect of *Lactobacillus mucosae* on in vitro rumen fermentation characteristics of dried brewers grain, methane production and bacterial diversity. *Asian-Australas. J. Anim. Sci.* **2014**, *27*, 1562–1570. [CrossRef] [PubMed]
40. Blummel, M.; Becker, K. The degradability characteristics of fifty-four roughages and roughage neutral detergent fibres as described by in vitro gas production and their relationship to voluntary feed intake. *Br. J. Nutr.* **1997**, *77*, 757–768. [CrossRef]
41. Chumpawadee, S.; Sommart, K.; Vongpralub, T.; Pattarajinda, V. Nutritional evaluation of non-forage high fibrous tropical feeds for ruminant using in vitro gas production technique. *Walailak J. Sci. Technol.* **2005**, *2*, 209–218.

42. Kricka, W.; Fitzpatrick, J.; Bond, U. Metabolic engineering of yeasts by heterologous enzyme production for degradation of cellulose and hemicellulose from biomass: A perspective. *Front. Microbiol.* **2014**, *5*, 174. [CrossRef] [PubMed]
43. Wanapat, M.; Polyorach, S.; Boonnop, K.; Mapato, C.; Cherdthong, A. Effects of treating rice straw with urea or urea and calcium hydroxide upon intake, digestibility, rumen fermentation and milk yield of dairy cows. *Livest. Sci.* **2009**, *125*, 238–243. [CrossRef]
44. Hart, F.J.; Wanapat, M. Physiology of digestion of urea-treated rice straw in swamp buffalo. *Asian-Australas. J. Anim. Sci.* **1992**, *5*, 617–622. [CrossRef]
45. Hungate, R.E. *The Rumen and Its Microbes*; Academic Press: New York, NY, USA, 1966.
46. Suriyapha, C.; Ampapon, T.; Viennasay, B.; Matra, M.; Wann, C.; Wanapat, M. Manipulating rumen fermentation, microbial protein synthesis, and mitigating methane production using bamboo grass pellet in swamp buffaloes. *Trop. Anim. Health Prod.* **2020**, *52*, 1609–1615. [CrossRef] [PubMed]
47. Zicarelli, F.; Calabro, S.; Piccolo, V.; D'Urso, S.; Tudisco, R.; Bovera, F.; Cutrignelli, M.I.; Infascelli, F. Diets with different forage/concentrate ratios for the Mediterranean Italian buffalo: In vivo and in vitro digestibility. *Asian-Australas. J. Anim. Sci.* **2008**, *21*, 75–82. [CrossRef]
48. Oeztuerk, H.; Schroeder, B.; Bayerbach, M.; Breves, G. Influence of living and autoclaved yeasts of *Saccharomyces boulardii* on in vitro ruminal microbial metabolism. *J. Dairy Sci.* **2005**, *88*, 2594–2600. [CrossRef]
49. Vyas, D.; Uwizeye, A.; Mohammed, R.; Yang, W.Z.; Walker, N.D.; Beauchemin, K.A. The effects of active dried and killed dried yeast on subacute ruminal acidosis, ruminal fermentation, and nutrient digestibility in beef heifers. *J. Anim. Sci.* **2014**, *92*, 724–732. [CrossRef]
50. Miller-Webster, T.; Hoover, W.H.; Holt, M.; Nocek, J.E. Influence of yeast culture on ruminal microbiological metabolism in continuous culture. *J. Dairy Sci.* **2002**, *85*, 2009–2014. [CrossRef]
51. Oeztuerk, H. Effect of live and autoclaved yeast cultures on ruminal fermentation in vitro. *J. Anim. Feed Sci.* **2009**, *18*, 142–150. [CrossRef]
52. Cunha, C.S.; Marcondes, M.I.; Silva, A.L.; Gionbelli, T.R.S.; Novaes, M.A.S.; Knupp, L.S.; Virginio, G.F., Jr.; Veloso, C.M. Do live or inactive yeasts improve cattle ruminal environment? *R. Bras. Zootec.* **2019**, *48*, e20180259. [CrossRef]
53. Matra, M.; Wanapat, M.; Cherdthong, A.; Foiklang, S.; Mapato, C. Dietary dragon fruit (*Hylocereus undatus*) peel powder improved in vitro rumen fermentation and gas production kinetics. *Trop. Anim. Health Prod.* **2019**, *51*, 1531–1538. [CrossRef]
54. Chaucheras-Durand, F.; Ameilbonne, A.; Auffret, P.; Bernard, M.; Mialon, M.M.; Duniere, L.; Forano, E. Supplementation of live yeast based feed additive in early life promotes rumen microbial colonization and fibrolytic potential in lambs. *Sci. Rep.* **2019**, *9*, 19216. [CrossRef] [PubMed]
55. Van Soest, P.J. *Nutritional Ecology of the Ruminant*; O&B Books Inc.: Corvallis, OR, USA, 1982; pp. 22–39.
56. Kang, S.; Wanapat, M.; Phesatcha, K.; Norrapoke, T.; Foiklang, S.; Ampapon, T.; Phesatcha, B. Using krabok (*Irvingia malayana*) seed oil and *Flimingia macrophylla* leaf meal as a rumen enhancer in an in vitro gas production system. *Anim. Prod. Sci.* **2017**, *57*, 327–333. [CrossRef]
57. Calsamiglia, S.; Cardozo, P.W.; Ferret, A.; Bach, A. Changes in rumen microbial fermentation are due to a combined effect of type of diet and pH. *J. Anim. Sci.* **2008**, *86*, 702–711. [CrossRef] [PubMed]

 *fermentation*

Article

Isocitric Acid Production from Ethanol Industry Waste by *Yarrowia lipolytica*

Svetlana V. Kamzolova *, Vladimir A. Samoilenko, Julia N. Lunina and Igor G. Morgunov

G.K. Skryabin Institute of Biochemistry and Physiology of Microorganisms, Pushchino Center for Biological Research of the Russian Academy of Sciences, Prospect Nauki 5, Pushchino, 142290 Moscow, Russia; samva@rambler.ru (V.A.S.); luninaju@rambler.ru (J.N.L.); morgunovs@rambler.ru (I.G.M.)
* Correspondence: kamzolova@rambler.ru; Tel.: +7-926-414-5620

Abstract: There is ever increasing evidence that isocitric acid can be used as a promising compound with powerful antioxidant activity to combat oxidative stress. This work demonstrates the possibility of using waste product from the alcohol industry (so-called ester-aldehyde fraction) for production of isocitric acid by yeasts. The potential producer of isocitric acid from this fraction, *Yarrowia lipolytica* VKM Y-2373, was selected by screening of various yeast cultures. The selected strain showed sufficient growth and good acid formation in media with growth-limiting concentrations of nitrogen, sulfur, phosphorus, and magnesium. A shortage of Fe^{2+} and Ca^{2+} ions suppressed both *Y. lipolytica* growth and formation of isocitric acid. The preferential synthesis of isocitric acid can be regulated by changing the nature and concentration of nitrogen source, pH of cultivation medium, and concentration of ester-aldehyde fraction. Experiments in this direction allowed us to obtain 65 g/L isocitric acid with a product yield (Y_{ICA}) of 0.65 g/g in four days of cultivation.

Keywords: microbial synthesis; yeast; isocitric acid; *Yarrowia lipolytica*; waste from alcohol industry; ester-aldehyde fraction; optimization

Citation: Kamzolova, S.V.; Samoilenko, V.A.; Lunina, J.N.; Morgunov, I.G. Isocitric Acid Production from Ethanol Industry Waste by *Yarrowia lipolytica*. *Fermentation* **2021**, *7*, 146. https://doi.org/10.3390/fermentation7030146

Academic Editor: Ronnie G. Willaert

Received: 15 July 2021
Accepted: 31 July 2021
Published: 4 August 2021

Publisher's Note: MDPI stays neutral with regard to jurisdictional claims in published maps and institutional affiliations.

Copyright: © 2021 by the authors. Licensee MDPI, Basel, Switzerland. This article is an open access article distributed under the terms and conditions of the Creative Commons Attribution (CC BY) license (https://creativecommons.org/licenses/by/4.0/).

1. Introduction

Over the past two decades, isocitric acid has attracted the attention of researchers as a promising compound for the pharmaceutical and food industry [1–11]. The antioxidant activity of microbially produced isocitric acid exceeds that of the classic antioxidant ascorbic acid [7]. Isocitric acid beneficially influences the spatial memory of intact laboratory rats and those intoxicated with heavy metals. It reduces the number of inadequate conditioned reactions and promotes the development of skills that meet new spatial conditions [8,10].

Despite many positive properties, isocitric acid is still rarely used due to its difficult production. Chemical synthesis gives a mixture of four stereoisomers of isocitric acid: threo-D_S-, threo-Ls-, erythro-Ds-, and erythro-Ls- [12], of which only threo-D_S- or (2*R*,3*S*)-isocitric acid (according to IUPAC Organic Nomenclature) is a metabolite of the tricarboxylic acid (TCA) cycle used by all aerobic organisms. It is this isomer that is commonly called isocitric acid (ICA). At the moment, the technology of separation of mixtures of stereoisomers has not been developed. In other words, chemical synthesis cannot be used for production of pharmacopoeial quality ICA. As for the production of naturally occurring ICA with the aid of the specially cultivated plant *Sedum spectabile*, the cost of ICA produced by this method is too high (USD 750 per 1g ICA) [8].

Alternatively, ICA can be produced using the yeast *Yarrowia lipolytica* [3,11]. The representatives of this species are nonpathogenic and have a status of GRAS (generally regarded as safe). They do not form mycotoxins and potentially allergenic spores during cultivation. Their metabolites are not toxic, mutagenic or clastogenic and hence can be used as food additives [13,14]. The microbial synthesis of ICA is relatively simple, fast, and cheap. As a result, the produced ICA becomes affordable and cost-effective for wide application.

The existing microbial methods of ICA production suggest the use of purified substrates as sources of carbon and energy. The range of such substrates includes ethanol [7,8,15,16], rapeseed oil [17–19], sunflower oil [1,3,4,20], glucose [21–24] and glycerol [21–25]. Besides, researchers used cheap wastes from biodiesel production [10,26,27]. In the works cited above, it is emphasized that the main condition of ICA production is the limitation of yeast growth by nitrogen source and the excess of carbon source. In all cases, ICA was synthesized together with citric acid (CA).

An urgent task is to expand the list of growth substrates used, increase the yield of organic acids, and reduce the cost of the final product. When developing the process of ICA production by yeast *Y. lipolytica*, we decided to use waste from the alcohol industry known as ester-aldehyde fraction (EAF).

EAF is formed as an ethanol distillation byproduct and contains some amount of ethanol and a range of volatile substances with a specific color and odor. These substances include aldehydes, methanol, esters, and carboxylic acids. Despite the low content of these substances, they are harmful, especially aldehydes, and cause a strong irritant effect on the mucosa of the eyes and upper respiratory tract. As a result, EAF is widely used only in the paint and varnish industry. At the same time, the main component of EAF, ethanol, is readily utilized by the yeast *Y. lipolytica* in concentrations up to 3 wt% [28] and may serve as an excellent substrate for the biosynthetic production of ICA [7,8,15,16]. To date, EAF has not been investigated as a potential source of carbon and energy for microbial ICA production.

The aim of this work was to study the possibility of using EAF for ICA production with the aid of yeasts, to select a promising producer, and to determine the cultivation conditions that provide an enhanced and directed synthesis of ICA.

2. Materials and Methods

Experiments were carried out with 35 wild strains of various yeast species and genera obtained from the All-Russian Collection of Microorganisms (VKM), as well as with the mutant and recombinant strains of *Y. lipolytica* from the culture collection of the laboratory of aerobic microbial metabolism of the Institute of Biochemistry and Physiology of Microorganisms of Russian Academy of Sciences. The mutant strain *Y. lipolytica* UV/NNG was derived from the wild strain *Y. lipolytica* VKM Y-2373 by ultraviolet irradiation (4 min) and treatment with N-methyl-N′-nitro-N-nitrosoguanidine (50 µg/mL) [17]. The recombinant strain *Y. lipolytica* ACO1 (no. 20) with the superexpressed gene of aconitate hydratase was derived from the wild strain *Y. lipolytica* 607 [18]. All strains under study were maintained at +4 °C on Reader agar with paraffin and re-cultured each 3 months.

All chemicals were of analytical grade (Mosreactiv, Russia). EAF represented a light-yellow liquid with a persistent unpleasant odor. EAF contained ethanol (90 vol%), aldehydes (0.5 g/L in terms of acetaldehyde), esters (0.4 g/L in terms of ethylacetate), and methanol (1 vol%).

To select ICA producers, yeast strains were cultivated at 29 °C on a shaker (130 rpm) in large tubes (2 cm in diameter and 20 cm long) with 5 mL of Reader medium with a 10-fold reduced content of ammonium sulfate. This medium contained (g/L): $(NH_4)_2SO_4$, 0.3; $MgSO_4 \cdot 7H_2O$, 0.7; $Ca(NO_3)_2 \cdot 4H_2O$, 0.4; NaCl, 0.5; KH_2PO_4, 1.0; and K_2HPO_4, 0.2. The medium was supplemented with yeast extract "Difco" (0.5 g/L) and Burkholder microelement solution of the following composition (mg/L): KJ, 0.1; B^{3+}, 0.01; Mn^{2+}, 0.01; Zn^{2+}, 0.01; Cu^{2+}, 0.01; Mo^{2+}, 0.01; and Fe^{2+}, 0.05. EAF was added in portions (2 g/L) to the concentration of 20 g/L. Cultivation lasted 4 days.

To study the effect of growth-limiting components of cultivation media, the strain *Y. lipolytica* VKM Y-2373 was cultivated at 29 °C on the shaker (130 rpm) in 750-mL Erlenmeyer flasks with 50 mL of a modified Reader medium with the concentrations of KH_2PO_4, K_2HPO_4, and sulfur were reduced 100-fold; the concentrations of $MgSO_4$ and calcium were reduced 500-fold; and the concentration of iron was reduced by 50 times in comparison with the original Reader medium. EAF was added in portions (2 g/L) to the concentration of 20 g/L. During cultivation, pH was maintained automatically at a level of 6.0 by adding the necessary amount of a sterile solution of NaOH. Cultivation lasted 4 days.

Experiments on the influence of nitrogen source were carried out in the flasks as described above. Various nitrogen sources were added in the concentration of 63 mg/L (in terms of nitrogen). Experiments on the effect of pH were described below in details in the Results section.

To study the effect of EAF concentration and dynamics of ICA formation, *Y. lipolytica* VKM Y-2373 was cultivated in a 10-l ANKUM-2M fermentor (IBP RAS, Pushchino, Russia) with the culture volume of 5 L. The cultivation temperature was 29 °C, the concentration of dissolved oxygen was 55–60% (of air saturation), pH was 6.0, and agitation was 800 rpm. The modified Reader medium contained (g/L): $(NH_4)_2SO_4$, 3.0; $MgSO_4 \cdot 7H_2O$, 1.4; $Ca(NO_3)_2$, 0.8; NaCl, 0.5; KH_2PO_4, 2.0; K_2HPO_4, 0.2; double volume of Burkholder solution. The concentration of Zn^{2+} and Fe^{2+} was increased to 0.3 and 1.2 mg/L, respectively. Yeast autolysate was added to the fermentor before inoculation at a concentration of 6.3 mL/L. EAF was added in portions (from 1 to 10 g/L) at the moments when oxygen concentration in the fermentor fell by 10% (of air saturation) from the basal level. Cultivation lasted 4 days.

Yeast growth was followed by measuring the biomass dry weight: 1–3 mL of the culture broth was filtered through a membrane filter; yeast cells were washed with H_2O and dried under vacuum at 110 °C to the constant weight.

Concentration of NH_4^+ was determined potentiometrically with an Ecotest-120 ionometer using an Ekom-NH4 electrode (Econix, Moscow, Russia).

Ethanol and impurities in the EAF were determined by gas-liquid chromatography on a Chrom-5 chromatograph (Laboratory Instruments, Praha, Czech Republic) with a flame-ionization detector using a glass column (200 × 0.3 mm) packed with 15% Reoplex-400 on Chromaton N-AW (0.16–0.20 mm) at a column temperature of 65 °C; argon was used as a carrier gas.

To analyze organic acids, the culture broth was centrifuged (8000× *g*, 20 °C, 3 min); 1 mL of the supernatant was diluted with an equal volume of 8% $HClO_4$ and the concentration of organic acids was measured on an HPLC chromatograph (LKB, Sweden) equipped with an Inertsil ODS-3 reversed-phase column (250 × 4 mm, Elsiko, Moscow, Russia) at 210 nm; 20 mM phosphoric acid was used as a mobile phase with a flow rate of 1.0 mL·min^{-1}; the column temperature was maintained at 35 °C. Quantitative determination of acids was carried out with the help of calibration curves constructed with the use of appropriate standards (Sigma-Aldrich, St. Louis, MO, USA). Moreover, diagnostic kits (Roche Diagnostics GmbH, Mannheim, Germany) were used for the assay of ICA and CA concentrations. The determination of ICA was based on the measurement of NADPH produced during the conversion of ICA to α-ketoglutarate, a reaction catalyzed by isocitrate dehydrogenase. The determination of CA was based on the measurement of NADH produced during the conversion of CA to oxaloacetate and its decarboxylation product, pyruvate, and subsequent conversion to L-malate and L-lactate. Reactions were catalyzed by citrate lyase, malate dehydrogenase and L-lactate dehydrogenase.

The calculation of production parameters, such as the product yield (Y_{ICA}), the specific growth rate (μ), the specific rate of ICA synthesis (q_{ICA}), and volume productivity (Q_{ICA}), have been described earlier [20,26].

The results presented in this paper are the means of experiments performed in triplicate or quadruplicate. Each measurement was repeated twice. Standard deviation did not exceed 10%. The product yield (Y_{ICA}), the specific rate of synthesis (q_{ICA}), and volume productivity (Q_{ICA}) were calculated based on average values of ICA.

3. Results

3.1. Selection of ICA Producer

The quantitative composition of the metabolites excreted from the cells of various yeast strains when their growth was limited by nitrogen shortage is shown in Table 1. As seen from this table, most of the strains studied (26 from 35) excreted ICA in amounts from 0.06 to 5.51 g/L when they grew on EAF under nitrogen deficiency. Four of these strains produced ICA in small amounts (below 0.1 g/L); 12 strains produced it in amounts from 0.1 to 0.5 g/L; 2 strains, from 0.5 to 1 g/L; 3 strains, from 1 to 2 g/L; and 2 strains, from 2 to 3 g/L. The maximum ICA-producing ability was exhibited by three strains, *Y. lipolytica* VKM Y-2373, *Y. lipolytica* UV/NNG, and *Y. lipolytica* ACO1 no. 20 (5.51, 4.61, and 3.73 g/L, respectively). Besides ICA, the yeast strains excreted other metabolites, such as citric acid (CA), acetic acid (AA), α-ketoglutaric acid (KGA), succinic acid (SA), malic acid (MA), and fumaric acid (FA). Most strains of the species *Y. lipolytica* excreted mainly ICA (from 32.3 to 53.5% of the total amount of acids), the maximum relative content of ICA (53.5% of the total amount of acids) being demonstrated by the wild strain *Y. lipolytica* VKM Y-2373. It is this strain that was chosen for further investigations as a promising producer of ICA from EAF.

Table 1. Acid production by yeast in the medium containing EAF as a carbon source under nitrogen limitation.

Strain	Acids (g/L)							ICA (% of Total Acids)
	ICA	AA	CA	KGA	SA	MA	FA	
Aciculoconidium aculeatum VKM Y-1301	0	0	0.11	0	0	0.11	0.06	-
Babjeviella inositovora VKM Y-2494	0.10	0.09	0.13	0.08	0	0.04	0.05	22.7
Blastobotrys adeninivorans VKM Y-2676	0	0.11	0.33	0.16	0	0.11	0.11	-
Candida intermedia	0.08	0.12	0.01	0.12	0	0.12	0.12	17.8
C. saitoana 127	0.10	0	0.45	0.07	0	0.1	0.07	13.9
C. utilis VKM Y-33	0	0.14	1.05	0.42	0.32	0.45	0.10	-
C. zeylanoides VKM Y-14	0.50	0.12	1.0	0.10	0.32	0.44	0.07	20.2
C. zeylanoides VKM Y-2324	0.38	0.12	0.55	0.13	0.25	0.22	0.06	23.0
C. valida VKM Y-1493	0.14	0.12	0	0	0.52	0.12	0.01	15.6
Diutina catenulata VKM Y-5	0.15	0.15	0.21	1.50	0.33	0.62	0.08	5.1
D. rugosa VKM Y-67	0	0.14	0.15	0	0.2	0	0.07	-
Kluyveromyces wickerhamii VKM Y-589	0	0.62	0.18	0.08	0.21	0.10	0.05	-
Kregervanrija fluxuum VKM Y-240	0.31	0.11	0	0.10	0.32	0.15	0.08	31.3
Meyerozyma guilliermondii	0.20	0.53	0.45	0.09	0.15	0.12	0.05	13.0
Pichia besseyi VKM Y-2084	0	0.12	0.43	0.45	0.40	0.32	0.03	-
P. media VKM Y-1381	0.20	0.10	0.40	0.10	0.33	0.10	0.04	16.3
P. membranifaciens VKM Y-292	0	0.13	0.15	0	0.13	0	0.07	-
Sugiyamaella paludigena VKM Y-2443	0.08	0.55	0	0.40	0.55	0.32	0.07	4.2
Torulaspora candida 420	0.10	0.45	0.32	0.10	0	0.10	0.08	9.3
T. globosa VKM Y-93	0.10	0.10	0	0.10	0.10	0.10	0.10	20.0
Wickerhamomyces anomalus VKM Y-118	0	0.15	0	0.10	0.43	0.12	0.07	-
Yarrowia lipolytica 12a	0.15	0.15	0	0.09	0.52	0.15	0.06	14.2
Y. lipolytica VKM Y-47	0	0.10	0.52	0.45	0.40	0.30	0.10	-
Y. lipolytica 68	0.06	0.10	0.07	0.10	0.10	0.10	0.10	11.3
Y. lipolytica 69	1.11	0.15	0.92	0.50	0.35	0.41	0.12	32.3
Y. lipolytica VKM Y-57	1.42	0.10	0.72	0.07	0.35	0.21	0.06	49.5
Y. lipolytica VKM Y-2412	1.75	0.10	1.50	0.09	0.24	0.10	0.20	46.3
Y. lipolytica 374/4	2.10	0.15	1.63	0.08	0.14	0.11	0.02	50.0
Y. lipolytica 571	0.55	0.10	0.30	0.10	0.10	0.10	0.10	44.0
Y. lipolytica 581	0.65	0.10	0.72	0.10	0.10	0.10	0.10	36.7
Y. lipolytica 585	0.06	0.08	0.10	0	0.30	0.10	0.08	9.4
Y. lipolytica 607	2.30	0.10	1.50	0.10	0.45	0.10	0.10	50.5
Y. lipolytica VKM-2373	5.51	0.5	4.24	0.10	0.13	0.12	0.08	53.5
Y. lipolytica UV/NNG	4.61	1.17	3.18	1.15	0.12	0.10	0.10	44.6
Y. lipolytica ACO1 no. 20	3.73	1.0	2.65	0.65	0.10	0.10	0.10	45.3

Acids: ICA—isocitric acid, AA—acetic acid, CA—citric acid, KGA—α-ketoglutaric acid, SA—succinic acid, MA—malic acid, FA—fumaric acid. All the data presented are the mean values of three experiments and two measurements for each experiment (SD < 10%).

3.2. Effect of Growth-Limiting Component of Cultivation Media

As seen from Table 2, the strain *Y. lipolytica* VKM Y-2373 growing in the complete medium with EAF did not produce acids. However, the limitation of its growth by either of the biogenic elements N, P, and S led to excretion of ICA and CA in the ratio varied from 1.7:1 to 1.3:1. The limitation of its growth by Mg increased the ICA:CA ratio to 3.4:1, however, the amount of excreted ICA was 2.7 times lower than under nitrogen deficiency. Under deficiency of Fe^{2+} (0.001 mg/L) or Ca^{2+} (0.136 mg/L), ICA and CA were not produced, although Fe^{2+} caused the formation of AA in a noticeable amount (1.25 g/L). The maximum ICA production (6.33 g/L) with a product yield (Y_{ICA}) (0.32 g/g) were observed in the case of nitrogen limitation. Moreover, yeast cells require nitrogen in greater amounts than other nutrient elements (P, S, Mg) and, hence, the deficiency of this element in the culture medium can easily be controlled. For this reason, further experiments were carried out with the limitation of yeast growth by a nitrogen source.

Table 2. Effect of limiting component on the growth of *Y. lipolytica* VKM Y-2373 and ICA production.

Parameters	Full Medium (mg/L) N—630, P—246, S—186, Mg—140, Ca—68, Fe—0.05	Limiting Component (mg/L)					
		N (63.0)	P (2.5)	S (1.9)	Mg (0.28)	Ca (0.136)	Fe (0.001)
Biomass (g/L)	8.9	2.33 ± 0.40	1.77 ± 0.15	1.67 ± 0.15	2.10 ± 0.20	1.03 ± 0.21	1.05 ± 0.25
ICA (g/L)	0	6.33 ± 0.32	5.27 ± 0.15	5.65 ± 0.25	2.37 ± 0.21	0	0
CA (g/L)	0	3.70 ± 0.20	3.43 ± 0.31	4.37 ± 0.21	0.70 ± 0.10	0	0
AA (g/L)	0	Tr	Tr	Tr	Tr	0	1.25 ± 0.32
ICA/CA ratio	-	1.7:1	1.5:1	1.3:1	3.4:1	-	-
Y_{ICA} (g/g)	-	0.32	0.26	0.30	0.11	-	-

"Tr" stands for trace amount.

3.3. Effect of Nitrogen Source

As seen from Table 3, the best source of nitrogen was found to be urea ((NH_2)$_2$CO) and ammonium sulfate ((NH_4)$_2$SO$_4$) (biomass of 2.35 and 2.33 g/L, respectively), but ammonium acetate (CH_3COONH_4) was unable to support the growth of *Y. lipolytica* VKM Y-2373 (biomass of 1.19 g/L).

Table 3. Effect of nitrogen source on the growth of *Y. lipolytica* VKM Y-2373 and ICA production.

Parameters	Nitrogen Concentration (63 mg/L)				
	$(NH_4)_2SO_4$	$(NH_2)_2CO$	NH_4CL	NH_4NO_3	CH_3COONH_4
Biomass (g/L)	2.33 ± 0.04	2.35 ± 0.31	1.63 ± 0.16	1.76 ± 0.14	1.19 ± 0.03
ICA (g/L)	6.33 ± 0.32	5.90 ± 0.15	5.01 ± 0.28	4.43 ± 0.39	2.03 ± 0.15
CA (g/L)	3.70 ± 0.20	3.52 ± 0.19	3.07 ± 0.21	2.93 ± 0.15	1.77 ± 0.21
AA (g/L)	0.5	Tr	0.5	Tr	Tr
ICA/CA ratio	1.7:1	1.7:1	1.6:1	1.5:1	1.2:1
Y_{ICA} (g/g)	0.32	0.30	0.25	0.22	0.10

"Tr" stands for trace amount.

Nitrogen source influenced not only yeast growth, but also the production of ICA. The production was at the maximum (6.33 g/L) with $(NH_4)_2SO_4$. With $(NH_2)_2CO$ the ICA production was slightly lower (5.9 g/L). With NH_4Cl and NH_4NO_3, the ICA production decreased by 21% and 30%, respectively, in comparison with the growth limitation by $(NH_4)_2SO_4$. CH_3COONH_4 was found to be the worst nitrogen source for ICA production

(2.03 g/L). The ICA to CA ratio varied from 1.2:1 to 1.7:1 depending on the nitrogen source used. The maximum shift toward the formation of ICA and the maximum product yield (Y_{ICA} = 0.32 g/g) was observed in the case of growth limitation by $(NH_4)_2SO_4$.

3.4. Effect of pH

The growth of *Y. lipolytica* VKM Y-2373 in media with EAF without titration was accompanied by a drastic fall of pH (down to 2) by the end of cultivation (Table 4) which was caused by consumption of the cation of physiologically acidic salt $(NH_4)_2SO_4$ and the accumulation of excreted organic acids in the medium. In this case, the production of ICA and CA was low (1.03 and 0.57 g/L, respectively). In contrast, the regular (daily) alkalization of the growth medium during yeast cultivation by the addition of 10 wt% NaOH led to a lower decrease in pH (down to 5) by the end of cultivation and the accumulation of ICA and CA in greater amounts (6.33 and 3.70 g/L, respectively). Still, better results were obtained when the medium was supplemented with 2 wt% $CaCO_3$ just after inoculation of the medium. This approach allowed us to maintain the pH of the medium at the initial level of 6 during the whole cultivation period. In this case, the production of ICA and CA increased to 8.3 and 5.53 g/L, respectively, although the accumulation of cell biomass even slightly decreased. The increase of $CaCO_3$ concentration to 3 wt% did not enhance the production of ICA.

Table 4. Effect of medium titration on *Y. lipolytica* VKM Y-2373 growth and ICA production.

Parameters	Without Titration	Titration		
		10 wt% NaOH	2 wt% CaCO$_3$	3 wt% CaCO$_3$
Initial pH	6.0	6.0	6.0	6.5
Final pH	2.0	5.0	6.0	6.5
Biomass (g/L)	1.50 ± 0.10	2.33 ± 0.04	1.90 ± 0.10	1.80 ± 0.10
ICA (g/L)	1.03 ± 0.15	6.33 ± 0.32	8.30 ± 0.30	9.00 ± 0.26
CA (g/L)	0.57 ± 0.12	3.70 ± 0.20	5.53 ± 0.15	6.03 ± 0.65
AA (g/L)	0.5	0.5	0.1	0.1
ICA/CA ratio	1.8:1	1.7:1	1.5:1	1.5:1
Y_{ICA} (g/g)	0.05	0.32	0.42	0.45

"Tr" stands for trace amount.

Further experiments were designed in order to understand in more detail the dependence of ICA biosynthesis on the pH of the cultivation medium. For this purpose, 1 M phosphate buffer with pH varied from 4 to 7 was added to the cultivation medium in an amount of 20 vol%. If the pH of the medium during cultivation shifted from the set value, it was corrected by the addition of either NaOH. NaOH, rather than other titrants, was chosen because it provided the maximum shift toward the formation of ICA.

The experiments showed that pH values from 4 to 6.5 are beneficial for yeast growth (Table 5). The range of pH values favorable for ICA production was narrower. At pH equal to 4.0, the production of ICA was low (3.04 g/L ICA, Y_{ICA} = 0.15 g/g). The maximum production of ICA (9.5 and 9.0 g/L, respectively), was observed at the pH values of 6.0 and 6.5.

Table 5. Effect of pH medium on the growth of *Y. lipolytica* VKM Y-2373 and ICA production.

Parameters	pH						
	4.0	**4.5**	**5.0**	**5.5**	**6.0**	**6.5**	**7.0**
Biomass (g/L)	2.76 ± 0.12	2.70 ± 0.26	2.60 ± 0.10	2.45 ± 0.10	2.30 ± 0.10	2.15 ± 0.10	2.00 ± 0.11
ICA (g/L)	3.04 ± 0.09	5.90 ± 0.20	6.59 ± 0.19	8.00 ± 0.23	9.50 ± 0.26	9.00 ± 0.10	7.31 ± 0.20
CA (g/L)	1.90 ± 0.17	3.67 ± 0.15	3.88 ± 0.11	4.53 ± 0.13	5.16 ± 0.15	5.05 ± 0.21	4.53 ± 0.13
AA (g/L)	0.1	0.1	0.1	0.1	0.1	0.1	0.1
ICA/CA ratio	1.6:1	1.6:1	1.7:1	1.8:1	1.8:1	1.8:1	1.6:1

3.5. Effect of EAF Concentration

EAF was added to the growth medium in the fermentor in the portions varied from 1 to 10 g/L. As seen from Table 6, the content of EAF in the medium weakly influenced the growth of *Y. lipolytica* VKM Y-2373, so that the accumulation of cell biomass was between 13 and 14.5 g/L. At the same time, the quantitative composition of excreted organic acids greatly depended on the EAF concentration. Indeed, at the EAF concentration equal to 1 g/L, the yeast strain synthesized ICA and CA in equal amounts (ICA:CA = 1:1) and the concentration of AA did not exceed 0.5 g/L. The increase in the concentration of EAF from 1 to 4 g/L promoted the production of ICA from 51.5 to 60.7 g/L and suppressed the production of CA from 51.3 to 31.3 g/L, so that the ICA:CA ratio became 2:1. The increase in the concentration of EAF from 4 to 6 g/L slightly suppressed the production of both ICA and CA (to 54.1 and 27.0 g/L, respectively). The concentration of EAF equal to 10 g/L greatly suppressed the formation of ICA (22.2 g/L) and CA (16.0 g/L) and promoted the accumulation of AA in the medium to 12 g/L. The maximum values of volume productivity (Q_{ICA} = 0.89 g ICA/L·h) and product yield (Y_{ICA} = 0.61 g/g) were observed when EAF was added to the medium in the concentration of 4 g/L.

Table 6. Effect of EAF on the growth of *Y. lipolytica* VKM Y-2373 and ICA production.

Parameters	EAF Content, Periodically Added to the Cultivation Medium (g/L)				
	1	**2**	**4**	**6**	**10**
Biomass (g/L)	13.0 ± 0.2	13.6 ± 1.2	13.3 ± 0.4	14.5 ± 0.2	14.1 ± 1.3
ICA (g/L)	51.5 ± 5.2	56.6 ± 4.4	60.7 ± 2.1	54.1 ± 3.5	22.2 ± 1.1
CA (g/L)	51.3 ± 2.1	33.4 ± 1.2	31.1 ± 2.2	27.0 ± 2.0	16.0 ± 1.1
AA (g/L)	0.5	0.5	0.5	0.5	12.0 ± 1.4
ICA/CA ratio	1:1	1.7:1	2:1	2:1	1.4:1
Y_{ICA} (g/g)	0.47	0.55	0.61	0.53	0.25
Q_{ICA} (g/L·h)	0.78	0.81	0.89	0.81	0.31

3.6. Dynamics of Yeast Growth and Acid Excretion

The dynamics of growth of *Y. lipolytica* VKM Y-2373 and acid excretion were studied using cultivation in the fermentor. Cell growth was limited by shortage of the nitrogen source (($NH_4)_2SO_4$ = 3 g/L). The growth substrate EAF was added in portions varied from 2 to 6 g/L. The concentration of dissolved oxygen comprised 60% (of air saturation); pH was maintained at the level of 6.0.

As seen from Figure 1, the growth curve of *Y. lipolytica* VKM Y-2373 had a clear-cut exponential phase (to 12 h of growth), a phase of growth retardation (from 12 to 24 h of growth), and then a stationary phase. The maximum specific growth rate (μ_{max} = 0.282 h^{-1}) was observed at 6 h of cultivation. The excretion of ICA and CA began during the transition of the yeast culture from exponential growth to retarded growth, when the specific growth rate (μ) fell by two times from its maximum value. In the late phase of retarded growth and stationary phase, the yeast strain continued to synthesize ICA and CA with a specific rate of synthesis (q_{ICA}) equal to 0.08–0.083 g/g·h; beginning from 60 h of growth, the rate of ICA synthesis decreased by 2 times. By the end of the cultivation period, the culture liquid contained 65.0 g/L ICA and 31.2 g/L CA with the ICA:CA ratio equal to 2.1:1. The maximum values of volume productivity (Q_{ICA}) and product yield (Y_{ICA}) were 0.95 g ICA/L·h and 0.65 g/g, respectively.

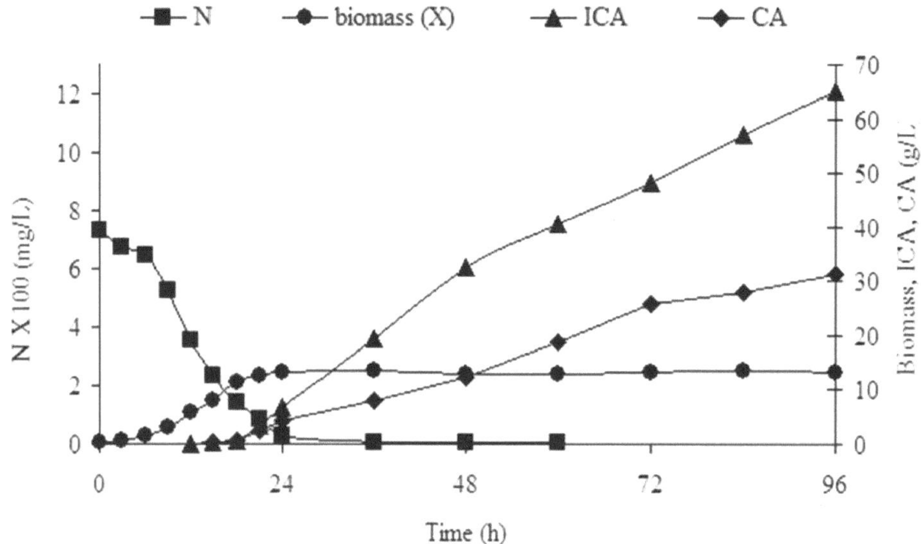

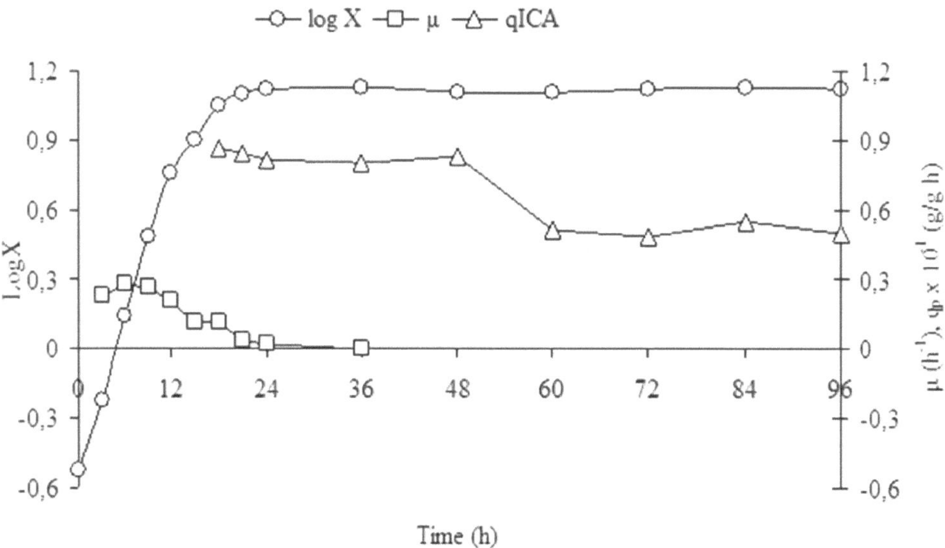

Figure 1. Time course of growth of *Y. lipolytica* VKM Y-2373, nitrogen consumption, citric acids production in the medium with EAF (**top**) and the calculated parameters of ICA production (**bottom**). N—nitrogen; ICA—isocitric acid; CA—citric acid; logX—logarithm of biomass; μ—the specific growth rate; q_{ICA}—the specific rate of ICA synthesis.

4. Discussion

The results obtained revealed the possibility to use a cheap hard-to-recycle waste—EAF for the microbial production of ICA by yeast strains.

ICA production from EAF was observed in 26 strains (Table 1), but the maximum ICA concentration and the relative content was indicated in the wild-type strain *Y. lipolytica* VKM Y-2373.

The following peculiarities of ICA production from EAF by the selected strain were revealed.

The strain *Y. lipolytica* VKM Y-2373 cultivated in the complete nutrient medium showed good growth but without excretion of organic acids. The basic condition of excretion of ICA and CA from EAF was the limitation of *Y. lipolytica* VKM Y-2373 growth by shortage of nitrogen source (as well as sources of phosphorus, sulfur, or magnesium). The limitation of the strain growth by Ca^{2+} ions did not cause acid excretion, while the Fe^{2+} ions deficiency stimulated the excretion of acetic acid, but not ICA or CA (Table 2). It should be noted that the limitation of culture growth by shortage of biogenic elements is a common approach to stimulation of the excretion of organic acids. For example, the excretion of CA by *Y. lipolytica* was stimulated by deficiency of N and P [29–31]; the excretion of itaconic acid by *Aspergillus terreus* was stimulated by P shortage [32,33]; and the excretion of gluconic acid by *Aspergillus niger* was stimulated by shortage of N and P [34]. The most part of studies concerned the biosynthesis of ICA and CA by *Y. lipolytica* was performed under conditions of nitrogen deficiency [15–17,22,23,35]. On the other hand, some authors reported that the addition of 0.7 g/L $(NH_4)_2SO_4$ to the medium with the glycerol-containing wastes from biodiesel production shifts the biosynthesis of citric acids toward the preferential synthesis of ICA, as is evident from the shift of the CA:ICA ratio from 1.2:1 to 1:11.5 [26].

Y. lipolytica VKM Y-2373 was found to be able assimilate all nitrogen sources investigated, including inorganic and organic forms. However, it should be noted that the most suitable nitrogen source is ammonium sulfate because it provides the maximum ICA production and its concentration can easily be controlled (Table 3). In experiments with the wide ICA-producing strains of *Y. lipolytica* grown in media with rapeseed oil, glycerol, glucose, and ethanol as the carbon sources, the authors usually employed NH_4Cl or $(NH_4)_2SO_4$ as the nitrogen source [22,35]. In experiments with the recombinant strains *Y. lipolytica* CIT 1 and CIT 2 with the super-expressed citrate synthase genes, the authors used NH_4Cl as the nitrogen source when the strains were grown on glycerol, while $(NH_4)_2SO_4$ when the strains were grown on vegetable oils [19].

The strain *Y. lipolytica* VKM Y-2373 is able to grow in a range of pH values from 2 to 6.5 (Table 4). The optimal pH for growth of this strain does not coinside with the optimal pH for ICA production (6.0–6.5) (Table 5). This data corresponds to our earlier data that ICA is optimally produced by this strain at pH = 6.0 when it is grown in media with rapeseed oil [17], ethanol [16], and biodiesel wastes [10]. Literature data clearly demonstrate the metabolic flexibility of the yeast species *Y. lipolytica* on pH, depending on the particular strain and the carbon source used. Moeller et al. (2007) reported that the production of ICA increases with the pH value and reaches the maximum at pH 7.5 [35]. At the same time, Papanikolaou et al. (2002) showed that ICA synthesis does not depend on pH [36]. There is evidence that the genetically modified strains of *Y. lipolytica* with the superexpressed *GUT1/GUT2* genes cultivated in media with biodiesel wastes efficiently produced ICA and CA even at pH = 3 and thus did not require the addition of a neutralizing agent [27]. This feature can significantly reduce the profitability of the process. It should be noted that the cultivation of the wide strains of *Y. lipolytica* at pH values of 2.5–3.5 generally caused the accumulation of polyols [37,38], while its cultivation at pH = 4.5–6.0 led to the accumulation of citric acids [37–39].

EAF in elevated concentrations may suppress the growth and acid formation by *Y. lipolytica* VKM Y-2373 (Table 6). For this reason, EAF should be added to the culture medium in small portions (2–6 g/L) rather than in bulk. The specified value of such portions for EAF is considerably lower than for glucose (120 g/L) [35]. Still greater portions of the carbon source glycerol (up to 150 g/L) can be used in the case of the transformant *Y. lipolytica* A101.1.31 with the super-expressed genes *CIT1* and *CIT2* encoding citrate synthase [19].

Under optimized cultivation conditions, *Y. lipolytica* VKM Y-2373 grown on EAF produced 65.0 g/L ICA with the ICA:CA ratio equal to 2.1:1 and product yield Y_{ICA} = 0.65 g/g. The same strain grown on pure ethanol produced 90.5 g/L ICA with a product yield of 0.77 g/g when cultivated in batch culture [16] and 109.6 g/L ICA with a product yield of 0.80 g/g when cultivated in repeated-batch culture [8]. Presumably, the reduced biosynthetic activity of *Y. lipolytica* VKM Y-2373 in media with EAF as the carbon source is associated with harmful impurities (aldehydes, esters, and methanol) in this substance. More efficient processes of ICA production are reported with the use of vegetable oils as the carbon source, which provide the accumulation of 68.4–93 g/L ICA with the product yield Y_{ICA} from 0.64 to 0.95 g/g [1,3,4,17,18,20]. The most efficient process of ICA production was realized using the yeast strain *Y. lipolytica* YALI0E34672g with the enhanced coexpression of the gene coding for the mitochondrial succinate–fumarate carrier YlSfc1 which controls ICA efflux from mitochondria and adenosine monophosphate deaminase YlAMPD genes together with inactivation of citrate mitochondrial carrier YlYHM2 gene. This strain provided the accumulation of 136.7 g/L ICA with a process selectivity of 88.1% [24].

5. Conclusions

It was shown for the first time that EAF could be successfully used for the synthesis of ICA by *Y. lipolytica* yeast. Although the ICA production from EAF (65 g/L; product yield (Y_{ICA}) of 0.65 g/g) was lower than that from pure ethanol (90.5 g/L; product yield (Y_{ICA}) of 0.77 g/g) [16], EAF can be considered as a promising substrate for microbiological ICA production taking into account its low cost. The optimal fermentation regime which ensures a good growth of selected strain *Y. lipolytica* VKM Y-2373 and directed synthesis of ICA was determined. It included the limitation of growth by nitrogen; $(NH_4)_2SO_4$ as nitrogen source; pH values from 4 to 6.5; and the addition of EAF by small portions (2–6 g/L).

Author Contributions: Conceptualization, S.V.K., V.A.S., I.G.M.; investigation, S.V.K., J.N.L.; methodology, S.V.K.; resources, V.A.S., I.G.M.; supervision, I.G.M.; writing—original draft, S.V.K. All authors have read and agreed to the published version of the manuscript.

Funding: This research received no external funding.

Institutional Review Board Statement: Not applicable.

Informed Consent Statement: Not applicable.

Data Availability Statement: Data is contained within the article.

Conflicts of Interest: The authors declare no conflict of interest.

References

1. Heretsch, P.; Thomas, F.; Aurich, A.; Krautscheid, H.; Sicker, D.; Giannis, A. Syntheses with a chiral building block from the citric acid cycle: (2R,3S)-isocitric acid by fermentation of sunflower oil. *Angew. Chem. Int. Ed. Engl.* **2008**, *47*, 1958–1960. [CrossRef] [PubMed]
2. Kulkarni, M.G.; Shaikh, Y.B.; Borhade, A.S.; Dhondge, A.P.; Chavhan, S.W.; Desai, M.P.; Birhade, D.R.; Dhatrak, N.R.; Gannimani, R. The efficient synthesis of (3R,3aS,6aR)-hexahydrofuro[2,3-b]furan-3-ol and its isomers. *Tetrahedron Asymmetry* **2010**, *21*, 2394–2398. [CrossRef]
3. Aurich, A.; Specht, R.; Müller, R.A.; Stottmeister, U.; Yovkova, V.; Otto, C.; Holz, M.; Barth, G.; Heretsch, P.; Thomas, F.A.; et al. Microbiologically produced carboxylic acids used as building blocks in organic synthesis. In *Reprogramming Microbial Metabolic Pathways. Subcellular Biochemistry*; Wang, X., Chen, J., Quinn, P., Eds.; Springer: Dordrecht, The Netherlands, 2012; Volume 64, pp. 391–423. [CrossRef]
4. Aurich, A.; Hofmann, J.; Oltrogge, R.; Wecks, M.; Glaser, R.; Blömer, L.; Mauersberger, S.; Roland, A.; Müller, R.A.; Sicker, D.; et al. Improved isolation of microbiologically produced (2R,3S)-isocitric acid by adsorption on activated carbon and recovery with methanol. *Org. Process. Res. Dev.* **2017**, *21*, 866–870. [CrossRef]

5.   Moore, G.L.; Stringham, R.W.; Teager, D.S.; Yue, T.Y. Practical synthesis of the bicyclic darunavir side chain: (3R,3aS,6aR)-hexahydrofuro[2,3-b]furan-3-ol from monopotassium isocitrate. *Org. Process. Res. Dev.* **2017**, *21*, 98–106. [CrossRef]

6.   Yang, J.; Kim, M.J.; Yoon, W.; Kim, E.Y.; Kim, H.; Lee, Y.; Min, B.; Kang, K.S.; Son, J.H.; Park, H.T.; et al. Isocitrate protects DJ-1 null dopaminergic cells from oxidative stress through NADP+-dependent isocitrate dehydrogenase (IDH). *PLoS Genet.* **2017**, *13*, e1006975. [CrossRef]

7.   Morgunov, I.G.; Karpukhina, O.V.; Kamzolova, S.V.; Samoilenko, V.A.; Inozemtsev, A.N. Investigation of the effect of biologically active threo-Ds-isocitric acid on oxidative stress in *Paramecium caudatum*. *Prep. Biochem. Biotechnol.* **2018**, *48*, 1–5. [CrossRef]

8.   Morgunov, I.G.; Kamzolova, S.V.; Karpukhina, O.V.; Bokieva, S.V.; Inozemtsev, A.N. Biosynthesis of isocitric acid in repeated-batch culture and testing of its stress-protective activity. *Appl. Microbiol. Biotechnol.* **2019**, *103*, 3549–3558. [CrossRef] [PubMed]

9.   Bullin, K.; Hennig, L.; Herold, R.; Krautscheid, H.; Richter, K.; Sicker, D. An optimized method for an (2R,3S)-isocitric acid building block. *Mon. Chem.* **2019**, *150*, 247–253. [CrossRef]

10.   Morgunov, I.G.; Kamzolova, S.V.; Karpukhina, O.V.; Bokieva, S.V.; Lunina, J.N.; Inozemtsev, A.N. Microbiological production of isocitric acid from biodiesel waste and its effect on spatial memory. *Microorganisms* **2020**, *8*, 462. [CrossRef] [PubMed]

11.   Fickers, P.; Cheng, H.; Sze, K.; Lin, C. Sugar alcohols and organic acids synthesis in *Yarrowia lipolytica*: Where Are We? *Microorganisms* **2020**, *8*, 574. [CrossRef]

12.   Vickery, H.B. A suggested new nomenclature for the isomers of isocitric acid. *J. Biol. Chem.* **1962**, *237*, 1739–1741. [CrossRef]

13.   Groenewald, M.; Boekhout, T.; Neuvéglise, C.; Gaillardin, C.; Van Dijck, P.W.; Wyss, M. *Yarrowia lipolytica*: Safety assessment of an oleaginous yeast with a great industrial potential. *Crit. Rev. Microbiol.* **2014**, *40*, 187–206. [CrossRef]

14.   Zinjarde, S.S. Food-related applications of *Yarrowia lipolytica*. *Food Chem.* **2014**, *152*, 1–10. [CrossRef]

15.   Finogenova, T.V.; Shishkanova, N.V.; Fausek, E.A.; Eremina, S.S. Biosynthesis of isocitric acid from ethanol by yeasts. *Appl. Microbiol. Biotechnol.* **1991**, *36*, 231–235. [CrossRef]

16.   Kamzolova, S.V.; Shamin, R.V.; Stepanova, N.N.; Morgunov, G.I.; Lunina, J.N.; Allayarov, R.K.; Samoilenko, V.A.; Morgunov, I.G. Fermentation conditions and media optimization for isocitric acid production from ethanol by *Yarrowia lipolytica*. *Biomed. Res. Int.* **2018**, e2543210. [CrossRef]

17.   Kamzolova, S.V.; Dedyukhina, E.G.; Samoilenko, V.A.; Lunina, J.N.; Puntus, I.F.; Allayarov, R.K.; Chiglintseva, M.N.; Mironov, A.A.; Morgunov, I.G. Isocitric acid production from rapeseed oil by *Yarrowia lipolytica* yeast. *Appl. Microbiol. Biotechnol.* **2013**, *97*, 9133–9144. [CrossRef]

18.   Laptev, I.A.; Filimonova, N.A.; Allayarov, R.K.; Kamzolova, S.V.; Samoilenko, V.A.; Sineoky, S.P.; Morgunov, I.G. New recombinant strains of the yeast *Yarrowia lipolytica* with overexpression of the aconitate hydratase gene for the obtainment of isocitric acid from rapeseed oil. *Appl. Biochem. Microbiol.* **2016**, *52*, 699–704. [CrossRef]

19.   Hapeta, P.; Rakicka-Pustułka, M.; Juszczyk, P.; Robak, M.; Rymowicz, W.; Lazar, Z. Overexpression of citrate synthase increases isocitric acid biosynthesis in the heast *Yarrowia lipolytica*. *Sustainability* **2020**, *12*, 7364. [CrossRef]

20.   Kamzolova, S.V.; Samoilenko, V.A.; Lunina, J.N.; Morgunov, I.G. Effects of medium components on isocitric acid production by *Yarrowia lipolytica* yeast. *Fermentation* **2020**, *6*, 112. [CrossRef]

21.   Finogenova, T.V.; Shishkanova, N.V.; Ermakova, I.T.; Kataeva, I.A. Properties of *Candida lipolytica* mutants with the modified glyoxylate cycle and their ability to produce citric and isocitric acid. II. Synthesis of citric and isocitric acid by mutants and peculiarities of their enzyme systems. *Appl. Microbiol. Biotechnol.* **1986**, *23*, 378–383. [CrossRef]

22.   Förster, A.; Jacobs, K.; Juretzek, T.; Mauersberger, S.; Barth, B. Overexpression of the ICL1 gene changes the product ratio of citric acid production by *Yarrowia lipolytica*. *Appl. Microbiol. Biotechnol.* **2007**, *77*, 861–869. [CrossRef] [PubMed]

23.   Holz, M.; Förster, A.; Mauersberger, S.; Barth, G. Aconitase overexpression changes the product ratio of citric acid production by *Yarrowia lipolytica*. *Appl. Microbiol. Biotechnol.* **2009**, *81*, 1087–1096. [CrossRef] [PubMed]

24.   Yuzbasheva, E.Y.; Scarcia, P.; Yuzbashev, T.V.; Messina, E.; Kosikhina, I.M.; Palmieri, L.; Shutov, A.V.; Taratynova, M.O.; Amaro, R.L.; Palmieri, F.; et al. Engineering *Yarrowia lipolytica* for the selective and high-level production of isocitric acid through manipulation of mitochondrial dicarboxylate-tricarboxylate carriers. *Metab. Eng.* **2021**, *65*, 156–166. [CrossRef] [PubMed]

25.   Ermakova, I.T.; Shishkanova, N.V.; Melnikova, O.F.; Finogenova, T.V. Properties of *Candida lipolytica* mutants with the modified glyoxylate cycle and their ability to produce citric and isocitric acid. I. Physiological, biochemical and cytological characteristics of mutants grown on glucose or hexadecane. *Appl. Microbiol. Biotechnol.* **1986**, *23*, 372–377. [CrossRef]

26.   Da Silva, L.V.; Tavares, C.B.; Amaral, P.F.F.; Coehlo, M.A.Z. Production of citric acid by *Yarrowia lipolytica* in different crude oil concentrations and in different nitrogen sources. *Chem. Eng. Trans.* **2012**, *27*, 199–204.

27.   Rzechonek, D.A.; Dobrowolski, A.; Rymowicz, W.; Mirończuk, A.M. Aseptic production of citric and isocitric acid from crude glycerol by genetically modified *Yarrowia lipolytica*. *Bioresour. Technol.* **2019**, *271*, 340–344. [CrossRef] [PubMed]

28.   Barth, G.; Gaillardin, C. Physiology and genetics of the dimorphic fungus *Yarrowia lipolytica*. *FEMS Microbiol. Rev.* **1997**, *19*, 219–237. [CrossRef]

29.   Rywińska, A.; Wojtatowicz, M.; Rymowicz, W. Citric acid biosynthesis by *Yarrowia lipolytica* A-101-1.31 under deficiency of various medium macrocomponents. *Electron. J. Pol. Agric. Univ.* **2006**, *9*, 15.

30.   Rywińska, A.; Juszczyk, P.; Wojtatowicz, M.; Rymowicz, W. Chemostat study of citric acid production from glycerol by *Yarrowia Lipolytica*. *J. Biotechnol.* **2011**, *152*, 54–57. [CrossRef]

31.   Kamzolova, S.V.; Morgunov, I.G. Metabolic peculiarities of the citric acid overproduction from glucose in yeasts *Yarrowia lipolytica*. *Bioresour. Technol.* **2017**, *243*, 433–440. [CrossRef]

32. Willke, T.; Vorlop, K.D. Biotechnological production of itaconic acid. *Appl. Microbiol. Biotechnol.* **2001**, *56*, 289–295. [CrossRef]
33. Karaffa, L.; Kubicek, C.P. Citric acid and itaconic acid accumulation: Variations of the same story? *Appl. Microbiol. Biotechnol.* **2019**, *103*, 2889–2902. [CrossRef] [PubMed]
34. Ramachandran, S.; Fontanille, P.; Pandey, A.; Larroche, C. Gluconic acid: Properties, applications and microbial production. *Food Technol. Biotech.* **2006**, *44*, 185–195.
35. Moeller, L.; Strehlitz, B.; Aurich, A.; Zehnsdorf, A.; Bley, T. Optimization of citric acid production from glucose by *Yarrowia Lipolytica*. *Eng. Life Sci.* **2007**, *7*, 504–511. [CrossRef]
36. Papanikolaou, S.; Muniglia, L.; Chevalot, I.; Aggelis, G.; Marc, I. *Yarrowia lipolytica* as a potential producer of citric acid from raw glycerol. *J. Appl. Microbiol.* **2002**, *92*, 737–744. [CrossRef] [PubMed]
37. Tomaszewska, L.; Rakicka, M.; Rymowicz, W.; Rywinska, A. A comparative study on glycerol metabolism to erythritol and citric acid in *Yarrowia lipolytica* yeast cells. *FEMS Yeast Res.* **2014**, *14*, 966–976. [CrossRef]
38. Egermeier, M.; Russmayer, H.; Sauer, M.; Marx, H. Metabolic flexibility of *Yarrowia lipolytica* growing on glycerol. *Front. Microbiol.* **2017**, *8*, 49. [CrossRef] [PubMed]
39. Papanikolaou, S.; Beopoulos, A.; Koletti, A.; Thevenieau, F.; Koutinas, A.A.; Nicaud, J.M.; Aggelis, G. Importance of the methyl-citrate cycle on glycerol metabolism in the yeast *Yarrowia lipolytica*. *J. Biotechnol.* **2013**, *168*, 303–314. [CrossRef]

Article

Physico-Chemical Characteristics and Amino Acid Content Evaluation of Citric Acid by-Product Produced by Microbial Fermentation as a Potential Use in Animal Feed

Mutyarsih Oryza.S [1], Sawitree Wongtangtintharn [1,*], Bundit Tengjaroenkul [2], Anusorn Cherdthong [1], Sirisak Tanpong [1], Pasakorn Bunchalee [3], Padsakorn Pootthachaya [1], Alissara Reungsang and Sineenart Polyorach [5]

[1] Department of Animal Science, Faculty of Agriculture, Khon Kaen University, Khon Kaen 40002, Thailand; mutyarsih.or@kkumail.com (M.O.); anusornc@kku.ac.th (A.C.); sirisak1985@hotmail.com (S.T.); padsakornp@kkumail.com (P.P.)

[2] Department of Veterinary Public Health, Faculty of Veterinary Medicine, Khon Kaen University, Khon Kaen 40002, Thailand; btengjar@kku.ac.th

[3] Department of Biology, Faculty of Science, Mahasarakham University, Mahasarakham 44150, Thailand; pasakorn.b@msu.ac.th

[4] Department of Biotechnology, Faculty of Technology, Khon Kaen University, Khon Kaen 40002, Thailand; alissara@kku.ac.th

[5] Department of Animal Production, Faculty of Agricultural Technology, King Mongkut's Institute of Technology, Ladkrabang, Bangkok 10520, Thailand; neenart324@hotmail.com

[*] Correspondence: sawiwo@kku.ac.th

Citation: Oryza.S, M.; Wongtangtintharn, S.; Tengjaroenkul, B.; Cherdthong, A.; Tanpong, S.; Bunchalee, P.; Pootthachaya, P.; Reungsang, A.; Polyorach, S. Physico-Chemical Characteristics and Amino Acid Content Evaluation of Citric Acid by-Product Produced by Microbial Fermentation as a Potential Use in Animal Feed. *Fermentation* **2021**, *7*, 149. https://doi.org/10.3390/fermentation7030149

Academic Editor: Ronnie G. Willaert

Received: 22 July 2021
Accepted: 9 August 2021
Published: 11 August 2021

Publisher's Note: MDPI stays neutral with regard to jurisdictional claims in published maps and institutional affiliations.

Copyright: © 2021 by the authors. Licensee MDPI, Basel, Switzerland. This article is an open access article distributed under the terms and conditions of the Creative Commons Attribution (CC BY) license (https://creativecommons.org/licenses/by/4.0/).

Abstract: The production of citric acid produces 70% waste product or by-product. This by-product is produced by microbial fermentation which could be used as an alternative raw material for animal feed because it still contains citric acid, which could help to reduce pathogenic bacteria. The objective of this study is to evaluate the physical and chemical value of citric acid by-product from rice (CABR) to compare the properties with those of rice bran and broken rice and to determine its potential as an alternative energy source in animal feed. The chemical composition of CABR was calculated using proximate analysis. The color of CABR was darker, and the bulk density value was 549.65 (g/L) ($p < 0.05$). With free flow, the angle of repose was 40°, and the particle size had less polygonal starch granules. CABR had a low pH of 4.77 and contained 19.80% crude protein, 11.97% crude fiber, and 4005.72 kcal/kg of energy. CABR had a higher crude protein value than broken rice and rice bran and a higher gross energy value than broken rice but less than rice bran. It also had a higher crude fiber value ($p > 0.05$). The results suggest that CABR could be utilized as an energy and protein source for animal feed formulations.

Keywords: cereal crops; organic acid; nutritive value; alternative feedstuff

1. Introduction

Citric acid is a source of organic acid and is produced by microbe fermentation. It has wide uses, but 75% of it is used in the beverage and food industries as an ingredient in carbonated drinks, followed by pharmaceuticals, cosmetics, and animal feed [1]. Globally, 1.7 million tons of citric acid are produced per year, and the amount is predicted to increase annually [2]. Corn and cassava are the main raw materials for citric acid production, and nowadays, there is a chance of producing citric acid from plants [3]. In particular, rice is also commonly used for citric acid production in Thailand. Rice is one of the most important cereals and a primary food for the majority of the world population, especially in Asian countries. Global rice production has increased by 2.5 percent per year on average over the last decade, reaching 744.4 million tons in 2014. The citric acid industry generates a lot of waste and by-product, which can lead to pollution and environmental issues if not effectively managed. Therefore, there is a need to develop economically and environmentally friendly methods for citric acid production. Converting the by-products as feed is the way to increase their value and decrease the environtmental problem [4]. Tanpong et al. [5] reported that the by-product of citric acid production contains cellulose, sugar, starch, and protein. Citric acid by-products from cassava contain 3.588 kcal/kg of energy and 6.11% crude protein and could be utilized as animal feed.

Feed is an essential factor in animal stock sectors. Feed is the most significant expenditure in the livestock industry and represents around 70% of the total production cost. Feed containing formulations with functional components are needed to improve livestock productivity, minimize mortality, and improve the feed conversion ratio [6]. As a general rule, feedstuff's physical and chemical properties are very influential in selecting ingredients in feed formulation.

The physical characteristics of alternative feeds are essential for planning feed rations. They affect the planning and design of feed storage on farms [7]. The physical properties include the shape, particle size, bulk density, angle of repose, color, and pH related to processing and handling in feed production [5]. By-products have advantages that are directly related to their low price as a feed additive, which can decrease animal feed costs when used as a feed replacement. Thus, the objective of this study was to evaluate the physical and chemical properties of citric acid by-product from rice (CABR) produced by microbial fermentation and its potential as an alternative energy source in animal feed.

2. Materials and Methods

2.1. Sample Collection

Samples were provided by PS Nutrition Company Limited, Sai Mai, Bangkok, Thailand. Citric acid production was carried out using rice extract media and inoculated with *Aspergillus niger*. Then, the waste products from the citric acid production from rice (CABR), broken rice (BR), and rice bran (RB) were used as samples. The total weight of each sample was 50 kg, which was collected by random sampling using a tapered bag trier. The samples were carefully handled whilst maintaining their original state for analysis in the Laboratory Department of Animal Science, Faculty of Agriculture, Khon Kaen University, Khon Kaen, Thailand. The methods of Association of American Feed Control Official (AAFCO) were followed [8].

2.2. Physical Characteristic Measurement

The physical characteristics of the samples were observed, such as the color, bulk density, angle of repose, and particle distribution. The procedures reported by Tanpong et al. [5] were used to measure the physical properties of each feedstuff. The particle size and distribution were calculated as follows:

$$\text{Retain (\%)} = (\text{the total sample weight in the sieve/total weight of sample}) \times 100$$

$$\text{Passing (\%)} \ 100 - \text{retain (\%)}$$

$$D_{gw=log^{-1}} \left[\sum_{i=1}^{n} \left(w_i log d_i \right) \div \sum_{i=1}^{n} w_i \right]$$

where W_i is the mass in each sieve (g) and d_i is the sieve size (mm), which is calculated as $(d \times)$. The geometric mean diameters or median size of particle (D_{gw}) followed the method from [9].

2.3. Microscopy Compound

According to the method reported by Vasconcelos et al. [9], the structure of morphological starch granules and plant cell walls of the samples was described and observed under a compound microscope (JNOEC, XS-212 201, Beijing, China) at 40× magnification and a stereo microscope (NIKON SMZ-1, Tokyo, Japan) at 20–60× magnification.

Scanning Electron Microscope (SEM) analysis was conducted using an SEM electron microscope (JEOL-JSM 6460 LV, Tokyo, Japan) at $50\times$, $500\times$, and $1000\times$ magnifications at an accelerating voltage of 20 kV.

2.4. Chemical Composition

Proximate analysis was performed using the methods of the AOAC [10]. We analyzed the moisture, ash, soluble ash, insoluble ash, crude protein, crude fiber, crude fat, and nitrogen-free extract. The gross energy (GE) was analyzed via an automatic Adiabatic bomb calorimeter (AC500 Isoperibol Clorimeter, LECO Corp., St. Joseph, MI, USA) following the method of the Leco company. The pH was measured with a pH meter after mixing 10 g of the sample in a beaker, adding 100 mL of distilled water, and stirring for 30 min.

2.5. Citric Acid Measurement

Following the method of Ezea et al. [11], the citric acid content in CABR was measured. The samples were treated by titration with 0.1 NaOH and phenolphthalein as an indicator. The citric acid content (%) was calculated with the following equation:

$$Citricacid(\%) = \frac{N \times W1 XTV \times DF}{W2 \times 10}$$

where N is the normality, $W1$ is the equivalent weight of citric acid, TV is the titrated value, DF is the dilute factor, and $W2$ is the weight of the sample.

2.6. Amino Acid Determination

The extraction of amino acids from CABR was performed according to Nimbalkar et al. [12]. Amino acid contents were measured according to the method of Thiele et al. [13] and Chumroenphat et al. [14] using the Liquid Chromatography with tandem mass spectrometry (LC-MS/MS) system. LC-MS/MS analysis was performed on a triple quadrupole tendem mass spectrometer (Shimadzu Corp., Kyoto, Japan) coupled with a 1290 Infinity LC system (Agilent Technologies, Santa Clara, CA, USA). The chromatographic separation of amino acids was carried out on an Atlantis Silica HILIC column (4.6 mm $\times$ 100 mm, 3 μm particle size) (Waters Corporation, Midford, MA, USA). Mobile phases were (A) 5% acetic acid in water and (B) 10% methanol in acetonitrile. The LC gradient was t(min)/B (%); 0/5, 25/50, 27/98, 29/98, 29.1/5, and 40/5 operated at a flow rate of 0.25 mL/min. The injection volume was 2 μL. The column was coupled with a mass spectrometer for quantification. The mass spectrometer was performed in multiple reaction monitoring mode (MRM) with argon. Amino acids were counted using internal standard calibration curves and external standard calibration curves. All data were demonstrated on a fresh weight (fw) basis as g/kg.

2.7. Aflatoxins and Fumonisin Measurement

The aflatoxins in the samples were detected and quantified with the in-house system of Central Laboratory (Thailand) Co., Ltd. (TE-CH-025) based on AOAC 991.31 and 994.08 [15]. Samples were blind-coded and processed at 2–8 °C. Before analysis, 50 g of ground sample was put in a clean disposable extraction bottle containing 250 mL of 70% methanol, and the bottle was shaken for 3 min to extract the sample. One minute was then allowed for the solids to fall to the bottom of the bottle, and they were filtered with filter paper.

The amount of fumonisin B1 and B2 were calculated using an in-house process using LC-MS/MS, which combines high-pressure liquid chromatography (HPLC) separation with mass spectrometry detection power. High-pressure liquid chromatography HPLC with tandem mass spectrometry (MS/MS) was used to make the determination. Samples were immediately transferred to sealed bags to prevent moisture changes and stored. The sample powders were dissolved in methanol and acetonitrile to prepare stock standard solutions. Appropriate amounts of sample standard solutions and aflatoxin standard solutions were combined and diluted to a volume with methanol to prepare mixed standard solutions of mycotoxins.

All solutions were stored in the dark at $-20\,^{\circ}C$ and prepared for sample pretreatment. The final sample concentration in the extract of the sample pretreatment was injected for LC-MS/MS analysis. The sensitivity of the method was estimated by the limit of detection (LOD). The LOD was determined as the lowest concentration giving a response of three times the average of the base-line noise obtained from non-contaminated aflatoxin peanut samples that had been spiked with a mixed standard stock solution containing the four investigated aflatoxins [16].

2.8. Statistical Analysis

The data were analyzed by using the procedure of the Statistical Analysis System Institute (SAS, 2015). All data were subjected to analysis of variance (ANOVA) with a completely randomized design (CRD). Differences among means with $p < 0.05$ were accepted as representing statistically significant differences, which were determined by Duncan's New Multiple Range Test (DMRT).

3. Results

3.1. Physical Characteristics

The physical characteristics of CABR are shown in Table 1, respectively compared with broken rice and rice bran. The bulk density of CABR was 549.65 g/L, which is 57.94% lower than that of broken rice (868.12 g/L) and 21.13% higher than that of rice bran (453.78 g/L). The results show that the angle of repose for CABR was 40.6°, which can be classified as fair to passable flow, whereas broken rice showed an angle of repose of approximately 39.45°, which could be classified as a fair to passable flow, and rice bran's angle of repose value was 50.6°, which classified as very poor (46–50°). Broken rice has bulky and more massive particles (99.92 g) in mesh 20 when compared with CABR (19.92 g) and rice bran (31.21 g). The color space value was analyzed by the CIELAB system and is shown in Table 1. The results show L* = 45.02, a* = 5.64, and b* = 13.88. for CABR L* of CABR which was lower than that of broken rice (78.37) and rice bran (76.90). The a* value of CABR was higher than that of broken rice (1.66) and rice bran (1.20), and CABR's b* value decreased from rice bran (18.89) and broken rice (17.92). The particle size and distribution of CABR are shown in Table 2. Compared with broken rice and rice bran, most CABR particle sizes were increased after processing. When comparing the passing percentage using sieve numbers 20 to 100, the result for CABR was higher than that of broken rice. The geometric mean diameter (the median particle size) was 232 μm, as a small particle size of the sample.

Table 1. The physical characteristics of citric acid by-product from rice (CABR), broken rice (BR), and rice bran (RB).

Parameter	CABR	BR	RB	SEM	*p*-Value
Bulk density (g/L)	549.65 [b]	868.12 [a]	453.78 [c]	5.867	0.0001
Angle of repose (°)	40.6 [b]	39.45 [a]	50.6 [c]	0.380	0.0001
Color					
L*	45.02 [b]	78.37 [a]	76.90 [c]	0.397	0.0001
A*	5.64 [a]	1.66 [b]	1.20 [c]	0.080	0.0001
B*	13.88 [c]	18.89 [a]	17.92 [b]	0.147	0.0001

[a,b,c] Means in the same row without a common letter are different at $p < 0.01$, SEM: standard error of mean.

Table 2. Particle size and distribution analyses of citric acid by-product from rice (CABR), broken rice (BR), and rice bran (RB).

Sieve Mesh no.	Sieve Size (μm)	Sample (g)			Retain (%)			SEM	Cumulative (%)			Passing (%)			D_{gw} [1] (μm)			SEM
		CABR	BR	RB	CABR	BR	RB		CABR	BR	RB	CABR	BR	RB	CABR	BR	RB	
20	850	19.92	99.92	31.21	20.09 c	99.78 a	31.09 b	1.56	20.09	99.78	31.09	79.91	0.22	68.91	232 c	600 a	338 b	5.915
40	425	22.82	0.19	40.3	23.01 b	0.20 c	40.15 a	0.87	43.01	99.98	71.25	56.9	0.02	28.75				
60	250	19.35	0.00	19.37	19.51 b	0.01 b	19.3	0.82	62.61	99.98	90.54	37.39	0.02	9.46				
80	180	14.26	0.00	7.58	14.38 a	0.00 c	7.55 b	0.92	76.98	99.99	98.1	23.01	0.01	1.9				
100	150	8.91	0.00	01.20	7.77 a	0.00 c	1.20 b	0.09	84.75	99.99	99.3	15.25	0.01	0.7				
Pan		15.12	0.01	0.70	15.25 a	0.01 c	0.70 b	0.20	100.00	100.00	100.00	0	0	0				
Total		100.42	100.14	100.37	100	100	100											

[1] Geometric mean diameter in μm by mass of sample, [a,b,c] means in the same row without a common letter are different at $p < 0.01$, SEM: standard error of mean.

3.2. Microscopy Compound

A stereo microscope (Figure 1), compound microscope (Figure 2), and scanning electron microscope (SEM) (Figure 3) were used to show differences in particle size and content of fiber as a starch between CABR, broken rice, and rice bran. CABR exhibited a darker color under the stereo microscope when compared to rice bran and broken rice, which supported the results of the physical analysis (Table 1) and particle size and distribution (Table 2). The ultrastructure morphology of feedstuff is characterized using SEM micrographs at $50\times$, $500\times$, and $1000\times$ magnification. The result showed CABR starch granules are polygonal in shape.

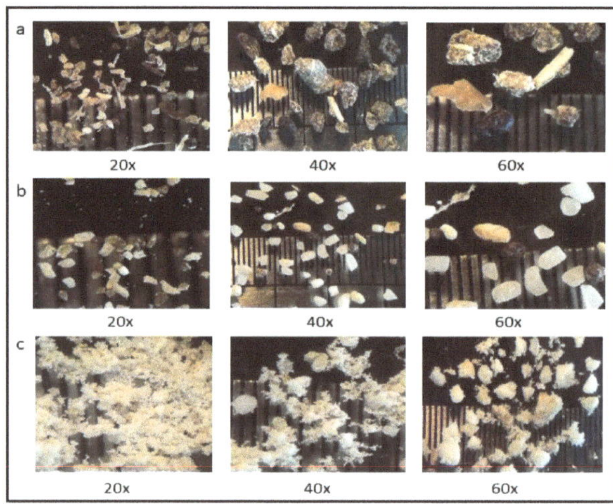

Figure 1. Stereoscopic micrographs of citric acid by-product from rice (CABR) (**a**), rice bran (**b**), and broken rice (**c**) at $\times 20$, $\times 40$, and $\times 60$ magnifications.

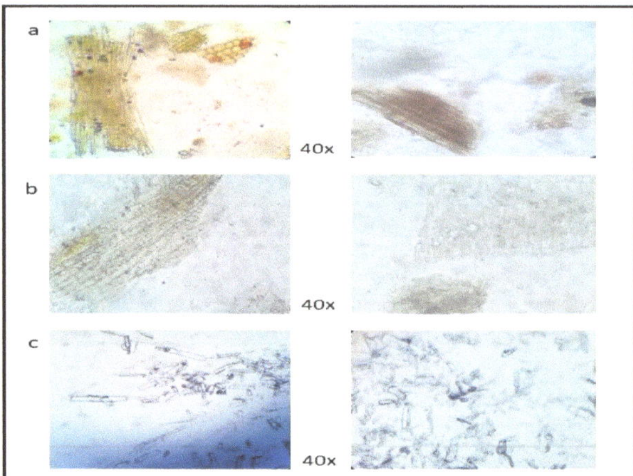

Figure 2. Compound micrographs of citric acid by-product from rice (CABR) (**a**), rice bran (**b**), and broken rice (**c**) at $\times 40$ magnifications.

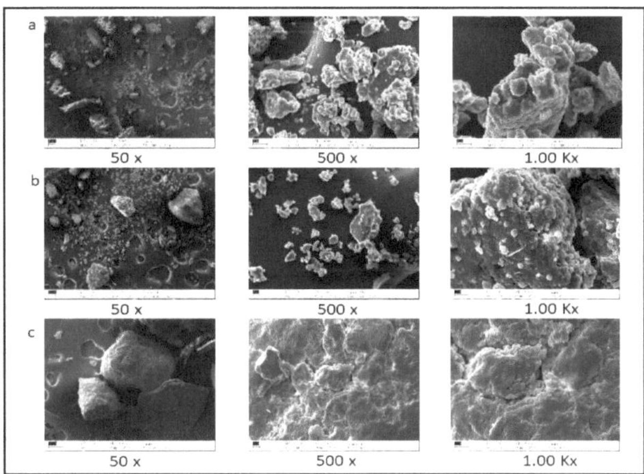

Figure 3. Scanning electron micrographs (SEM) of citric acid by-product from rice (CABR) (**a**), rice bran (**b**), and broken rice (**c**) at ×50, ×500, and ×1000 magnifications.

3.3. Chemical Composition

The chemical properties of CABR were determined by proximate analysis and are shown in Table 3. The results revealed that CABR contained 8.26% moisture, 9.35% ash, 5.20% soluble ash, 4.15 % insoluble ash, 3.98% ether extract, 0.43% of calcium, 0.07% of phosphorus, 19.80% of crude protein, and 4005.72 kcal/kg of gross energy. CABR contains crude protein higher than broken rice and rice bran (Table 3). Moreover, CABR contains low pH and contains citric acid 3.3%.

Table 3. Nutritive values and chemical composition of citric acid by-product from rice (CABR), broken rice (BR), and rice bran (RB).

Parameter	CABR	BR	RB	SEM	*p*-Value
Moisture (%)	8.26 [b]	9.35 [a]	8.43 [b]	0.114	0.0012
Ash (%)	9.35 [a]	0.49 [c]	0.49 [b]	0.154	0.0001
Soluble ash (%)	5.20 [b]	0.49 [c]	6.88 [a]	0.148	0.0002
Insoluble ash (%)	4.15 [a]	0.00 [c]	0.39 [b]	0.013	0.0001
Ether extract (%)	3.98 [b]	1.41 [c]	14.28 [a]	0.161	0.0001
Crude fiber (%)	11.97 [a]	0.10 [b]	0.76 [b]	0.297	0.0002
Nitrogen-free extract (%)	46.64 [c]	82.20 [a]	55.89 [b]	0.428	0.0001
Ca (%)	0.43 [b]	0.76 [a]	0.76 [a]	0.050	0.0001
Phosphorus (%)	0.07 [a]	0.04 [b]	0.01 [c]	0.003	0.0001
Crude protein (%)	19.80 [a]	6.47 [c]	13.37 [b]	0.159	0.0001
Gross energy (kcal/kg)	4,005.72 [b]	3,780.52 [c]	4,287.11 [a]	7.430	0.0001
pH	4.77 [b]	6.46 [a]	6.51 [a]	0.053	0.0001
Citric acid content (%)	3.3	-	-		

[a,b,c] Means in the same row without a common letter are different at $p < 0.05$, SEM: standard error of mean.

3.4. Amino Acid Composition

The amino acid composition of CABR, broken rice, and rice bran are shown in Table 4. The table shows the dispensable and indispensable amino acid content of each sample. CABR contained the highest value of aspartic acid, glutamic acid, threonine, alanine, and valine compared with broken rice and rice bran. CABR also contains a low value of methionine and lysine.

Table 4. Amino acid composition of citric acid by-product from rice (CABR), broken rice (BR), and rice bran (RB).

Amino Acid (g/kg)	CABR	BR	RB	SEM	*p*-Value
	Essential amino acid				
Leucine	3.72 [b]	2.72 [c]	4.42 [a]	17.23	0.0051
Valine	2.70 [a]	1.80 [b]	1.53 [c]	7.59	0.0012
Isoleucine	2.49 [b]	1.71 [c]	3.48 [a]	11.34	0.0013
Phenylamine	1.53 [b]	1.20 [b]	3.63 [a]	33.20	0.0098
Threonine	1.44 [a]	0.87 [b]	0.82 [b]	4.24	0.0012
Histidine	0.92 [b]	4.03 [a]	0.93 [b]	36.47	0.0052
Tryptophan	0.39 [c]	1.13 [a]	0.61 [b]	18.13	0.0565
Lysine	0.31 [c]	0.87 [b]	1.88 [a]	7.24	0.0005
Methionine	0.05 [c]	0.96 [a]	0.37 [b]	50.41	0.3216
	Non-essential amino acid				
Tyrosine	5.50 [c]	6.81 [b]	12.99 [a]	30.37	0.0003
Glycine	0.77 [a]	0.77	0.28 [b]	4.83	0.0003
Proline	2.96 [b]	1.49 [c]	7.58 [a]	24.04	0.0003
Alanine	9.07 [a]	4.26 [b]	2.47 [c]	12.72	0.0001
Cysteine	0.05 [b]	0.06 [a]	0.07 [a]	0.51	0.1828
Arginine	3.77 [b]	3.05 [b]	5.82 [a]	53.64	0.0293
Aspartic acid	0.81 [a]	0.38 [b]	0.13 [c]	4.28	0.0013
Glutamic acid	1.99 [a]	0.67 [c]	0.99 [b]	7.37	0.0008
Serine	0.07 [c]	0.09 [b]	0.11 [a]	1.08	0.0738
Asparagine	0.32 [b]	5.07 [a]	0.66 [b]	33.23	0.0013
Glutamine	0.31 [c]	0.90 [b]	1.97 [a]	6.90	0.0004

[a,b,c] Means in the same row without a common letter are different at *p* < 0.05, SEM: standard error of mean.

3.5. Mycotoxins Contamination

The mycotoxin contamination of CABR is shown in Table 5. The CABR was not contaminated with aflatoxin (B1, B2, G1 and G2) and fumonisin (B1 and B2) with LOD of 0.8 and 100.000 µg/kg for aflatoxin and fumonisin, respectively, which could be safe for animal consumption.

Table 5. Observation and LOD (limit of detection) of mycotoxin contamination in citric acid by-products from rice (CABR).

Parameter	Observation	LOD
Aflatoxin B1	ND	0.8 µg/kg
Aflatoxin B2	ND	0.8 µg/kg
Aflatoxin G1	ND	0.8 µg/kg
Aflatoxin G2	ND	0.8 µg/kg
Fumonisin B1	ND	100.00 µg/kg
Fumonisin B2	ND	100.00 µg/kg

ND (not detected).

4. Discussion

The bulk density of CABR was lower than that of broken rice and higher than that of rice bran. Tanpong et al. [5] found that the bulk density of citric acid by-product from cassava was 601.00 g/L and was higher than that of cassava root meal (64.18%). Bulk density varies with the particle size and compaction (packing) of the feed. Increasing the value of the bulk density was influenced by the moisture of raw material. The earlier study reported that the cassava chips' bulk density was affected by the moisture content [17]. The differences among the sizes of each sample affected the bulk density value. Broken rice has a massive particle size and contains the highest bulk density followed by CABR and rice bran. The larger the particle size of raw materials, the greater the bulk density [5]. The bulk physical property of an alternative feed is essential to plan designing, transporting, and storing the feed. Different processes during harvest and manufacturing of a product impact the end products' physical properties or by-products used as animal feed [18].

The angle of repose of CABR is shown in Table 1. The result shows that the angle of repose of CABR classified as fair to passable flow. Baker [19] reported that the ability of powders to flow is referred to as flowability. Powder flowability is influenced by both the physical properties of the material and the specific processing conditions in the handling system. The particle size, density, surface features of materials, and the water and fat content of feed are all elements that influence the angle of repose. For feedstuffs with large particle size, the angle of repose will be small. Moreover, the average angle of repose depends on the moisture of the material. The raw material characteristics indicate flow behavior and affect the feed mixture. Fitzpatrick et al. [20] reported that feed powder properties affect the behavior during storage, handling, and processing. Therefore, it is helpful to predict storage capacity, including friction against a corroded bin wall, and to indicate the moisture content of raw material [21].

The color space value was analyzed by the CIELAB system and is shown in Table 1. The a* value of CABR was higher than that of broken rice and rice bran, and CABR's b* value decreased compared to rice bran and broken rice. Several researchers report that the fermentation process includes a browning reaction, which explains why the by-product becomes dark during fermentation, while lightness is related to rice's structure [5,22–24].

CABR had a small particle size that could pass through each sieve number better than broken rice. Comparative retention percentages retained from broken rice and rice bran with CABR in Table 1 indicate that most of the particle sizes of CABR increased after processing. The geometric mean diameter (the median particle size) of CABR shows that CABR has as a small particle size. According to Vu et al. [25], a smaller particle of the material dissolves faster than a larger one. The particle size distribution influences many material properties and indicates quality. The particle size and shape influence flow and compaction properties. A smaller particle can have lower nutrient digestibility, and larger particles can reduce animal feed intake [26]. More spherical particles enable a faster typical flow than smaller or high aspect-ratio particles. Smaller particles dissolve more quickly and lead to higher suspension viscosities than larger particles. The particle size of animal feed impacts its utilization and production. Larger particles can reduce animal feed intake by limiting surface area per unit and allowing enzyme digestion of nutrients. Smaller particles make it more difficult to separate ingredients, but they also increase viscosity in the digestive tract. Kiarie et al. [27] reveal that finer feed particles provide optimal usage of nutrients and improve animal performance due to a higher surface area that allows greater contact with digestive enzymes.

CABR exhibited a darker color under the stereo microscope when compared to rice bran and broken rice, which supported the results of the physical analysis (Table 1) and particle size and distribution (Table 2) due to the citric acid production fermentation process. However, the starch granules and cell walls of the samples were shown under the compound microscope. CABR contained a few residual starch granules, most likely due to the acidity of the citric acid, which may weaken the interaction between starch polymer chains. Mohammed [28] mentioned that after modifying the raw materials with citric acid, the starch granules were melted and fused to form a continuous phase, most likely because the acidity of the citric acid may weaken the interaction between the starch polymer chains of the starch by-products. This microscopic technology analysis could become the primary front-line protection in the program of feed quality control.

The chemical properties of CABR were shown in Table 3. CABR contains crude protein higher than rice bran at approximately 67.53% (Table 3). Tanpong et al. [29] reported that citric acid from cassava had a crude protein content of 6.11%, and when compared with the current study, the crude protein percentages of CABR were higher at 30.86% (Table 3). CABR also contained a remaining citric acid content of 3.3%, which could be beneficial in terms of microorganisms and improving animal growth. However, CABR showed the highest proportions of crude fiber, which can be a problem when fed to monogastric animals, such as poultry species, where only a small percentage of crude fiber is broken down in the gastrointestinal tract.

Rice normally has a high level of amino acids and protein when compared to other cereal varieties [30]. In the current study, CABR contained the highest value of valine, glutamic acid, alanine, threonine, and aspartic acid compared with broken rice and rice bran. CABR also includes a low value of methionine and lysine; therefore, this should also be considered when using CABR in animal diets. In protein nutrition, the balance of amino acids in the diet is essential. Amino acid antagonism occurs when chemically or structurally-related amino acids are imbalanced. Imbalanced amino acids might affect the performance of animals [31]. This report of complete amino acid profile in citric acid by-product from rice could be useful for formulating balanced animals diets.

The CABR is not contaminated with aflatoxin and fumonisin, which means it can be safe for animal consumption. A previous study in citric acid by-products from cassava showed that it contains 73.37 ppb of aflatoxin [29]. However, the aflatoxin in CABR is not detected in the current study, making this raw material safer to use as an animal feed. The aflatoxin contamination in feed could cause a reduction in immune response that may cause several diseases. Cereal grains, primarily corn, are widely used as an energy source in animal feed for different species. They were originating from tropical and subtropical regions that contain high amounts of aflatoxins. Contamination of various feeds containing mycotoxin continues to be a safety issue worldwide because it harms animal health [32]. Thus, mycotoxin contamination in feedstuff should be considered to ensure a safety product for animal and humans.

5. Conclusions

CABR's physical properties consist of small particle size, dark color, and having fewer starch granules. CABR contains 19.80% of crude protein, 11.97% of crude fiber, and 4,005.72 Kcal/kg of energy. The results imply that CABR could be utilized as an energy source for animal feed substitution. CABR has low pH and contains 3.3% citric acid, which could help inhibit bacteria in the GI tracts of animals and improve their growth performance. CABR is also not contaminated with aflatoxin or fumonisin, which means it could be safe feed for animals. The physical and chemical data from this research could be used as guidelines for handling raw materials before use in a feed formulation.

Author Contributions: Conceptualization, M.O., S.W., B.T., A.C., S.T., P.B., P.P., A.R. and S.P.; methodology, M.O., S.T., P.P., S.W. and B.T.; investigation, S.W. and B.T.; writing—original/draft preparation, M.O. and S.W.; supervision, M.O., S.W., B.T. and A.C.; formal analysis, M.O., S.W., B.T., S.T. and P.P.; visualization, M.O. and S.W.; funding acquisition, S.W., B.T. and A.C. All authors have read and agreed to the published version of the manuscript.

Funding: This research was funded by the Program Management Unit for Human & Institutional Development, Research, and Innovation (Grant no. B05F630053), and Thailand Research Fund (TRF) Senior Research Scholar (Grant No. RTA6280001).

Institutional Review Board Statement: Not applicable.

Informed Consent Statement: Not applicable.

Data Availability Statement: The data presented in this study are available on request from the corresponding author.

Acknowledgments: The authors would like to express their genuine gratitude to the Research and Graduate Studies Division, KKU, and PS Nutrition Company Limited for providing the offices of this exploration.

Conflicts of Interest: The authors declare no conflict of interest.

References

1. Show, P.L.; Oladele, K.O.; Siew, Q.Y.; Aziz Zakry, F.A.; Lan, J.C.W.; Ling, T.C. Overview of citric acid production from *Aspergillus niger*. *Front. Life Sci.* **2015**, *8*, 271–283. [CrossRef]
2. Dhillon, G.; Brar, S.; Dhillon, S.; Verma, M. Bioproduction and extraction optimization of citric acid from *Aspergillus niger* by rotating drum type solid-state bioreactor. *Ind. Crops Prod.* **2013**, *41*, 78–84. [CrossRef]
3. Alsudani, A.A.; Al-Shibli, M.K. Citric acid production from some local isolates of the fungus *Aspergillus niger* by rice husks filtrate medium. *Int. J. Recent Sci. Res.* **2015**, *6*, 5625–5633.
4. Khalil, M.; Berawi, M.A.; Heryanto, R.; Rizalie, A. Waste to energy technology: The potential of sustainable biogas production from animal waste in Indonesia. *Renew. Sustain. Energy Rev.* **2019**, *105*, 323–331. [CrossRef]
5. Tanpong, S.; Cherdthong, A.; Tengjaroenkul, B.; Tengjaroenkul, U.; Wongtangtintharn, S. Evaluation of physical and chemical properties of citric acid industrial waste. *Trop. Anim. Health Prod.* **2019**, *105*, 323–331. [CrossRef] [PubMed]
6. Makkar, H. Animal nutrition in a 360-degree view and a framework for future R&D work: Towards sustainable livestock production. *Anim. Prod. Sci.* **2016**, *56*, 1561–1568. [CrossRef]
7. Syamsu, J.A.; Yusuf, M.; Abdullah, A. Evaluation of physical properties of feedstuffs in supporting the development of feed mill at farmers group scale. *J. Adv. Agric. Technol.* **2015**, *2*, 147–150. [CrossRef]
8. AAFCO. *Association of American Feed Control Officials*; Association of American Feed Control Officials Inspection and Sampling Committee: Champaign, IL, USA, 2014.
9. Vasconcelos, L.M.; Brito, A.C.; Carmo, C.D.; Oliveira, P.H.G.A.; Oliveira, E.J. Phenotypic diversity of starch granules in cassava germplasm. *Genet. Mol. Res.* **2017**, *16*, 1–15. [CrossRef]
10. AOAC. *Association of Official Analytical Chemists; Official Methods of Analysis*, 15th ed.; Association of Analytical Chemists: Arlington, TX, USA, 1990.
11. Ezea, I.; Chiejina, N.V.; Ogbonna, J.C. Biological production of citric acid in solid state cultures of *Aspergillus niger*. *ChemXpress* **2016**, *8*, 201–207.
12. Nimbalkar, M.S.; Pai, S.R.; Pawar, N.V.; Oulkar, D.; Dixit, G.B. Free amino acid profiling in grain Amaranth using LC–MS/MS. *Food Chem.* **2012**, *134*, 2565–2569. [CrossRef]
13. Thiele, B.; Hupert, M.; Santiago-Schübel, B.; Oldiges, M.; Hofmann, D. Direct analysis of underivatized amino acids in plant extracts by LC-MS/MS (Improved Method). *Methods Mol. Biol.* **2019**, *2030*, 403–414. [CrossRef] [PubMed]
14. Chumroenphat, T.; Somboonwatthanakul, I.; Saensouk, S. Changes in curcuminoids and chemical components of turmeric (Curcuma longa L) under freeze-drying and low-temperature drying methods. *Food Chem.* **2021**, *339*, 128121. [CrossRef] [PubMed]
15. Weaver, C.; Trucksess, M. Determination of Aflatoxins in Botanical Roots by a Modification of AOAC Official MethodSM 991.31: Single-Laboratory Validation. *J. AOAC Int.* **2010**, *93*, 184–189. [CrossRef]
16. Sirhan, A.Y.; Tan, G.H.; Al-Shunnaq, A.; Abdulra'uf, L.; Wong, R.C.S. QuEChERS-HPLC method for aflatoxin detection of domestic and imported food in Jordan. *J. Liq. Chromatogr. Relat. Technol.* **2014**, *37*, 321–342. [CrossRef]
17. Nwabanne, J.T. Drying characteristics and engineering properties of fermented ground cassava. *African J. Biotechnol.* **2009**, *8*, 873–876. [CrossRef]
18. Ndou, S.P.; Gous, R.M.; Chimonyo, M. Prediction of scaled feed intake in weaner pigs using physico-chemical properties of fibrous feeds. *Br. J. Nutr.* **2013**, *110*, 774–780. [CrossRef] [PubMed]
19. Baker, S.; Herrman, T. *MF2051 Evaluating Particle Size*; Kansas State University: Manhattan, Kansas, 1976; pp. 1–6.
20. Fitzpatrick, J.; Iqbal, T.; Delaney, C.; Twomey, T.; Keogh, M.K. Effect of powder properties on the flowability of milk powders with different fat contents. *J. Food Eng.* **2004**, *64*, 435–444. [CrossRef]
21. Amidon, G.; Meyer, P.J.; Mudie, D. Particle, Powder, and Compact Characterization. In *Developing Solid Oral Dosage Forms: Pharmaceutical Theory and Practice*, 2nd ed.; Elsevier: Amsterdam, The Netherlands, 2017; pp. 271–293. ISBN 9780128024478.
22. Papagianni, M.; Boonpooh, Y.; Mattey, M.; Kristiansen, B. Substrate inhibition kinetics of Saccharomyces cerevisiae in fed-batch cultures operated at constant glucose and maltose concentration levels. *J. Ind. Microbiol. Biotechnol.* **2007**, *34*, 301–309. [CrossRef]
23. Max, B.; Salgado, J.; Rodríguez, N.; Cortes, S.; Converti, A.; Domínguez, J. Biotechnological production of citric acid. *Braz. J. Microbiol.* **2010**, *41*, 862–875. [CrossRef]
24. Angumeenal, A.R.; Venkappayya, D. An overview of citric acid production. *LWT-Food Sci. Technol.* **2013**, *50*, 367–370. [CrossRef]
25. Vu, T.V.; Delgado-Saborit, J.M.; Harrison, R.M. Review: Particle number size distributions from seven major sources and implications for source apportionment studies. *Atmos. Environ.* **2015**, *122*, 114–132. [CrossRef]
26. Nutrition, C.A. Impacts of feed texture and particle size on broiler and layer feeding patterns. In Proceedings of the XXIV World's Poultry Congress, Bahia, Brazil, 9 August 2012; pp. 1–8.
27. Kiarie, E.G.; Mills, A. Role of Feed Processing on Gut Health and Function in Pigs and Poultry: Conundrum of Optimal Particle Size and Hydrothermal Regimens. *Front. Vet. Sci.* **2019**, *6*, 1–13. [CrossRef] [PubMed]
28. Mohammed, H. 2 Effect of utilization organic acid supplement on broiler (ROS-308). *Int. J. Adv. Res. Biol. Sci.* **2016**, *3*, 76–81.
29. Tanpong, S. Utilization of Citric Acid by Product from Industrial Waste Used as Poultry Feed. Ph.D. Thesis, Khon Kaen University, Khon Kaen, Thailand, 2019.
30. Juliano, B. *Rice in human nutrition*; International Rice Research Institute, Food and Agriculture Organization of the United Nations: Rome, Italy, 1993; pp. 31–34.

31. Wu, G.; Bazer, F.; Dai, Z.; Li, D.; Wang, J.; Wu, Z. Amino acid nutrition in animals: Protein synthesis and beyond. *Annu. Rev. Anim. Biosci.* **2014**, *2*, 387–417. [CrossRef] [PubMed]
32. Dhanasekaran, D.; Panneerselvam, A.; Thajuddin, N. Evaluation of aflatoxicosis in hens fed with commercial poultry feed. *Turkish J. Vet. Anim. Sci.* **2009**, *33*, 385–391. [CrossRef]

 *fermentation*

MDPI

Article

Different Gene Expression Patterns of Hexose Transporter Genes Modulate Fermentation Performance of Four *Saccharomyces cerevisiae* Strains

Chiara Nadai [1,2,†], Giulia Crosato [2,†], Alessio Giacomini [1,2,*] and Viviana Corich [1,2]

1 Interdepartmental Centre for Research in Viticulture and Enology (CIRVE), University of Padova, Via XXVIII Aprile, 14-31015 Conegliano, Italy; chiara.nadai@unipd.it (C.N.); viviana.corich@unipd.it (V.C.)
2 Department of Agronomy Food Natural Resources Animals and Environment (DAFNAE), University of Padova, Viale dell'Università, 16-35020 Legnaro, Italy; giuliacrosato@yahoo.it
* Correspondence: alessio.giacomini@unipd.it; Tel.: +39-049-8272925
† These authors contributed equally to this work.

Abstract: In *Saccharomyces cerevisiae*, the fermentation rate and the ability to complete the sugar transformation process depend on the glucose and fructose transporter set-up. Hexose transport mainly occurs via facilitated diffusion carriers and these are encoded by the *HXT* gene family and *GAL2*. In addition, *FSY1*, coding a fructose/H+ symporter, was identified in some wine strains. This little-known transporter could be relevant in the last part of the fermentation process when fructose is the most abundant sugar. In this work, we investigated the gene expression of the hexose transporters during late fermentation phase, by means of qPCR. Four *S. cerevisiae* strains (P301.9, R31.3, R008, isolated from vineyard, and the commercial EC1118) were considered and the transporter gene expression levels were determined to evaluate how the strain gene expression pattern modulated the late fermentation process. The very low global gene expression and the poor fermentation performance of R008 suggested that the overall expression level is a determinant to obtain the total sugar consumption. Each strain showed a specific gene expression profile that was strongly variable. This led to rethinking the importance of the *HXT3* gene that was previously considered to play a major role in sugar transport. In vineyard strains, other transporter genes, such as *HXT6/7*, *HXT8*, and *FSY1*, showed higher expression levels, and the resulting gene expression patterns properly supported the late fermentation process.

Keywords: *HXT*; fructose symporter; late fermentation

Citation: Nadai, C.; Crosato, G.; Giacomini, A.; Corich, V. Different Gene Expression Patterns of Hexose Transporter Genes Modulate Fermentation Performance of Four *Saccharomyces cerevisiae* Strains. *Fermentation* **2021**, *7*, 164. https://doi.org/10.3390/fermentation7030164

Academic Editor: Ronnie G. Willaert

Received: 5 August 2021
Accepted: 17 August 2021
Published: 23 August 2021

Publisher's Note: MDPI stays neutral with regard to jurisdictional claims in published maps and institutional affiliations.

Copyright: © 2021 by the authors. Licensee MDPI, Basel, Switzerland. This article is an open access article distributed under the terms and conditions of the Creative Commons Attribution (CC BY) license (https://creativecommons.org/licenses/by/4.0/).

1. Introduction

In wine alcoholic fermentation, glucose and fructose present in grape must are co-fermented by yeasts to ethanol and carbon dioxide. Grape must usually contains equal or very similar amounts of both sugars [1]. Glucose is known to be the preferred carbon source for *Saccharomyces cerevisiae*. Although fructose is used concomitantly with glucose, the latter is the first sugar to be depleted from the medium during fermentation [2,3]. Consequently, fructose becomes the main sugar present during the late stages of alcoholic fermentation, and wine yeasts have to ferment this non-preferred sugar in the presence of large amounts of ethanol. The stress associated with these conditions, together with nutritional imbalances, may result in sluggish or stuck fermentations [1,4]. Moreover, it has been reported that stuck fermentations are frequently characterized by an unusually high fructose-to-glucose ratio [2]. The ability of wine yeasts to ferment fructose is therefore critically important for the maintenance of a high rate of fermentation at the end of the process and for fermentation of the must to dryness. The reasons for the difference between the glucose fermentation rate and the fructose fermentation rate are unclear, but one of the first steps in hexose metabolism is generally thought to be involved [1].

The first essential step towards the utilization of hexose sugars is their uptake by yeast cells. In yeast, hexose uptake may proceed through facilitated diffusion carriers and energy-dependent active proton-sugar symporters [3,5,6]. Hexose transport in *S. cerevisiae* occurs via facilitated diffusion carriers and these are encoded by several genes, including the *HXT* genes, the *GAL2* gene encoding a galactose transporter, and *SNF3* and *RGT2* encoding two glucose sensors [5,6]. Among the 17 *HXT* genes in *S. cerevisiae*, only seven of them, Hxt1p–Hxt7p, are required for growth on glucose or fructose [3,7]. Although all the hexose transporters in *S. cerevisiae* can also transport fructose, glucose is the preferential sugar for Hxt carriers [3]. The catabolic hexose transporters exhibit different affinities for their substrates; furthermore, the expression of their corresponding genes is controlled by the glucose sensors, according to the availability of carbon sources [8]. Expression of the *HXTs*, all of which exhibit different levels of glucose affinity, is differentially regulated depending on extracellular glucose concentrations [9].

The complete genome sequence of the commercial wine yeast strain EC1118 [10] and the vineyard strain P301 [11] revealed the presence of a new gene (named EC-1O4_6634g in the former study and P301_O3_0021 in the latter study) highly similar to the *S. pastorianus* *FSY1* gene [10]. This gene was designated *FSY1* as well, and Galeote and colleagues [3] demonstrated that, in *S. cerevisiae*, it encoded a high-affinity fructose/H+ symporter. In the EC1118 genome, *FSY1* is in the C region that resulted from a horizontal gene transfer (HGT) from *Torulaspora microellipsoides*, a distant yeast species identified in the wine environment [12]. This region includes the *FOT* genes that confer a strong competitive advantage during grape must fermentation by increasing the number and diversity of oligopeptides that yeast can utilize as a source of nitrogen, thereby improving biomass formation, fermentation efficiency, and cell viability. Thus, the acquisition of the C region genes is related to better fermentation performance in the nitrogen-limited wine fermentation environment [13].

Moreover, *FSY1* expression is repressed by high concentrations of glucose or fructose and is induced by ethanol as the sole carbon source [3]. This observation leads to suppose that this transporter is active in the last part of the fermentation process, where ethanol concentration is high and fructose residue is more abundant than glucose. Although the presence of *HXT* gene family and *GAL2* gene is well documented in the *Saccharomyces* genus, studies on *FSY1* gene diffusion among *Saccharomyces* strain genomes is lacking, as well as the *FSY1* gene expression level during the fermentation process.

Previously, sugar transporter genes expression studies were mainly focused on a single strain in the fermentation process [14–17]. Therefore, in this work, the gene expression patterns of four *S. cerevisiae* strains (P301.9, R31.3, R008, and EC1118) have been studied at two time points of the late fermentation phase. The aim was to investigate how the strain gene expression pattern modulated the late fermentation process.

2. Materials and Methods

2.1. Yeast Strains

In the present work, four *Saccharomyces cerevisiae* strains were used: The industrial strain EC1118 (Lallemand Inc.—Montreal, QC, Canada) and three natural strains isolated in a vineyard: P301.9, R008, and R31.3 (University of Padova, Padova, Italy).

2.2. Fermentation Trial and Samplings

Fermentations were performed at 25 °C in synthetic grape must MS300 [18] modified for the carbon source: 100 g/L of glucose and 100 g/L of fructose were added instead of 200 g/L of glucose. Three independent biological replicates were carried out for each strain in 1 L bioreactors (Multifors, Infors HT, Basel, CH, Switzerland). These instruments are equipped with sensors to monitor temperature and a flow meter to determine CO_2 outflow (red-y mod. GSM-A95A-BN00, Infors HT, Basel, CH, Switzerland) (range 1–20 mL/min). Strict anaerobiosis was not imposed, but fermentation conditions were largely anaerobic due to the design of the bioreactor and the effect of CO_2 production. CO_2 production was monitored by the flow meter every 5 min to determine the rate of CO_2 production. For each strain, approximately 3×10^6 cells/mL have been inoculated in 1 L of MS300 must. For each strain, cells were collected for Real Time-PCR assay at two sampling points during the late fermentation phase, when 45 and 60 g/L of CO_2 were produced. After sampling, the cells were centrifuged to remove the growth media and immediately frozen at $-80\ °C$. At the same sampling points, 50 mL of fermented media have been collected and frozen for chemical analysis.

2.3. Chemical Analysis

HPLC analysis was performed to determine the concentrations of residual sugars and ethanol, as described by Lemos Junior [19]. Ten microliters of sample was analyzed by the Waters 1525 HPLC binary pump (Waters, Milford, MA, USA) equipped with a 300×7.8 mm stainless-steel column packed with Aminex HPX_87H HPLC column (Bio-Rad, Hercules, CA, USA) and a Waters 2414 Refractive Index Detector (Waters, Milford, MA, USA). The analyses were performed isocratically at 0.6 mL/min and 65 °C with a cation-exchange column (300 by 7.8 mm [inner diameter]; Aminex HPX-87H) and a Cation H+ Microguard cartridge (Bio-Rad Laboratories, Hercules, CA, USA), using 0.01 N H_2SO_4 as the mobile phase. Ammonia and amino nitrogen were measured by means of specific enzymatic kits (Steroglass, Perugia, Italy) according to the manufacturer's instructions.

2.4. RNA Extraction and Reverse Transcription

Total RNA was extracted using the TRIzol® Plus RNA Purification Kit (Ambion, Austin, TX, USA). The concentration, purity, and integrity of RNA samples were determined by spectrophotometric analysis using the SPARK® multimode microplate reader (Tecan Trading AG, Switzerland), considering the absorbance ratio at 260/280 nm and at 230/260 nm. The quality and integrity of RNAs were confirmed by electrophoresis on 1.5% agarose gels under denaturing conditions (2% formaldehyde, v/v, 20 mM MOPS, 5 mM sodium acetate, 1 mM EDTA, pH 7.0). To obtain DNA-free RNA, the total RNA previously extracted (1 µg) was treated with DNase I (Fermentas, Waltham, MA, USA) according to the manufacturer's instructions. cDNA was synthesized using Revert Aid Reverse Transcriptase (Fermentas, Waltham, MA, USA) according to the manufacturer's instructions using both polyT (16) primers (MWG-biotech; 0.5 µg/µL) and random hexamers (Promega; 0.5 µg/µL). Samples were stored at -20 °C until Real-Time PCR was run.

The quality control of cDNAs was checked by end-point PCR in a PTC200 thermal cycler (MJ Research Inc., Waltham, MA, USA). Amplification of the gene *APE2* (F—TGCGCATCAATGTAATGTGGAAGCAGAGTA, R—TGAAATCAGGTTCCACGGTTAAA TCGTAGTGT) was performed on cDNAs both for checking the reverse-transcription efficiency and for excluding genomic DNA contamination. Amplified samples were run on 1.5% agarose gel containing 1X GelRedTM Nucleic Acid Gel Stain (Biotium, Fremont, CA, USA). Run was performed on a horizontal electrophoresis apparatus with TBE 0.5x as the running buffer (44.5 mM Tris, 44.5 mM boric acid, 1 mM EDTA) and the bands were visualized by UV trans-illumination.

2.5. Primer Design

PCR primers of the investigated genes for real-time assays are listed in Table 1. They were designed using Primer-BLAST (http://www.ncbi.nlm.nih.gov/tools/primer-blast/, accessed on 5 May 2017). The yeast database was used to check primer specificity on sequences of other yeast species. Special attention was given to primer length (15–25 bp), annealing temperature (58–62 °C), nucleotides composition, 3′-end stability, and amplicon size (80–200 bp). All primers were synthesized and OPC purified by Metabion International AG (Metabion International AG, Planegg, Germany). After synthesis, the primer specificity was tested by end-point PCR and gel electrophoresis.

Table 1. List of the investigated genes and details of primers and amplicons for each gene.

Gene	Molecular Function (SGD Curated)	Primer Sequence [5′ → 3′]	Amplicon Length	Efficiency %	R^2
HXT1	HeXose Transporter (low-affinity glucose transporter)	F: 5′-GAA GCT GGC AGA ATC GAC GA-3′ R: 5′-TAT GGA TGG TCA GGT GGG CA-3′	71 bp	102.9	0.996
HXT2	HeXose Transporter (high-affinity glucose transporter)	F: 5′-TGA ACT CCC AGC AAA GCC AA-3′ R: 5′-TCC CAA CCA AAG ACA AAC CCA-3′	90 bp	104.9	0.992
HXT3	HeXose Transporter (low affinity glucose transporter)	F: 5′-CAC GTT ATT TGG TTG AAG CTG GT-3′ R: 5′-GTT GAA TGA ATG GAT GGT CTG GG-3′	93 bp	101.3	0.998
HXT4	HeXose Transporter (high-affinity glucose transporter)	F: 5′-TGG TGG TAT GAC ATT CGT TCC-3′ R: 5′-CGC TGA CCT TAT TTG AAA GAG CA-3′	101 bp	104.7	0.989
HXT5	HeXose Transporter (moderate affinity for glucose)	F: 5′-TCC AAA TCG CCT CCA TTG ACA-3′ R: 5′-AAT ACC ACC AAC GCC CAG TC-3′	77 bp	102.5	0.983
HXT6/7	HeXose Transporter (high-affinity glucose transporter)	F: 5′-GAC TTT GGA AGA AGT CAA CAC CA-3′ R: 5′-TTC TTC AGC GTC GTA GTT GGC-3′	106 bp	100.0	0.998
HXT8	Protein of unknown function with similarity to hexose transporters	F: 5′-AAT TCT GTC CAG TGG CGT GT-3′ R: 5′-CGG AAC AAA CGT CAT ACC ACC-3′	81 bp	95.2	0.991
FSY1	Fructose/H^+ symporter (Galeote et al., 2010)	F: 5′-CGA TGT TAA AGG CGG GTG GA-3′ R: 5′-AAC GTG GTG ACT CGG GTA AG-3′	98 bp	95.1	0.989
GAL2	GALactose metabolism (also able to transport glucose)	F: 5′-GGG TCT GAA GGC TCC CAA AG-3′ R: 5′-ACA AAC AAA GCA AGG AAA CGG T-3′	85 bp	104.7	0.981

For each different primer pair, the efficiency of RT-PCR (E), slope values, and correlation coefficients (R^2) were determined using serial 1:5 dilutions of the template cDNA on the CFX96 cycler—Real-Time PCR Detection System (Bio-Rad Laboratories, Hercules, CA, USA). Efficiency was considered adequate when ranging from 95% to 105%, and R^2 was considered acceptable when greater than 0.98.

2.6. Real-Time PCR

Real-Time PCR was carried out on a CFX96 Cycler-Real Time PCR Detection System (Bio-Rad Laboratories, Inc., Hercules, CA, USA), in white-walled PCR plates (96 wells). A ready-to-use master-mix containing a fast proof-reading Polymerase, dNTPs, stabilizers, $MgCl_2$, and Eva Green dye was used according to the manufacturer's instructions (Bio-Rad Laboratories, Hercules, CA, USA). Reactions were prepared in a total volume of 15 μL containing 400 nM each primer (MWG), 1× Sso Fast Eva Green Supermix 2× (Bio-Rad Laboratories, Hercules, CA, USA), and 5 μL cDNA. The cycle conditions were set as follows: Initial template denaturation at 98 °C for 30 s, followed by 40 cycles of denaturation at 98 °C for 2 s, and combined primer annealing/elongation at 60 °C for 10 s. The amount of fluorescence for each sample, given by the incorporation of Eva Green into dsDNA, was measured at the end of each cycle and analyzed via CFX-Manager Software v2.0 (Bio-Rad Laboratories, Inc., Hercules, CA, USA). Melting curves of PCR amplicons were obtained using temperatures ranging from 65 °C to 95 °C. Data acquisition was performed for every 0.5 °C temperature increase with a 1-s step. For each target gene, each sample was analyzed in triplicate and no-template controls for each primer pair were included in all plates. Gene expression analysis was performed using the CFX-Manager Software v2.0, adopting the $2^{-\Delta\Delta CT}$ method. Four housekeeping genes have been used in Real-Time PCR gene expression analysis: The ALG9 and UBC6 primers designed by Teste [20]; and the FBA1 and PFK1 primers designed by Cankorur-Cetinkaya [21] and Nadai [22], respectively (Table 2). In this study, the lettering "total expression" indicates the sum of the normalized expression values of the genes, for each strain, relative to one of the two sampling points (45 or 60 g/L of produced CO_2).

Table 2. List of the reference genes and details of primers and amplicons for each gene.

Gene	Reference	Primer Sequence [5′ → 3′]	Amplicon Length	Efficiency %	R^2
ALG9	[20]	F: 5′-CAC GGA TAG TGG CTT TGG TGA ACA ATT AC-3′ R: 5′-TAT GAT TAT CTG GCA GCA GGA AAG AAC TTG GG-3′	156 bp	95.7	0.996
UBC6	[20]	F: 5′-GAT ACT TGG AAT CCT GGC TGG TCT GTC TC-3′ R: 5′-AAA GGG TCT TCT GTT TCA TCA CCT GTA TTT GC-3′	272 bp	99.0	0.985
FBA1	[21]	F: 5′-GGT TTG TAC GCT GGT GAC ATC GC-3′ R: 5′-CCG GAA CCA CCG TGG AAG ACC A-3′	125 bp	102.4	0.998
PFK1	[22]	F: 5′-GAG GTT GAT GCT TCT GGG TTC CGT-3′ R: 5′-TGT GGC GGT TTC GTT GGT GTC G-3′	138 bp	97.7	0.998

2.7. Statistical Analysis

The statistical data analysis was performed using XLSTAT software, vers. 2016.02 (Addinsoft, Paris, France). Data were subjected to Student's *t*-test or simple analysis of variance (one-way ANOVA) followed by the Tukey's *post-hoc* test. The analysis was carried out comparing the averages of three independent replicates, and differences were considered statistically significant for *p*-value lower than 0.05.

3. Results

3.1. Yeasts Fermentation Process

Fermentations in synthetic must were run using three *S. cerevisiae* strains, P301.9, R31.3, and R008, isolated from a vineyard and the commercial starter EC1118. The aim was to check strain behavior during the alcoholic fermentation focusing on the last part of the fermentation, starting form half of the total expected CO_2 (45 g/L) to the end of the process (late fermentation phase).

The fermentation profiles of the four strains were determined monitoring the fermentation rate (dCO_2/dt) (Figure 1).

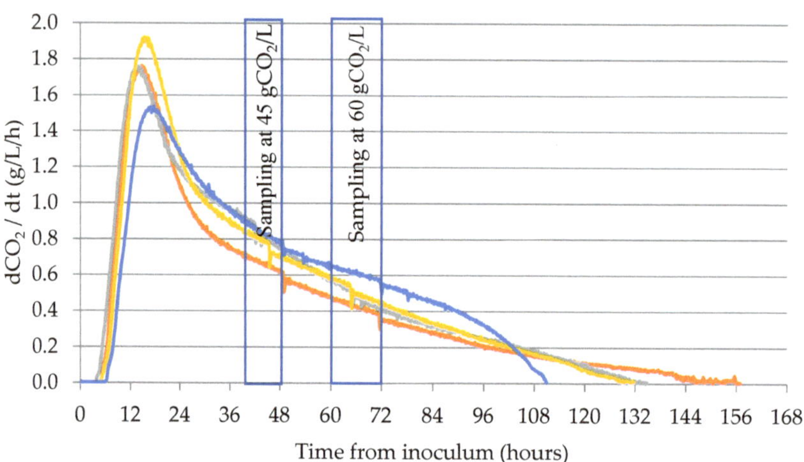

Figure 1. CO_2 production kinetics of strains R008 (—), R31.3 (—), P301.9 (—), and EC1118 (—). Data of dCO_2/dt are the average of three biological replicates.

During fermentation process, the fermentation rate peaked (Vmax) just before the entry into the yeast stationary growth phase and declined thereafter until the end of the fermentation [23]. Strain P301.9 showed the highest Vmax value (1.93 $gCO_2/L/h$), reached after 15.58 h from inoculum, while the lowest was obtained for EC1118 (1.53 $gCO_2/L/h$) reached in 16.75 h from inoculum. Intermediate values were registered for strains R008 and R31.3 (Table 3).

Table 3. Parameters of the fermentation kinetics.

Strain	CO$_2$ Production Peak (g/L/hours)	Peak Time (hours)	Sampling Time 45 g/L (hours)	Sampling Time 60 g/L (hours)	Fermentation Time (hours)
EC1118	1.53 ± 0.05 [A]	16.75 ± 0.85 [B]	50.3 ± 2.7 [A]	73.9 ± 3.7 [B]	110.2 ± 1.7 [A]
R008	1.76 ± 0.00 [B]	14.91 ± 0.33 [A]	46.3 ± 4.9 [A]	67.5 ± 3.1 [AB]	156.9 ± 1.0 [C]
R31.3	1.76 ± 0.05 [B]	14.08 ± 1.08 [A]	41.8 ± 1.2 [A]	64.6 ± 1.2 [A]	134.9 ± 4.2 [B]
P301.9	1.93 ± 0.05 [C]	15.58 ± 0.84 [AB]	43.2 ± 4.1 [A]	63.5 ± 1.5 [A]	131.7 ± 4.2 [B]

Data are expressed as the average of three biological replicates ± standard deviation. Uppercase letters indicate significant differences obtained from Tuckey post-hoc test after ANOVA analysis of variance between the strains (*alpha* = 0.05).

During the late fermentation phase, strains R008, R31.3, and P301.9 showed a marked decrease of the fermentation rate that is more evident in the last part of the fermentation. This trend was responsible for the increase of the fermentation time, with respect to the commercial strain EC1118. Therefore, EC1118 was the first strain to complete the fermentation, followed by P301.9, R31.3, and R008.

Despite the different fermentation trends, EC1118, P301.9, and R31.3 completed the sugar transformation, while R008 left a fructose residue of 7.62 g/L (Table 4). The ethanol concentration at the end of the fermentation reflected the sugar consumption. No significant differences in ethanol production were found among strains EC1118, P301.9, and R31.3. Strain R008 did not reach the same ethanol level, due to the sugar residues (Table 4).

Table 4. Ethanol and sugar residues at the end of the fermentations.

	Ethanol (% vol.)	Glucose (g/L)	Fructose (g/L)
EC1118	11.49 ± 0.04 [B]	nd	nd
P301.9	11.45 ± 0.04 [B]	nd	nd
R31.3	11.38 ± 0.26 [B]	nd	nd
R008	10.64 ± 0.07 [A]	nd	7.62 ± 0.04

Data are expressed as the average of three biological replicates $\pm$ standard deviation. Uppercase letters indicate significant differences obtained from Tuckey post-hoc test after ANOVA analysis of variance between the strains (*alpha* = 0.05). nd: not detected.

During the late fermentation phase, two samplings were performed at 45 g/L (half of the CO_2 produced) and 60 g/L of CO_2 produced. Due to differences in strain fermentation rates, these values were reached at different times.

The mean total sugar residue of the four strains was 79.69 ± 4.17 g/L at 45 g/L of CO_2 produced and 40.80 ± 5.96 g/L at 60 g/L of CO_2 produced (Supplementary Table S1). ANOVA analysis of variance found no significant differences among the strains' sugar residue at both sampling points (*alpha* = 0.05). This result indicates that the sampling time have been correctly chosen in order to harvest cells in the same physiological state.

At 45 g/L of CO_2 produced, strain EC1118 showed the lowest ratio between fructose and the total sugar residue (60.30%), corresponding to the most balanced intake of fructose and glucose with respect to the other strains, which have shown significantly higher ratios. The same pattern has been observed at 60 g/L of produced CO_2. For all strains, this ratio increased from 45 to 60 g/L of produced CO_2, during the late fermentation phase, confirming the glucophilic aptitude of *S. cerevisiae* (Table 5). Strain R31.3 showed the highest increase of the ratio between fructose and the total sugar residue from 45 to 60 g/L of produced CO_2 (12.27%), followed by P301.9 (10.11%), R008 (8.92%), and EC1118 (6.57%).

Table 5. Fructose/total sugars ration and amino nitrogen residue at 45 and 60 g/l of produced CO_2.

	Fructose/Total Sugars (%)			NH$_2$ (mg/L)		
	45 g/L	60 g/L	*p*	45 g/L	60 g/L	
EC1118	60.30 ± 0.43 [A]	66.87 ± 0.70 [A]	<0.0001 ***	61.17 ± 2.35	59.87 ± 3.10	Ns
R008	66.05 ± 0.29 [B]	74.97 ± 1.65 [B]	0.012 *	55.75 ± 1.95	63.10 ± 1.70	*
R31.3	67.01 ± 1.08 [B]	79.27 ± 1.84 [B]	0.016 *	51.93 ± 7.35	56.57 ± 6.33	*
P301.9	66.56 ± 0.74 [B]	76.67 ± 1.72 [B]	0.002 **	38.03 ± 4.29	39.20 ± 3.21	Ns

Data are expressed as the average of three biological replicates $\pm$ standard deviation. Uppercase letters indicate significant differences obtained from Tukey post-hoc test after ANOVA analysis of variance between the strains (*alpha* = 0.05). (***) $p < 0.001$, (**) $p < 0.01$, (*) $p < 0.05$, Ns (not significative) between the percentage of fructose on total sugar residue at 45 and 60 g/L of produced CO_2 by Student's *t* test.

At the two sampling points, assimilable nitrogen (ammonium and amino acids) residues were measured. At the beginning of the fermentation process, the assimilable nitrogen present in the synthetic must (300 mg/L) was at a high level and provided the suitable nitrogen amount required by the yeast strains to complete the fermentation when 200 g/L of sugar are present in the must [24]. Generally, ammonium nitrogen is completely consumed during the first part of the fermentation (up to Vmax) that corresponds to the last part of the exponential growth [23]. The amino-nitrogen residues are reported in Table 5 for each sampling point.

The amino-nitrogen residues reflected strain fermentation rates during the exponential growth phase. The three strains that showed high Vmax values left lower amino-nitrogen residues than EC1118. Among the three strains, different nitrogen consumptions were observed. R31.3 and P301.9, which showed a similar fermentation trend, revealed different nitrogen consumptions. Between the two sampling points, no further nitrogen consumption was apparently registered. This could be due to the nitrogen release by cell lysis that occurred during the stationary phase [14].

3.2. Expression of Hexose Transporters Genes during Late Fermentation

During the late fermentation phase, two cell samplings were performed, at 45 and 60 g/L of CO_2 produced, to analyze the gene expression of hexose transporters that are known to be active during the fermentation process, namely *HXT 1-8*, *GAL2*, and *FSY1*. All the investigated genes were present in the strain genome, but *FSY1* was not found in strain R008 [11].

Comparing strains gene expression (Figure 2), each strain evidenced a specific pattern both at 45 and at 60 g/L of CO_2 produced.

45 g/L of CO₂ produced

60 g/L of CO₂ produced

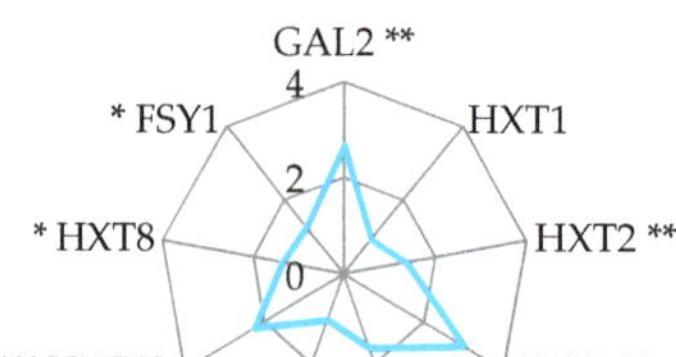

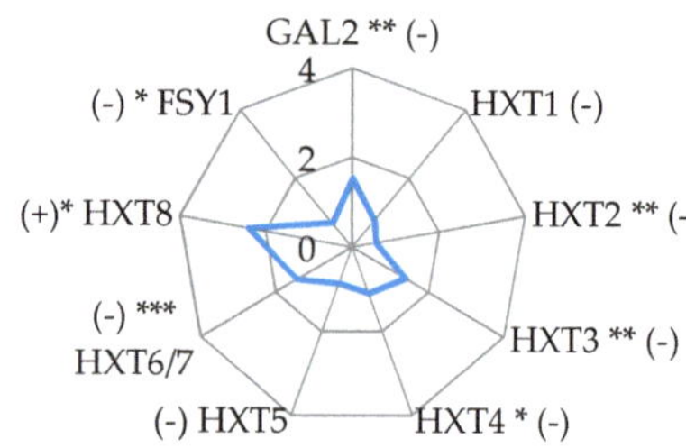

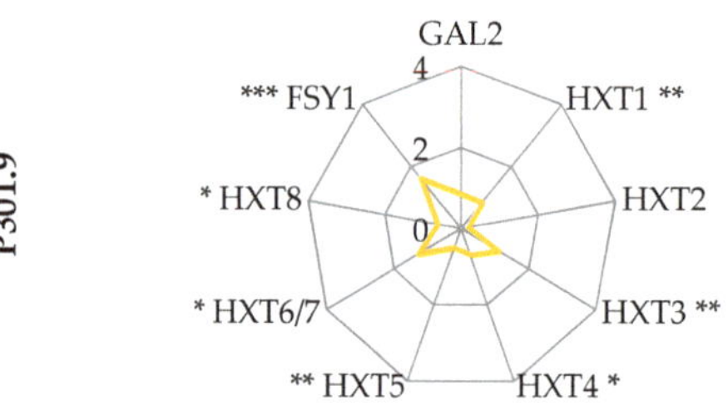

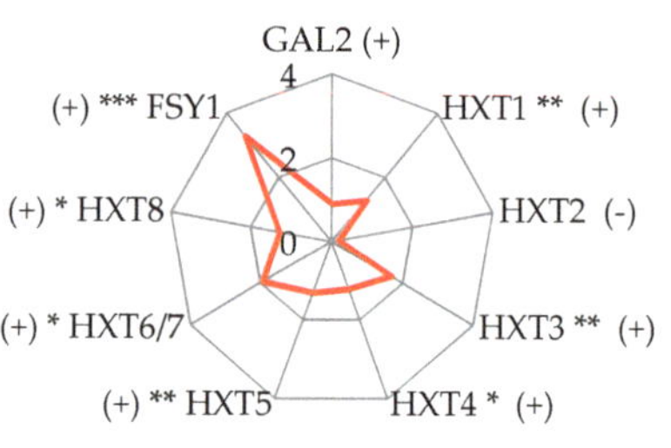

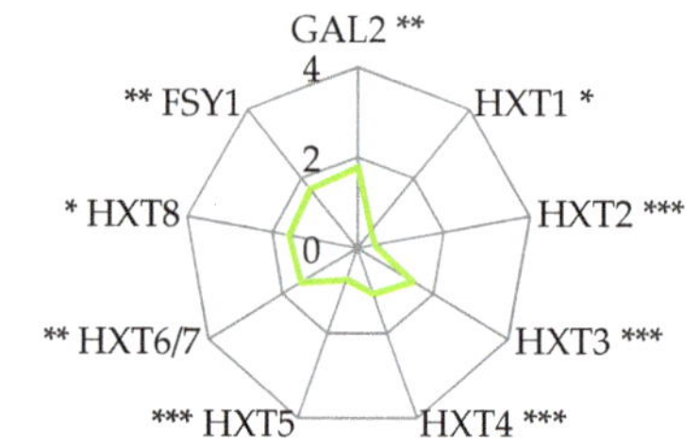

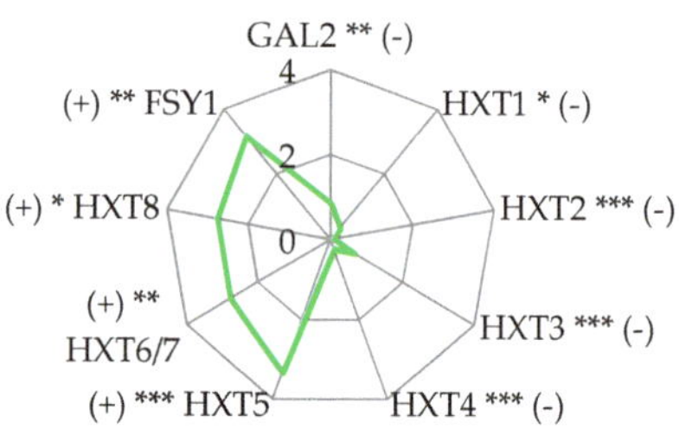

Figure 2. *Cont.*

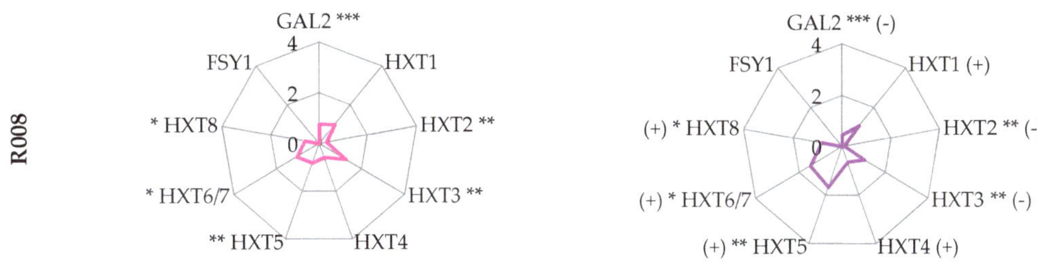

Figure 2. Gene expression pattern of hexose transporters. Letters indicate significant differences in the gene expression values obtained from ANOVA analysis of variance followed by the Tukey post-hoc test between the strains, at 45 or 60 g/L of CO_2 (*alpha* = 0.05). (***) $p < 0.001$, (**) $p < 0.01$, (*) $p < 0.05$ between gene expression values at 45 and 60 g/L of produced CO_2 by Student's *t* test for each yeast strain.

The global expression rate was calculated by summing the expression rate of each single gene analyzed. At 45 g/L of CO_2 produced, EC1118 showed the highest expression rate (15.48) and strain R008 showed the lowest (6.35). EC1118 was the only strain that decreased the global expression rate at 60 g/L of CO_2 produced (10.76), reaching a global expression rate value lower than P301.9 and R31.3 (13.08 and 14.25, respectively). R008, which left a sugar residue at the end of the fermentation, showed the lowest level of the global expression rate at both sampling times. Regarding the good fermenting yeasts (EC1118, P301.9, and R31.3), in the two vineyard strains, the expression of the tested genes increased at 60 g/L of CO_2 produced. On the contrary, in EC1118, most of the tested genes showed a decrease of expression.

When the expression of each gene was compared among the four strains (Table 6), different expression levels were always detected suggesting a strain-specific expression pattern at the two sampling points and trend.

HXT3 was the most expressed gene at 45 g/L (in EC1118 relative expression 3.004). This gene encodes a low-affinity transporter, and its expression requires the presence of glucose, but it is only weakly dependent on sugar concentration [25]. It has been demonstrated that, among the *HXT* family, *HXT3* has the highest capacity to support fermentation. In fact, Hxt3p is the only carrier that is expressed throughout the fermentation, consistent with the fact that it plays a key role in the process [26]. The robust expression and high stability of Hxt3p during the stationary phase is consistent with its capacity to maintain a high fermentation rate during starvation when expressed alone [7].

Transport-kinetic data support the idea that Hxt3p is the primary low-affinity transporter during stationary phase. All these previous findings were related to strain V5 isolated from Champagne [26], such as EC1118. In the other vineyard strains, the expression level is nearly half, and at 65 g/L, it decreased in EC1118, R008, and R31.3 but increased in P301.9. In the two vineyard strains that completed fermentation (P301.9 and R008), *HXT3* was not the most expressed strain both at 45 and 60 g/L of produced CO_2. This finding caused us to rethink the role of the Hxt3 transporter during fermentation. The main contribution of *HXT3* to maintain a high fermentation ratio during the late fermentation phase, as suggested by the literature, could be limited to specific yeast genotypes.

The second most expressed gene at 45 g/L of produced CO_2 was *GAL2*. Substrate-inducible and glucose repressible galactose transporter Gal2p, which is more than 60% identical to the Hxt transporters, mediates the transport of galactose by the mechanism of facilitated diffusion. It has been demonstrated that the Gal2p also mediates the transport of glucose in the *HXT1-7* null mutant [27]. This finding suggested that *GAL2* can play a role when glucose concentration is low such as in the late fermentation phase, although the expression level is reduced at 60 g/L of CO_2 produced with the exception of P301.9.

Table 6. Normalized expression of hexose transporter genes.

Gene	Strain	45 g/L of CO_2 Produced		60 g/L of CO_2 Produced	
GAL2	EC1118	2.660 ± 0.655	C	1.533 ± 0.434	B
	R008	0.775 ± 0.184	A	0.465 ± 0.082	A
	R31.3	1.801 ± 0.519	B	0.851 ± 0.521	A
	P301.9	0.862 ± 0.236	A	0.914 ± 0.131	AB
HXT1	EC1118	0.925 ± 0.239	B	0.771 ± 0.130	B
	R008	0.996 ± 0.125	B	1.072 ± 0.116	BC
	R31.3	0.475 ± 0.045	A	0.356 ± 0.075	A
	P301.9	0.849 ± 0.111	B	1.300 ± 0.254	C
HXT2	EC1118	1.364 ± 0.565	B	0.536 ± 0.385	B
	R008	0.304 ± 0.198	A	0.089 ± 0.017	A
	R31.3	0.399 ± 0.044	A	0.084 ± 0.045	A
	P301.9	0.181 ± 0.052	A	0.167 ± 0.035	A
HXT3	EC1118	3.004 ± 0.977	B	1.385 ± 0.134	BC
	R008	1.257 ± 0.106	A	1.031 ± 0.123	AB
	R31.3	1.454 ± 0.395	A	0.648 ± 0.101	A
	P301.9	1.117 ± 0.183	A	1.662 ± 0.399	C
HXT4	EC1118	1.642 ± 0.455	C	1.090 ± 0.224	C
	R008	0.578 ± 0.234	A	0.640 ± 0.126	B
	R31.3	1.065 ± 0.230	B	0.240 ± 0.128	A
	P301.9	0.686 ± 0.194	AB	1.209 ± 0.474	C
HXT5	EC1118	1.028 ± 0.466	B	0.859 ± 0.257	A
	R008	0.806 ± 0.281	AB	1.673 ± 0.754	A
	R31.3	0.731 ± 0.158	AB	3.341 ± 1.546	B
	P301.9	0.505 ± 0.116	B	1.301 ± 0.338	A
HXT6/7	EC1118	2.252 ± 0.181	C	1.439 ± 0.050	A
	R008	1.045 ± 0.129	A	1.502 ± 0.499	A
	R31.3	1.511 ± 0.447	B	2.782 ± 0.938	B
	P301.9	1.267 ± 0.277	AB	1.931 ± 0.613	AB
HXT8	EC1118	1.354 ± 0.496	B	2.430 ± 0.870	B
	R008	0.593 ± 0.097	A	0.862 ± 0.314	A
	R31.3	1.613 ± 0.390	B	2.782 ± 1.106	B
	P301.9	0.606 ± 0.213	A	1.289 ± 0.692	A
P301_O30021	EC1118	1.254 ± 0.496	A	0.712 ± 0.256	A
	R31.3	1.709 ± 0.308	A	3.164 ± 1.018	B
	P301.9	1.617 ± 0.203	A	3.301 ± 0.727	B

Uppercase letters indicate significant differences in the gene expression values obtained from ANOVA analysis of variance followed by the Tukey post-hoc test between the strains, at 45 or 60 g/L of CO_2 (*alpha* = 0.05).

The *HXT6* and *HXT7* genes encode the most closely related (they differ only in two amino acid residues), high-affinity transporters. Due to this high sequence similarity, the same primer couple was used for qPCR quantification. They are highly expressed at very low glucose concentrations, non-fermentable carbon sources (ethanol, glycerol), maltose and/or galactose, and repressed by moderate and high glucose concentrations [25]. Previous findings evidenced that the two genes displayed a similar expression profile during alcoholic fermentation. In strain V5, they displayed a strong burst of expression following the entry into the stationary phase and the respective proteins remained abundant through the late fermentation phase [26]. The EC1118 expression value at 45 g/L and 60 g/L of CO_2 produced confirmed the previous finding. The vineyard strains evidence an opposite trend. At 60 g/L, an increase in the expression level was observed for all the strains. These findings suggest that *HXT6* and *HXT7* have a major role in maintaining the high fermentation rate during the late fermentation phase.

The expression pattern and function of the hexose transporter encoded by *HXT8* is still poorly defined [28]. It has 70% similarity with *HXT6* [29]. It appears to be unable to support growth on glucose of the *HXT1-7* null strain [30]. However, when overexpressed in an *HXT1–17 GAL2* null strain, it was able to restore growth on glucose, fructose, and mannose, confirming that it is a functional hexose transporter [31]. The expression of Hxt8p is induced by low and repressed by high levels of glucose [32]. This finding was confirmed by the gene expression values displayed by the four strains. All yeast showed an increase in the expression level from 45 to 60 g/L of CO_2 produced, particularly in EC1118 and R31.3 that registered the highest expression values.

Hxt5p has a moderate affinity for glucose and a low affinity for fructose and for mannose [33]. It is apparently regulated by the growth rate of the cells rather than by the external glucose concentration [34]. In addition, it is upregulated upon nitrogen and carbon starvation [35]. Therefore, it could be of interest during the late fermentation phase. The expression level in EC1118 is stable at 45 and 60 g/L, and in vineyard strains, a marked increase was evidenced, particularly in the case of R31.3 where the expression level rose almost 5-fold.

The *HXT2* and *HXT4* genes, as well as *HXT5*, have been classified as genes encoding transporters with moderate affinity for glucose (Km values around 10 mM) [27,36] that are maximally expressed at a low concentration of glucose and repressed in the presence of high glucose concentrations [25,37]. Despite moderate affinity, the *HXT2* expression pattern showed a decreasing expression level from 45 to 60 g/L, and specially for vineyard strains was very low. These results were consistent with previous findings that evidenced a strong activation of the *HXT2* gene only during the first hours of fermentation, returning to very low levels during the growth phase. This transient induction of *HXT2* is somewhat contradictory to the known regulation of this gene, as it has been shown to be repressed by high glucose concentrations [38]. *HXT2* appeared to be able to bypass glucose repression during the lag phase but returned to a repressed state as growth resumed. Conversely, they did not observe an induction of *HXT2* at the end of fermentation when the hexose concentration became low and sub-repressive [26].

On the contrary, *HXT4* expression showed a strain-mediated trend: Decreasing in EC1118 and markedly so in R31.3, constant in R008, and increasing in P301.9.

HXT1, as well as *HXT3*, is a low-affinity carrier and it is strongly expressed at the beginning of fermentation, while its expression level decreased rapidly during the growth phase, consistent with the *HXT1* gene being induced by high sugar concentrations [26]. R31.3 seemed to confirm this previous finding as it showed a very low expression level decreasing from 45 to 60 g/L produced CO_2. At the two sampling points, similar expression levels were found in EC1118 and R008. Surprisingly, P301.9 showed a notably increased value at 60 g/L of CO_2 suggesting a role of *HXT1* in supporting the fermentation ratio during the late fermentation phase. The *FSY1* gene was firstly isolated in *Saccharomyces pastorianus* and its protein was the first fructose permease identified in a yeast species [39]. The EC1118 complete genome sequence revealed the presence of a gene sequenced responsible for a protein highly similar to that encoded by the *S. pastorianus*' specific fructose symporter *FSY1* gene [10]. The functional analysis of the EC1118 *FSY1* gene demonstrated that it encodes for a high-affinity fructose/H+ symporter [3]. Since hexoses are transported by facilitated diffusion via hexose carriers (Hxt), which prefer glucose to fructose, the utilization of fructose by wine yeast is critically important in the late phase of the fermentation where fructose is the most present sugar. Previous findings demonstrated that, in EC1118, *FSY1* was repressed by high concentrations of glucose or fructose and was highly expressed on ethanol as the sole carbon source [3]. Interestingly, the *FSY1* expression pattern was very different between EC1118 and the other strains that possessed this gene (R31.3 and P301.9). Indeed, they showed a marked increase of the expression level at 60 g/L of CO_2 produced. This finding suggested that due to the expression level, this gene could have a different impact on strain fermentation rate during late fermentation.

The expression percentage of each transporter-coding gene is reported in Figure 3.

	EC1118		P301.9		R31.3		R008	
	45 g/L	*60 g/L*	*45 g/L*	*60 g/L*	*45 g/L*	*60 g/L*	*45 g/L*	*60 g/L*
GAL2	17.2	14.3	11.2	7.0	16.7	6.0	12.2	6.3
HXT1	6.0	7.2	11.0	9.9	4.4	2.5	15.7	14.6
HXT2	8.8	5.0	2.4	1.3	3.7	0.6	4.8	1.2
HXT3	19.4	12.9	14.5	12.7	13.5	4.5	19.8	14.1
HXT4	10.6	10.1	8.9	9.2	9.9	1.7	9.1	8.7
HXT5	6.6	8.0	6.6	9.9	6.8	23.5	12.7	22.8
HXT6/7	14.5	13.4	16.5	14.8	14.0	19.5	16.4	20.5
HXT8	8.7	22.6	7.9	9.9	15.0	19.5	9.3	11.8
FSY1	8.1	6.6	21.0	25.2	15.9	22.2		
Total (%)	*100*	*100*	*100*	*100*	*100*	*100*	*100*	*100*
Variation (%)	*-30.5*		*+70.0*		*+32.4*		*+15.4*	

Figure 3. Different utilization of the tested hexose transporter genes among yeast strains. Data per each gene are expressed as percentage calculated on the total expression, for each strain and time point (45–60 g/L of produced CO_2). The difference between the total expression values from 45 to 60 g/L of produced CO_2 is reported as percentage of variation.

In EC1118, the decrease in the expression level from 45 to 60 g/L of produced CO_2 is 30,5%, while in the other strains the increase is from +15.4 in R008 to +70% in P301.9. At 45 g/L in EC1118, the expression of *HXT3*, *GAL2*, and *HXT6/7* accounted for more than 50% of the overall gene expression and at 60 g/L, *HXT8* was the most expressed gene. At 45 g/L in P301.9, *FSY1* is the most expressed gene followed by *HXT6/7* and *HXT3*. At 60 g/L, all these genes showed an increase in the expression level that is more evident in *FSY1*. In R31.3 at 45 g/L, *GAL2* is the most expressed gene and *HXT6/7*, *HXT8*, and *FSY1* showed similar expression percentages. At 60 g/L, they accounted for more than 50% of the overall gene expression the showed the highest increase in total expression (+76%), followed by R31.3 and R008. At 45 g/L of produced CO_2, strain EC1118 showed the highest total expression followed by R31.3, P301.9, and R008, while at 60 g/L of produced CO_2, strain R31.3 showed the highest total expression followed by P301.9, EC1118, and R008 (Figure 3).

At 45 g/L of produced CO_2, the most expressed genes in EC1118 were *HXT3*, *GAL2*, and *HXT6/7*. The expression pattern of *HXT3* and *HXT6/7* confirmed previous findings that assessed *HXT3* as the most highly expressed *HXT* gene during alcoholic fermentation and *HXT6/7*, expressed throughout fermentation [40]. *HXT3* and *HXT8* (together with *HXT1*) were the most expressed gene in R008, although at lower levels than that of EC1118. Interestingly, in P301.9, the most expressed gene was *FSY1* followed by *HXT6/7* and *HXT3*; in R31.3, *FSY1* showed an expression level similar to the most expressed genes, *HXT8* and *GAL2*. At 60 g/L of produced CO_2 in EC1118, a strong increase of *HXT8* was registered, while in P301.9, *FSY1* reached the 25% of the total gene expression. In R31.3, the *GAL2* gene expression strongly dropped down, while *FSY1*, *HXT8*, *HXT6/7*, and *HXT5* increased. Although the general low gene expression level in R008 caused an increase in the expression of *HXT5* and *HXT6/7*, *HXT8* was found (Figure 3).

4. Discussion

The hexose transporters' activity during late fermentation phase is crucial to ensure the complete transformation of the sugar into ethanol. In fact, a high fructose/glucose ratio may cause sluggish and stuck fermentations with high levels of fructose residue, which is a major problem in the wine industry [3,41].

The transport of glucose and fructose in the yeast *Saccharomyces cerevisiae* plays a crucial role in controlling the rate of wine fermentation, and the yeast fermentation performance is strongly influenced by the hexose carrier set-up [7,26,42]. *Saccharomyces cerevisiae* is the organism provided with the highest number of hexose transporter genes. In its genome, twenty genes encode hexose transporter-like proteins (*HXT1* to *HXT17*, *GAL2*, *SNF3*, and *RGT2*) [43].

The molecular function of these hexose transporters is redundant, as none of these transporters are essential for growth on glucose. The cooperation among the involved transporters determines the effectiveness of the sugar-uptake systems. In *S. cerevisiae*, two uptake systems have been described: The first, a constitutive low-affinity system involving *HXT2*, *HXT6*, and *HXT7*, and the second, a glucose-repressed high-affinity system responsible for the high-affinity transporters, *HXT1*, *HXT3*, or *HXT4* [27]. In contrast, the expression of *HXT5* [33] is influenced by the growth rate of cells and not by the extracellular glucose concentration. Strains lacking *HXT1* through *HXT7* genes are unable to grow on glucose, fructose, or mannose and have no glycolytic flux [8,27]. Despite extensive research, the functions of Hxt8-Hxt17 have remained poorly defined. *HXT8*, induced by low levels of glucose, appears to function as a glucose transporter since it can partially complement the glucose growth defect of the *hxt* null mutant [30]. Only recently, Treu and colleagues [44] found that the *HXT8* gene was expressed during the first stage of synthetic grape must fermentation, although at a lower level than *HXT1–7* genes. *HXT12* was found to be a pseudogene and only recently Jordan and co-workers [45] described Hxt13p, Hxt15p, Hxt16p, and Hxt17p as a novel type of polyol transporters, not involved in glucose and fructose transport. Another gene with similarity to the *HXT* family is present in the yeast genome: *GAL2*. It encodes a galactose transporter that is also a high-affinity glucose transporter [36]. The *FSY1* gene, encoding for a fructose/H+ symporter, previously identified in *S. pastorianus* [39], has been discovered and characterized in the EC1118 *Saccharomyces cerevisiae* strain [3,5]. In 2014, Treu and colleagues [11] found the fructose transporter in a vineyard strain (P301) with the same origin as the strains tested in this study. Following P301 transcriptome analysis under fermentation conditions reported that *FSY1* gene was expressed at lower levels with respect to other hexoses transporter genes [44]. Notwithstanding, *FSY1* gene is potentially very interesting as in the late fermentation phase, fructose is the main sugar residue due to yeast glucophilic behavior.

With the aim to evaluate the role of the different hexose transporter gene expressions in the strain ability to complete sugar fermentation, four *S. cerevisiae* strains were considered: he industrial wine EC1118 and the vineyard P301.9, R31.3, and R008. Previous results found that the strain R008 genome does not contain the *FSY1* gene [11].

Kinetics of carbon dioxide production differed among the tested strains, evidencing optimal (EC1118), intermediate (P301.9 and R31.3), and poor (R008) fermentation trends. During early fermentation (up to the fermentation rate peak, Vmax) that was consistent with the yeast exponential growth phase [23], vineyard strains showed better fermentation performance than EC1118, particularly P301.9. During late fermentation (after the fermentation rate peak to the end of the process), an opposite trend was observed. EC1118 kept the fermentation rate at a higher level than the other strains and completed the fermentation earlier. Strain R008 showed the lowest fermentation rate during the late fermentation phase, leaving a fructose residue and producing about 1% less ethanol. The fermentation trends were consistent with the fructose/total sugar ratio and amino acid consumption found at 45 g/L. EC1118 that showed the lowest fermentation level during early fermentation consumed the lowest amount of amino acid and fructose. P301.9 showed the highest Vmax value and consumed the highest number of amino acids, although the fructose/total sugar ratio showed no significant difference with the other vineyard strains. No significant differences were found between amino acid residues at 45 and 60 g/L in EC1118 and P301.9, the best fermenting strains. On the opposite, R31.3 and R008 showed a significant increase in amino nitrogen residues. This could be due to higher cell mortality level at 60 g/L that could also be responsible for the longer fermentation time.

All the *HXT* genes involved in glucose/fructose transport (*HXT1-8*) and *GAL2* and *FSY1* genes were tested by means of Real-Time PCR, and their relative gene expression has been determined. The results evidenced that all the genes tested were expressed during late fermentation regardless of their attitude to be inhibited by high or low glucose regulation pathways.

High-affinity transporter gene *HXT6-7*, intermediate affinity *HXT2* and *HXT4*, *HXT8*, and *GAL2* are induced by a low glucose concentration, therefore an increase in their expression is expected. On the contrary, low-affinity transporter genes *HXT1* and *HXT3*, induced by high glucose, were supposed to decrease their expression. EC1118 showed a decrease expression for all the genes, except for *HXT8*. In P301.9, all gene expressions increased, except for *HXT2* that remained unchanged. R31.3 showed a similar trend to EC1118, except for *HXT6/7* and *HXT8* that increased their expression level. Despite the very low total expression level, R008 showed an intermediate trend as *HXT6/7* and *HXT8* increased as in P301.9 and *HXT2*, *HXT3*, and *GAL2* decreased as in EC1118. These results suggested that the "low-glucose" signal that controls high- and intermediate-affinity transporters gene expression turned up at higher sugars concentration in EC1118 cells than the others. On the contrary, the "high-glucose" signal that controls low-affinity transporters gene expression turned up at higher sugar concentrations in P301.9 cells than the others.

HXT5 gene expression was maximally induced upon glucose and nitrogen depletion [34]. In this work, the *HXT5* gene expression increased in R31.3, P301.9, and R008 from 45 to 60 g/L of produced CO_2, while EC1118 remained low and constant. These findings are consistent with strain nitrogen consumption values as the vineyard strains consumed more nitrogen that EC1118 during early fermentation.

Previous works demonstrated that in EC1118, *FSY1* was repressed by high concentrations of glucose or fructose and was highly expressed on ethanol as the sole carbon source [3]. Surprisingly, in this study, *FSY1* was only poorly expressed at 45 and at 60 g/L in EC1118, evidencing a decrease in the expression from 45 to 60 g/L. On the contrary, in the two other strains containing this gene, the expression level was very high, and a marked increase was evident from 45 to 60 g/L. Therefore, the *FSY1* gene seems to play an important role in the sugar uptake of these yeast strains. EC1118 possessed the lowest "fructose/total residual sugars" ratio values (F/T ratio) at both sampling times. These data suggested that EC1118 is more capable of handling the F/T ratio than the other strains during early fermentation, leaving less fructose during late fermentation. In this condition, P301.9 and R31.3 took advantage of the high-affinity fructose/H+ symporter FSY1 to complete sugar consumption. Therefore, the high level of expression assured the fructose consumption during the late fermentation phase. This is not the case for R008, as they do not possess the *FSY1* gene. The lack of this gene, together with a low gene expression of all the genes of the HXT transport system, could be the cause of the poor fermentation performance during late fermentation.

5. Conclusions

In the fermentation trial, all the hexose transporter genes were expressed at both 45 and 60 g/L CO_2. This indicated that the corresponding sugar concentration range supported both the high and low glucose-dependent transporters.

The very low total gene expression of the transporters and the poor fermentation performance of R008 suggested that the overall expression level is a determinant to maintain a high fermentation rate during the late fermentation phase.

Each yeast showed a specific gene expression profile that was strongly variable among the strains. This led to rethinking the importance of the *HXT3* that was previously considered to play a major role in sugar transport during the overall fermentation process. The main contribution of *HXT3* to maintain a high fermentation ratio during late fermentation phase [7,40] could be limited to specific yeast genotypes. In different genetic contexts, other genes such as *HXT6/7*, *HXT8*, and *FSY1* were the most expressed and therefore responsible for sugar transport in the late fermentation phase.

Finally, two trends emerged from the data collected: EC1118, that reduced the gene expression of "low/high glucose"-induced genes, and P301.9, showing a gene expression increase. These findings suggest a strain-specific response of the "high/low glucose"-dependent genes that control most of the sugar transport in the yeast cell.

Supplementary Materials: The following are available online at https://www.mdpi.com/article/10.3390/fermentation7030164/s1, Table S1: Residual sugars at 45 and 60 g/L of produced CO_2.

Author Contributions: Conceptualization: V.C.; formal analysis: C.N. and G.C.; investigation: C.N. and G.C.; resources: V.C. and A.G., writing—original draft preparation: V.C., G.C. and C.N.; supervision: V.C.; project administration: V.C.; funding acquisition: A.G. and V.C. All authors have read and agreed to the published version of the manuscript.

Funding: This work was funded by Regione Veneto through its Programme FSE 2014–2020 (Grant No. 2105/32/2121/2015). G.C. was financially supported by University of Padova PhD Grant.

Institutional Review Board Statement: Not applicable.

Informed Consent Statement: Not applicable.

Data Availability Statement: Not applicable.

Conflicts of Interest: The authors declare no conflict of interest.

References

1. Guillaume, C.; Delobel, P.; Sablayrolles, J.-M.; Blondin, B. Molecular Basis of Fructose Utilization by the Wine Yeast *Saccharomyces cerevisiae*: A Mutated *Hxt3* Allele Enhances Fructose Fermentation. *Appl. Environ. Microbiol.* **2007**, *73*, 2432–2439. [CrossRef]
2. Berthels, N.; Corderootero, R.; Bauer, F.; Thevelein, J.; Pretorius, I. Discrepancy in Glucose and Fructose Utilisation during Fermentation by Wine Yeast Strains. *FEMS Yeast Res.* **2004**, *4*, 683–689. [CrossRef]
3. Galeote, V.; Novo, M.; Salema-Oom, M.; Brion, C.; Valério, E.; Gonçalves, P.; Dequin, S. *FSY1*, a Horizontally Transferred Gene in the *Saccharomyces cerevisiae* EC1118 Wine Yeast Strain, Encodes a High-Affinity Fructose/H+ Symporter. *Microbiology* **2010**, *156*, 3754–3761. [CrossRef]
4. Blateyron, L.; Sablayrolles, J.M. Stuck and Slow Fermentations in Enology: Statistical Study of Causes and Effectiveness of Combined Additions of Oxygen and Diammonium Phosphate. *J. Biosci. Bioeng.* **2001**, *91*, 184–189. [CrossRef]
5. Anjos, J.; Rodrigues de Sousa, H.; Roca, C.; Cássio, F.; Luttik, M.; Pronk, J.T.; Salema-Oom, M.; Gonçalves, P. Fsy1, the Sole Hexose-Proton Transporter Characterized in *Saccharomyces* Yeasts, Exhibits a Variable Fructose:H+ Stoichiometry. *Biochim. Biophys. Acta Biomembr.* **2013**, *1828*, 201–207. [CrossRef] [PubMed]
6. de Sousa, H.R.; Spencer-Martins, I.; Gonçalves, P. Differential Regulation by Glucose and Fructose of a Gene Encoding a Specific Fructose/H+ Symporter In *Saccharomyces* Sensu Stricto Yeasts. *Yeast* **2004**, *21*, 519–530. [CrossRef] [PubMed]
7. Luyten, K.; Riou, C.; Blondin, B. The Hexose Transporters of *Saccharomyces cerevisiae* Play Different Roles during Enological Fermentation. *Yeast* **2002**, *19*, 713–726. [CrossRef]
8. Boles, E.; Hollenberg, C.P. The Molecular Genetics of Hexose Transport in Yeasts. *FEMS Microbiol. Rev.* **1997**, *21*, 85–111. [CrossRef]
9. Kim, D.; Song, J.-Y.; Hahn, J.-S. Improvement of Glucose Uptake Rate and Production of Target Chemicals by Overexpressing Hexose Transporters and Transcriptional Activator Gcr1 in *Saccharomyces cerevisiae*. *Appl. Environ. Microbiol.* **2015**, *81*, 8392–8401. [CrossRef]
10. Novo, M.; Bigey, F.; Beyne, E.; Galeote, V.; Gavory, F.; Mallet, S.; Cambon, B.; Legras, J.-L.; Wincker, P.; Casaregola, S.; et al. Eukaryote-to-Eukaryote Gene Transfer Events Revealed by the Genome Sequence of the Wine Yeast *Saccharomyces cerevisiae* EC1118. *Proc. Natl. Acad. Sci. USA* **2009**, *106*, 16333–16338. [CrossRef]
11. Treu, L.; Toniolo, C.; Nadai, C.; Sardu, A.; Giacomini, A.; Corich, V.; Campanaro, S. The Impact of Genomic Variability on Gene Expression in Environmental *Saccharomyces cerevisiae* Strains: Genome Analysis of Environmental Yeast Strains. *Environ. Microbiol.* **2014**, *16*, 1378–1397. [CrossRef]
12. Marsit, S.; Mena, A.; Bigey, F.; Sauvage, F.X.; Couloux, A.; Guy, J.; Legras, J.L.; Barrio, E.; Dequin, S.; Galeote, V. Evolutionary Advantage Conferred by an Eukaryote-to-Eukaryote Gene Transfer Event in Wine Yeasts. *Mol. Biol. Evol.* **2015**, *32*, 1695–1707. [CrossRef]

13. Devia, J.; Bastías, C.; Kessi-Pérez, E.I.; Villarroel, C.A.; De Chiara, M.; Cubillos, F.A.; Liti, G.; Martínez, C.; Salinas, F. Transcriptional Activity and Protein Levels of Horizontally Acquired Genes in Yeast Reveal Hallmarks of Adaptation to Fermentative Environments. *Front. Genet.* **2020**, *11*, 293. [CrossRef] [PubMed]

14. Alexandre, H.; Ansanay-Galeote, V.; Dequin, S.; Blondin, B. Global Gene Expression during Short-Term Ethanol Stress in *Saccharomyces cerevisiae*. *FEBS Lett.* **2001**, *498*, 98–103. [CrossRef]

15. Boer, V.M.; de Winde, J.H.; Pronk, J.T.; Piper, M.D.W. The Genome-Wide Transcriptional Responses of *Saccharomyces cerevisiae* Grown on Glucose in Aerobic Chemostat Cultures Limited for Carbon, Nitrogen, Phosphorus, or Sulfur. *J. Biol. Chem.* **2003**, *278*, 3265–3274. [CrossRef] [PubMed]

16. Gibson, B.R.; Boulton, C.A.; Box, W.G.; Graham, N.S.; Lawrence, S.J.; Linforth, R.S.T.; Smart, K.A. Carbohydrate Utilization and the Lager Yeast Transcriptome during Brewery Fermentation. *Yeast* **2008**, *25*, 549–562. [CrossRef] [PubMed]

17. Wu, J.; Zhang, N.; Hayes, A.; Panoutsopoulou, K.; Oliver, S.G. Global Analysis of Nutrient Control of Gene Expression in *Saccharomyces cerevisiae* during Growth and Starvation. *Proc. Nat. Acad. Sci. USA* **2004**, *101*, 3148–3153. [CrossRef] [PubMed]

18. Bely, M.; Sablayrolles, J.; Barre, P. Description of Alcoholic Fermentation Kinetics: Its Variability and Significance. *Am. J. Enol. Viticult.* **1990**, *40*, 319–324.

19. Lemos Junior, W.J.F.; Nadai, C.; Crepalde, L.T.; de Oliveira, V.S.; de Matos, A.D.; Giacomini, A.; Corich, V. Potential Use of *Starmerella bacillaris* as Fermentation Starter for the Production of Low-Alcohol Beverages Obtained from Unripe Grapes. *Int. J. Food Microbiol.* **2019**, *303*, 1–8. [CrossRef]

20. Teste, M.-A.; Duquenne, M.; François, J.M.; Parrou, J.-L. Validation of Reference Genes for Quantitative Expression Analysis by Real-Time RT-PCR in *Saccharomyces cerevisiae*. *BMC Mol. Biol.* **2009**, *10*, 99. [CrossRef]

21. Cankorur-Cetinkaya, A.; Dereli, E.; Eraslan, S.; Karabekmez, E.; Dikicioglu, D.; Kirdar, B. A Novel Strategy for Selection and Validation of Reference Genes in Dynamic Multidimensional Experimental Design in Yeast. *PLoS ONE* **2012**, *7*, e38351. [CrossRef]

22. Nadai, C.; Campanaro, S.; Giacomini, A.; Corich, V. Selection and Validation of Reference Genes for Quantitative Real-Time PCR Studies during *Saccharomyces cerevisiae* Alcoholic Fermentation in the Presence of Sulfite. *Int. J. Food Microbiol.* **2015**, *215*, 49–56. [CrossRef]

23. Rossignol, T.; Dulau, L.; Julien, A.; Blondin, B. Genome-Wide Monitoring of Wine Yeast Gene Expression during Alcoholic Fermentation. *Yeast* **2003**, *20*, 1369–1385. [CrossRef] [PubMed]

24. Vendramini, C.; Beltran, G.; Nadai, C.; Giacomini, A.; Mas, A.; Corich, V. The Role of Nitrogen Uptake on the Competition Ability of Three Vineyard *Saccharomyces cerevisiae* Strains. *Int. J. Food Microbiol.* **2017**, *258*, 1–11. [CrossRef] [PubMed]

25. Zaman, S.; Lippman, S.I.; Schneper, L.; Slonim, N.; Broach, J.R. Glucose Regulates Transcription in Yeast through a Network of Signaling Pathways. *Mol. Syst. Biol.* **2009**, *5*, 245. [CrossRef] [PubMed]

26. Perez, M.; Luyten, K.; Michel, R.; Riou, C.; Blondin, B. Analysis of Hexose Carrier Expression during Wine Fermentation: Both Low- and High-Affinity Hxt Transporters Are Expressed. *FEMS Yeast Res.* **2005**, *5*, 351–361. [CrossRef] [PubMed]

27. Reifenberger, E.; Boles, E.; Ciriacy, M. Kinetic Characterization of Individual Hexose Transporters of *Saccharomyces cerevisiae* and Their Relation to the Triggering Mechanisms of Glucose Repression. *Eur. J. Biochem.* **1997**, *245*, 324–333. [CrossRef]

28. Horák, J. Regulations of Sugar Transporters: Insights from Yeast. *Curr. Genet.* **2013**, *59*, 1–31. [CrossRef]

29. Leandro, M.J.; Sychrová, H.; Prista, C.; Loureiro-Dias, M.C. ZrFsy1, a High-Affinity Fructose/H+ Symporter from Fructophilic Yeast *Zygosaccharomyces rouxii*. *PLoS ONE* **2013**, *8*, e68165. [CrossRef]

30. Leandro, M.J.; Fonseca, C.; Gonçalves, P. Hexose and Pentose Transport in Ascomycetous Yeasts: An Overview. *FEMS Yeast Res.* **2009**, *9*, 511–525. [CrossRef] [PubMed]

31. Wieczorke, R.; Krampe, S.; Weierstall, T.; Freidel, K.; Hollenberg, C.P.; Boles, E. Concurrent Knock-out of at Least 20 Transporter Genes Is Required to Block Uptake of Hexoses in *Saccharomyces cerevisiae*. *FEBS Lett.* **1999**, *464*, 123–128. [CrossRef]

32. Özcan, S.; Johnston, M. Function and Regulation of Yeast Hexose Transporters. *Microbiol. Mol. Biol. Rev.* **1999**, *63*, 554–569. [CrossRef] [PubMed]

33. Diderich, J.A.; Merijn Schuurmans, J.; Van Gaalen, M.C.; Kruckeberg, A.L.; Van Dam, K. Functional Analysis of the Hexose Transporter Homologue *HXT5* in *Saccharomyces cerevisiae*. *Yeast* **2001**, *18*, 1515–1524. [CrossRef] [PubMed]

34. Verwaal, R.; Paalman, J.W.G.; Hogenkamp, A.; Verkleij, A.J.; Verrips, C.T.; Boonstra, J. *HXT5* Expression is Determined by Growth Rates in *Saccharomyces cerevisiae*. *Yeast* **2002**, *19*, 1029–1038. [CrossRef] [PubMed]

35. Buziol, S.; Becker, J.; Baumeister, A.; Jung, S.; Mauch, K.; Reuss, M.; Boles, E. Determination of in Vivo Kinetics of the Starvation-Induced Hxt5 Glucose Transporter Of. *FEMS Yeast Res.* **2002**, *2*, 283–291. [CrossRef]

36. Maier, A.; Völker, B.; Boles, E.; Fuhrmann, G.F. Characterisation of Glucose Transport in *Saccharomyces cerevisiae* with Plasma Membrane Vesicles (Countertransport) and Intact Cells (Initial Uptake) with Single Hxt1, Hxt2, Hxt3, Hxt4, Hxt6, Hxt7 or Gal2 Transporters. *FEMS Yeast Res.* **2002**, *2*, 539–550. [CrossRef] [PubMed]

37. Ozcan, S.; Johnston, M. Two Different Repressors Collaborate to Restrict Expression of the Yeast Glucose Transporter Genes *HXT2* and *HXT4* to Low Levels of Glucose. *Mol. Cell Biol.* **1996**, *16*, 5536–5545. [CrossRef]

38. Ozcan, S.; Johnston, M. Three Different Regulatory Mechanisms Enable Yeast Hexose Transporter (HXT) Genes to Be Induced by Different Levels of Glucose. *Mol. Cell Biol.* **1995**, *15*, 1564–1572. [CrossRef]

39. Gonçalves, P.; Rodrigues de Sousa, H.; Spencer-Martins, I. *FSY1*, a Novel Gene Encoding a Specific Fructose/H+ Symporter in the Type Strain of *Saccharomyces carlsbergensis*. *J. Bacteriol.* **2000**, *182*, 5628–5630. [CrossRef]

40. Varela, C.; Cárdenas, J.; Melo, F.; Agosin, E. Quantitative Analysis of Wine Yeast Gene Expression Profiles under Winemaking Conditions: Gene Expression 'in' Wine Fermentation. *Yeast* **2005**, *22*, 369–383. [CrossRef]
41. Bisson, L. Stuck and Sluggish Fermentations. *Am. J. Enol. Vitic.* **1999**, *50*, 107–119.
42. Pritchard, L.; Kell, D.B. Schemes of Flux Control in a Model of *Saccharomyces cerevisiae* Glycolysis: Flux Control in Yeast Glycolysis. *Europ. J. Biochem.* **2002**, *269*, 3894–3904. [CrossRef] [PubMed]
43. Kayikci, Ö.; Nielsen, J. Glucose Repression in *Saccharomyces Cerevisiae*. *FEMS Yeast Res.* **2015**, *15*, fov068. [CrossRef]
44. Treu, L.; Campanaro, S.; Nadai, C.; Toniolo, C.; Nardi, T.; Giacomini, A.; Valle, G.; Blondin, B.; Corich, V. Oxidative Stress Response and Nitrogen Utilization Are Strongly Variable in *Saccharomyces cerevisiae* Wine Strains with Different Fermentation Performances. *Appl. Microbiol. Biotechnol.* **2014**, *98*, 4119–4135. [CrossRef]
45. Jordan, P.; Choe, J.-Y.; Boles, E.; Oreb, M. Hxt13, Hxt15, Hxt16 and Hxt17 from *Saccharomyces cerevisiae* Represent a Novel Type of Polyol Transporters. *Sci. Rep.* **2016**, *6*, 23502. [CrossRef] [PubMed]

 *fermentation*

Article

Addition of Active Dry Yeast Could Enhance Feed Intake and Rumen Bacterial Population While Reducing Protozoa and Methanogen Population in Beef Cattle

Kampanat Phesatcha [1],*, Krittika Chunwijitra [2], Burarat Phesatcha [3], Metha Wanapat [4] and Anusorn Cherdthong [4],*

[1] Department of Animal Science, Faculty of Agriculture and Technology, Nakhon Phanom University, Nakhon Phanom 48000, Thailand
[2] Department of Food Technology, Faculty of Agriculture and Technology, Nakhon Phanom University, Nakhon Phanom 48000, Thailand; krittika@npu.ac.th
[3] Department of Agricultural Technology and Environment, Faculty of Sciences and Liberal Arts, Rajamangala University of Technology Isan, Nakhon Ratchasima 30000, Thailand; Burarat_kat@hotmail.co.th
[4] Tropical Feed Resources Research and Development Center (TROFREC), Department of Animal Science, Faculty of Agriculture, Khon Kaen University, Khon Kaen 40002, Thailand; metha@kku.ac.th
* Correspondence: kampanatmon@gmail.com (K.P.); anusornc@kku.ac.th (A.C.); Tel.: +66-4320-2362 (A.C.)

Citation: Phesatcha, K.; Chunwijitra, K.; Phesatcha, B.; Wanapat, M.; Cherdthong, A. Addition of Active Dry Yeast Could Enhance Feed Intake and Rumen Bacterial Population While Reducing Protozoa and Methanogen Population in Beef Cattle. *Fermentation* **2021**, *7*, 172. https://doi.org/10.3390/fermentation7030172

Academic Editor: Ronnie G. Willaert

Received: 11 August 2021
Accepted: 28 August 2021
Published: 30 August 2021

Publisher's Note: MDPI stays neutral with regard to jurisdictional claims in published maps and institutional affiliations.

Copyright: © 2021 by the authors. Licensee MDPI, Basel, Switzerland. This article is an open access article distributed under the terms and conditions of the Creative Commons Attribution (CC BY) license (https://creativecommons.org/licenses/by/4.0/).

Abstract: Urea–lime-treated rice straw fed to Thai native beef cattle was supplemented with dry yeast (DY) (*Saccharomyces cerevisiae*) to assess total feed intake, nutrient digestibility, rumen microorganisms, and methane (CH_4) production. Sixteen Thai native beef cattle at 115 ± 10 kg live weight were divided into four groups that received DY supplementation at 0, 1, 2, and 3 g/hd/d using a randomized completely block design. All animals were fed concentrate mixture at 0.5% of body weight, with urea–lime-treated rice straw fed ad libitum. Supplementation with DY enhanced total feed intake and digestibility of neutral detergent fiber and acid detergent fiber ($p < 0.05$), but dry matter, organic matter and crude protein were similar among treatments ($p > 0.05$). Total volatile fatty acid (VFA) and propionic acid (C3) increased ($p < 0.05$) with 3 g/hd/d DY supplementation, while acetic acid (C2) and butyric acid (C4) decreased. Protozoal population and CH_4 production in the rumen decreased as DY increased ($p < 0.05$). Populations of *F. succinogenes* and *R. flavefaciens* increased ($p < 0.05$), whereas methanogen population decreased with DY addition at 3 g/hd/d, while *R. albus* was stable ($p > 0.05$) throughout the treatments. Thus, addition of DY to cattle feed increased feed intake, rumen fermentation, and cellulolytic bacterial populations.

Keywords: beef cattle; digestibility; ruminal fermentation; yeast

1. Introduction

To develop more effective ruminant production systems, ruminants must have a high fermentation capacity. The ability of the microbial ecology to digest organic substances into milk and meat precursors is required for increased production [1]. Many feed additives, such as direct-fed microorganisms, are employed to promote livestock productivity. Yeast-derived products, such as *Saccharomyces cerevisiae*, stand out in this group because they are beneficial to animal health and ruminal enhancer [2]. In the rumen, yeast can utilize the remaining dissolved oxygen, sparing anaerobic microbes from the damaging effects of oxygen. Yeasts can increase rumen maturity and regulate ruminal pH by competing with lactic generating bacteria, minimizing the danger of acidosis [3]. Yeast products lower rumen pH by encouraging microorganisms that convert lactate into short-chain fatty acids [4]. Yeast improves cattle feed digestion and metabolism in a variety of ways, including increasing nutritional digestibility, optimizing volatile fatty acid proportions, decreasing ammonia–nitrogen, lowering pH fluctuation, and stimulating microbial communities in the rumen [5]. Furthermore, yeast provides several growth factors, pro-vitamins, and other stimulants to rumen microbes while decreasing rumen redox potential and stimulating the growth of ruminal bacteria, primarily cellulolytic bacteria, which enhance fiber degradation [6]. Moreover, yeast enhanced total volatile fatty acids (VFAs) while decreasing acetate proportion. Supplementation of yeast can help cellulolytic bacteria and increase the digestibility of Nellore cattle [7]. This could be due to increasing cellulolytic and lactate-utilizing bacterial populations modifying lactate-to-propionate fermentative pathways [8]. Feed intake, milk yield, weight gain, digestibility of nutrients, cellulolytic bacteria numbers, and volatile fatty acid patterns have all been shown to improve with yeast supplementation in the ruminant diet [9,10]. Furthermore, there is a limitation of data because only a few researchers have studied the rumen microorganism and methane emission in cattle fed urea–lime-treated rice straw. The aim of this study was to study supplementation with *Saccharomyces cerevisiae* on feed intake, nutritional digestibility, rumen fermentation, rumen microorganism, and methane production in Thai native beef calves fed urea–lime-treated rice straw as a basal roughage.

2. Materials and Methods

The study design and plan strictly followed the norms of the Animal Ethics Committee of Nakhon Phanom University, Mueang Nakhon Phanom, Thailand (permission No. AENPU A2/2560). This study primarily involved laboratory analysis of ruminant feeds, for which permission to collect rumen fluid from animals was granted in accordance with the Thailand Ethics of Animal Experimentation of the National Research Council.

2.1. Animals, Feed, and Experimental Design

Sixteen Thai native beef cattle with 115 ± 10 kg live weight were blocked into four groups to receive active *Saccharomyces cerevisiae* (dry form) supplementation at 0, 1, 2, and 3 g/hd/day. The yeast strain *S. cerevisiae* in this study was obtained from the Renu Nakhon district, Nakhon Phanom Province, Thailand. All animals received a concentrated mixture at 0.5% body weight, while urea–lime-treated rice straw, water, and mineral blocks were available ad libitum.

Ingredient compositions of concentrate mixture and nutrient composition are presented in Table 1. The rice straw-treated urea–calcium hydroxide was made by adding 2 kg of urea and 2 kg of $Ca(OH)_2$ in 100 L to 100 kg of rice straw. The quantity of urea and calcium hydroxide solution was sprayed onto rice straw bales and then covered with a sheet of plastic for at least 10 days before feeding to the beef cattle [11].

Table 1. Compositions of concentrate mixtures and urea–calcium hydroxide-treated rice straw.

Items	Concentrate	Urea–Calcium Hydroxide-Treated Rice Straw
Concentrate ingredients, % dry matter basis		
Cassava chip	63.5	
Coconut meal	10.5	
Palm kernel meal	7.5	
Rice bran	11.0	
Urea	3.0	
Molasses	2.0	
Mineral mixture	1.0	
Salt	1.0	
Sulfur	0.5	
Chemical composition		
Dry matter, %	89.7	50.7
	% of dry matter	
Organic matter	94.6	90.1
Ash	5.4	9.9
Crude protein	14.0	4.3
Neutral detergent fiber	25.6	70.2
Acid detergent fiber	15.8	48.7
Total digestible nutrients (TDN) *	79.8	50.4

* Calculated value %TDN = [%digestible CP + Crude fiber (CF) + Nitrogen free extract (NFE) + (2.25 × %Digestible Ether extract (EE)].

The animals were dewormed and given a 14-day acclimation period prior to the experiment. The feeding trial lasted 90 days, with the digestibility test taking place in the final week. The individual feed was made and fed to the cattle in the morning and evening. To measure daily feed intake, the amount of feed supplied and denied was recorded every day, and to determine weight change, weighing was performed every two weeks before feeding time.

2.2. Samples Collection and Chemical Analyses of Samples

Representative feed, fecal, and urine samples were collected throughout the last 7 days, weighed, and oven-dried at 60 °C for 48 h. The composite samples were dried at 60 °C before being processed (1 mm screen using Cyclotech Mill, Tecator, 1093, Hoganas, Sweden) and tested for DM, CP, and ash [12]. The acid detergent fiber (ADF) was calculated and expressed including residual ash. Determined neutral detergent fiber (NDF) in samples with the addition of alpha-amylase but without sodium sulfite, and the findings are given inclusive of residual ash according to Van Soest et al. [13]. Nutrient digestibility was calculated using acid insoluble ash [14], and DM, OM, NDF, and ADF digestibility were determined from the ratio of AIA in feed and feces, and digestibility of nitrogen was determined from ratios of AIA and N in feed and feces.

On the final day of the trial, rumen fluid and blood samples were collected at 0 and 4 h following the morning feed. Each time, a stomach tube connected to a vacuum pump was utilized to collect approximately 200 mL of rumen fluid from the rumen. Rumen fluid pH and temperature were immediately measured, and 50 mL of rumen fluid was collected and mixed with 5 mL of 1M H_2SO_4 to stop microbial activity fermentation before centrifugation at $16{,}000 \times g$ for 15 min. A total of 20 cc of supernatant was taken and frozen at -20 °C before being analyzed in the laboratory for ammonia–nitrogen (NH_3–N) using micro-Kjeldahl methods [12].

High Performance Liquid Chromatography was used to examine rumen fluid samples for VFAs (HPLC; Model Water 600; UV detector, Millipore Corp., Milford, MA, USA). Rumen CH_4 production was approximated using equation of Moss et al. [15]. VFA proportions are as follows: production of $CH_4 = 0.45$ (acetate, C_2) + 0.275 (propionate, C_3) + 0.4 (butyrate, C_4).

The second portion was fixed with 10% formalin for the determined protozoal population using the direct count microscopic method as described by Galyean [16].

The community DNA was isolated from rumen fluid and digesta. QIAgen DNA Mini Stool Kit columns were used to purify the DNA (QIAGEN, Valencia, California, USA). Real-time PCR was used to determine the relative populations of total bacteria, rumen bacteria for fiber degradation (*Ruminococcus albus*, *Ruminococcus flavefaciens*, *Butyrivibrio fibrisolvens*, and *Fibrobacter succinogenes*), and methanogen. Total DNA was extracted from the samples using the method described by Stevenson et al. [17]. Extracted DNA was utilized as a template in real-time PCR experiments with specified primers to measure the microbial population of *R. albus*, and *R. flavefaciens* [18], *B. fibrisolvens* [17], *F. succinogenes* [19], and methanogen [20]. The DNA standards for real-time PCR amplification and detection were determined using a Chromo 4™ system (Bio-Rad, California, USA). The data of microbial population were transferred to log10 prior to statistical analysis.

At the same time as the rumen fluid was collected, a 10 mL blood sample was taken from the jugular vein into a tube containing 0.1 g of ethylenediaminetetraacetic acid (EDTA). All tubes were centrifuged at $3000 \times g$ for 15 min to obtain plasma and then stored at -20 °C for further analyses of blood urea nitrogen (BUN) [21].

2.3. Statistical Methods

The data were analyzed using the MIXED procedure in SAS software [22]. The mathematical model assumption used was:

$$Yi = \mu + T_i + \beta_i + \varepsilon_i$$

where Y_i is the dependent variable, μ is the overall mean, T_i is the ith treatment effect (supplementation of *Saccharomyces cerevisiae* at 0, 1, 2, 3 g/hd/day), βi is the ith block effect, and $_i$ is the residual error of the ith observation. Differences among means with $p < 0.05$ were represented as statistically significant differences. Orthogonal polynomials for diet responses were determined by linear and quadratic effects.

3. Results

3.1. Feed Intake and Digestibility

DY supplementation enhanced total feed intake of urea–lime-treated rice straw by Thai native beef cattle ($p < 0.05$) but did not alter digestibility DM, OM, or CP ($p > 0.05$). Supplementation at 3 g/hd/d increased digestibility of fiber (NDF, ADF) ($p < 0.05$; Table 2).

Table 2. Effect of yeast supplementation on voluntary feed intake and nutrient digestibility in Thai native beef cattle.

Items	Yeast Supplementation (g/day)				SEM	Contrast	
	0	1	2	3		Linear	Quadratic
Dry matter intake							
Roughage intake							
kg/day	1.9 [a]	2.0 [a]	2.3 [b]	2.5 [c]	0.18	0.04	0.43
g/kg BW$^{0.75}$	65.9 [a]	66.0 [a]	68.1 [b]	70.6 [c]	0.76	0.04	0.52
Concentrate intake							
kg/day	0.7	0.7	0.7	0.7	0.31	0.17	0.41
g/kg BW$^{0.75}$	17.8	18.9	17.7	19.0	1.67	0.15	0.32
Total feed intake							
kg/day	2.6 [a]	2.7 [a]	3.0 [b]	3.3 [c]	0.09	0.04	0.05
g/kg BW$^{0.75}$	84.7 [a]	84.9 [a]	85.7 [b]	89.6 [c]	1.24	0.04	0.05
Nutrient digestibility, %							
Dry matter	57.5	58.9	60.2	60.0	0.17	0.14	0.47
Organic matter	62.8	62.7	62.4	63.4	0.05	0.25	0.32
Crude protein	58.6	59.1	59.7	60.2	0.09	0.17	0.21
Neutral detergent fiber	50.1 [a]	51.9 [a]	53.2 [b]	55.2 [c]	0.08	0.03	0.04
Acid detergent fiber	41.4 [a]	42.3 [a]	44.9 [b]	46.9 [c]	0.06	0.02	0.03

[a,b,c] means within a row with different superscripts differ significantly ($p < 0.05$); SEM = standard error of the mean.

3.2. Rumen Fermentation, and Blood Urea Nitrogen

Table 3 shows the effect of DY on rumen fermentation and BUN. The ruminal pH (6.6–6.8) and ruminal temperature (39.0–39.5 °C) remained stable ($p > 0.05$). The concentration of NH_3–N increased in the DY supplementation groups and was highest at 3 g/hd/d but did not affect the concentration of BUN ($p > 0.05$).

Table 3. Effect of yeast supplementation on fermentation characteristics and blood urea nitrogen in Thai native beef cattle.

Items	Yeast Supplementation (g/day)				SEM	Contrast	
	0	1	2	3		Linear	Quadratic
Ruminal pH	6.8	6.8	6.7	6.6	0.09	0.09	0.15
Temperature, °C	39.5	39.0	39.4	39.5	0.22	0.34	0.46
NH_3–N, mg/dL	12.1 [a]	12.9 [a]	13.3 [b]	15.6 [c]	0.30	0.02	0.03
BUN, mg/dL	9.1	9.5	10.7	11.3	0.06	0.52	0.62
Total VFAs, mmol/L	90.1 [a]	92.8 [a]	96.5 [b]	100.3 [c]	0.15	0.03	0.04
VFAs, mol/100mol							
Acetic acid (C_2)	68.1 [c]	66.6 [b]	66.4 [b]	64.8 [a]	0.18	0.02	0.04
Propionic acid (C_3)	20.9 [a]	22.6 [b]	24.1 [c]	26.0 [d]	0.16	0.02	0.03
Butyric acid (C_4)	11.0 [b]	10.8 [b]	9.5 [a]	9.2 [a]	0.07	0.03	0.05
C_2: C_3	3.3 [c]	2.9 [b]	2.8 [b]	2.5 [a]	0.31	0.04	0.07
CH_4 (mM)	29.3 [c]	28.1 [b]	27.1 [b]	25.7 [a]	0.25	0.04	0.05

[a,b,c,d] means within a row with different superscripts differ significantly ($p < 0.05$); SEM = standard error of the mean; NH_3–N = ammonia–nitrogen; BUN = blood urea nitrogen; VFAs = volatile fatty acids; CH_4 = methane production = 0.45 (C_2) − 0.275 (C_3) + 0.4 (C_4) calculated according to Moss et al. [15].

3.3. Volatile Fatty Acid (VFA) Profiles and Methane (CH_4) Production

Concentrations of total volatile fatty acid (TVFA) and propionic acid (C3) increased ($p < 0.05$) with DY supplementation, particularly for DY at 3 g/hd/d. However, acetic acid (C2) and butyric acid (C4) concentrations, C2: C3 ratio and CH_4 production reduced with the addition of DY at 3 g/hd/d.

3.4. Microbial Population

Protozoal population significantly reduced ($p < 0.05$) with LY addition at 3 g/hd/d. The bacteria, *F. succinogenes*, *B. fibrisolvens* and *R. flavefaciens* increased, whereas the methanogenic population decreased with DY addition at 3 g/hd/d. *R. albus* was stable ($p > 0.05$) throughout all treatments (Table 4).

Table 4. Effect of yeast supplementation on microbial population in Thai native beef cattle.

Items	Yeast Supplementation (g/day)				SEM	Contrast	
	0	1	2	3		Linear	Quadratic
Direct count, cell/mL							
Protozoa, $\times 10^6$ cell/mL	8.1 [d]	6.9 [c]	5.2 [b]	3.5 [a]	0.19	0.04	0.05
Real-time PCR, copies/mL rumen content							
F. succinogenes, $\times 10^6$	3.2 [a]	3.6 [a]	4.8 [b]	5.9 [c]	0.07	0.04	0.07
R. flavefaciens, $\times 10^5$	2.1 [a]	2.4 [a]	3.9 [b]	4.8 [c]	0.31	0.04	0.06
R. albus, $\times 10^6$	5.0	4.9	5.2	5.5	0.16	0.06	0.08
B. fibrisolvens, $\times 10^5$	2.5 [a]	3.1 [a]	4.4 [b]	6.8 [c]	0.21	0.04	0.05
Methanogens, $\times 10^2$	6.6 [a]	5.8 [a]	4.7 [b]	3.4 [c]	0.09	0.04	0.05

[a,b,c,d] means within a row with different superscripts differ significantly ($p < 0.05$); SEM = standard error of the mean.

4. Discussion

4.1. Feed Intake and Nutrient Digestibility

Total feed intake increased with DY supplementation, with the highest found at 3 g/hd/d ($p < 0.05$), concurring with Crossland et al. [6], who found that adding yeast to cattle diet increased dry matter intake. Supplementation of DY at 3 g/hd/d also increased fiber digestibility (NDF, ADF) ($p < 0.05$) due to the ability of yeast to scavenge excess oxygen in the rumen, lower the redox potential, and enhance the degradability of NDF and ADF. Yeast provides an ecological setting that encourages the proliferation and activity of microbes, especially cellulolytic bacteria that enhance NDF and ADF breakdown. Guedes et al. [23] discovered that feeding cattle with yeast improved NDF degradation of maize silage. By contrast, Mir and Mir [24] found that supplementing cattle feed with live yeast did not impact DM and NDF degradation in the rumen. Satori et al. [10] stated that the highest total intake and average daily gain were observed in cattle supplemented with yeast at below 6 g/d.

4.2. Rumen Ecology and Blood Urea-Nitrogen

Ruminal pH and temperature values for all DY supplementations were reported in the optimal range by Phesatcha et al. [11]. In general, rumen pH stability benefits acid-sensitive cellulolytic bacteria and is extremely beneficial to beef cattle, especially fattening cattle. Monnerat et al. [25] and Ghasemi et al. [26] reported that adding yeast to high concentrate cattle feed did not affect rumen pH.

The NH_3–N concentration increased with DY supplementation at 3 g/hd/d. The concentration of BUN was similar among treatments and ranged between 9.1and 11.3 mg/dl, and in the normal range as reported by Wanapat and Pimpa [27]. By contrast, Li et al. [28] found that the addition of yeast to cattle feed decreased BUN.

4.3. Ruminal Volatile Fatty Acid (VFA) Profiles and Methane (CH$_4$) Production

In this study, beef calves fed DY had higher total VFA levels and higher C3 levels, while the C2, C4, and C2 to C3 ratio were lower than those no supplemented group. This was due to an increase in the lactate-utilizing bacteria *Selenomonas ruminantium* and *Megasphaera elsdenii* that convert lactate to C3, with their growth stimulated by yeast supplementation [10,29]. When compared to the control, yeast supplementation increased total VFA, C3, and valeric acid but decreased C2 and the C2 to C3 ratio [30,31]. Dawson et al. [32] found that for yeast supplements containing in vitro total VFA, the molar proportion of C3 increased while C2 decreased. Variable effects of yeast on rumen fermentation efficiency can be attributed to dose, diet type, different yeast strains, animal physiological stage, and feeding systems [7,33].

Major alterations of CH$_4$ in ruminants are produced through propionate fermentation, and CH$_4$ production decreased with yeast supplementation. This result concurred with Phesatcha et al. [11] and Wang et al. [34], who found that CH$_4$ production decreased with yeast supplementation, while Munoz et al. [35] observed that DY supplementation increased CH$_4$ production in lactating dairy cows and Bayat et al. [36] determined that yeast did not influence CH$_4$ emissions. Diverse effects of yeast supplementation on CH$_4$ synthesis were attributed to varying yeast strains, dosages, and diets utilized in the trials [28]. Yeast can be used to minimize CH$_4$ emissions and was shown to lower methane production in the rumen by encouraging acetogens to use more hydrogen in the process of acetate formation by Darabighane et al. [37]. In this study, the methanogen population was reduced with yeast supplementation and was lowest at 3 g/hd/d.

The protozoal population was reduced with DY supplementation. Microbial populations studied using real-time PCR revealed that *R. flavefaciens*, *B. fibrisolvens*, and *F. succinogenes* increased, while methanogen population decreased with yeast supplementation at 3 g/hd/d. This result concurred with Sousa et al. [7], who reported that the addition of yeast significantly increased the relative population of *R. flavefaciens*. The addition of DY stimulated the growth of cellulolytic bacterial populations (*R. flavefaciens* and *F. succinogenes*), while suppressing growth of the lactate-producing bacterium (*Streptococcus bovis*), thereby improving the consistency of rumen fermentation [38]. Enhanced fiber degradation increased total cellulolytic bacteria in the rumen. Ding et al. [39] found that the addition of yeast increased bacteria, fungi, protozoa, lactate-utilizing bacteria, and rate of fiber decomposition. Growth factors induced by organic acids and vitamins provided by yeast may enhance cellulolytic bacterial and fungal colonization in the rumen. Yeast promoted microbial proliferation, specifically lactic acid-utilizing bacteria, and reduced acidosis [3,9]. Furthermore, as a facultative anaerobe organism, yeast gathers available oxygen on the surface of freshly swallowed meals to sustain metabolic activity, thereby lowering rumen redox potential. Removal of oxygen improves growth conditions for strict anaerobic cellulolytic bacteria, increasing their adherence to fodder particles and shortening the cellulolytic process [33]. Jiang et al. [29] examined the ruminal microbiota of cows fed with different amounts of yeast. They found that the number of *Butyrivibrio fibrisolvens*, an important hemicellulolytic species, was reduced in cows supplemented with a high dose of yeast. In our study, methane production and methanogen population reduced with yeast supplementation, and the lowest values were found at 3 g/hd/d. By contrast, Lu et al. [40] reported that adding yeast at 6 and 12 g/d decreased methane production without affecting the number or diversity of methanogens.

5. Conclusions

The addition of DY at 3 g/hd/d enhanced total feed intake, rumen fermentation, and total bacteria populations while reducing protozoal population and CH$_4$ production in beef cattle fed with urea–lime-treated rice straw. However, there are certain drawbacks related to the fattening beef cattle influenced by DY addition, which requires further study.

Author Contributions: Conceptualization, K.P.; methodology, K.P., K.C. and B.P., validation, K.P. and A.C.; formal analysis, K.P., B.P. and K.C.; investigation, K.P.; resources, K.P. and B.P.; data curation, K.P.; manuscript drafting, K.P.; manuscript editing and finalizing, K.P., A.C. and M.W.; visualization, K.P.; supervision, A.C. and M.W.; project administration, K.P.; funding acquisition, K.P.; All authors have read and agreed to the published version of the manuscript.

Funding: The authors express their most sincere gratitude to the Nakhon Phanom University Fund through the middle class researcher scholarships (MR3/2563), the Research Program on the Research and Development of Winged Bean Root Utilization as Ruminant Feed; the Increase Production Efficiency and Meat Quality of Native Beef and Buffalo Research Group; Research and Graduate Studies, Khon Kaen University (KKU).

Institutional Review Board Statement: The study was conducted under approval procedure no. AENPU A2/2560 of animal Ethics and Care issue by Nakhon Phanom University.

Informed Consent Statement: Not applicable.

Data Availability Statement: Not applicable.

Acknowledgments: Thanks are extended to the Tropical Feed Resources Research and Development Center (TROFREC), Department of Animal Science, Faculty of Agriculture, Khon Kaen University, Khon Kaen, Thailand, for their support with the use of their facilities.

Conflicts of Interest: We declare that no conflict of interest exists among the authors.

References

1. Elghandour, M.M.Y.; Khusro, A.; Adegbeye, M.J.; Tan, Z.; Abu Hafsa, S.H.; Greiner, R.; Ugbogu, E.A.; Anele, U.Y.; Salem, A.Z.M. Dynamic role of single-celled fungi in ruminal microbial ecology and activities. *J. Appl. Microbiol.* **2020**, *128*, 950–965. [CrossRef]
2. Faccio-Demarco, C.; Mumbach, T.; Oliveira-de-Freitas, V.; Fraga e Silva-Raimondo, R.; Medeiros-Gonçalves, F.; Nunes-Corrêa, M.; Burkert-Del Pino, F.A.; Mendonça-Nunes-Ribeiro Filho, H.; Cassal-Brauner, C. Effect of yeast products supplementation during transition period on metabolic profile and milk production in dairy cows. *Trop. Anim. Health Prod.* **2019**, *51*, 2193–2201. [CrossRef]
3. Chaucheyras-Durand, F.; Ameilbonne, A.; Bichat, A.; Mosoni, P.; Ossa, F.; Forano, E. Live yeasts enhance fibre degradation in the cow rumen through an increase in plant substrate colonization by fibrolytic bacteria and fungi. *J. Appl. Microbiol.* **2016**, *120*, 560–570. [CrossRef] [PubMed]
4. Desnoyers, M.; Giger-Reverdin, S.; Bertin, G.; Duvaux-Ponter, C.; Sauvant, D. Meta-analysis of the influence of *Saccharomyces cerevisiae* supplementation on ruminal parameters and milk production of ruminants. *J. Dairy Sci.* **2009**, *92*, 1620–1632. [CrossRef]
5. Cagle, C.M.; Fonseca, M.A.; Callaway, T.R.; Runyan, C.A.; Cravey, M.D.; Tedeschi, L.O. Evaluation of the effects of live yeast on rumen parameters and in situ digestibility of dry matter and neutral detergent fiber in beef cattle fed growing and finishing diets. *Appl. Anim. Sci.* **2020**, *36*, 36–47. [CrossRef]
6. Crossland, W.L.; Cagle, C.M.; Sawyer, J.E.; Callaway, T.R.; Tedeschi, L.O. Evaluation of active dried yeast in the diets of feedlot steers. II. Effects on rumen pH and liver health of feedlot steers. *J. Anim. Sci.* **2019**, *97*, 1347–1363. [CrossRef]
7. Sousa, D.O.; Oliveira, C.A.; Velasquez, A.V.; Souza, J.M.; Chevaux, E.; Mari, L.J.; Silva, L.F.P. Live yeast supplementation improves rumen fibre degradation in cattle grazing tropical pastures throughout the year. *Anim. Feed Sci. Technol.* **2018**, *236*, 149–158. [CrossRef]
8. Opsi, F.; Fortina, R.; Tassone, S.; Bodas, R.; López, S. Effects of inactivated and live cells of Saccharomyces cerevisiae on in vitro ruminal fermentation of diets with different forage:concentrate ratio. *J. Agric. Sci.* **2012**, *150*, 271–283. [CrossRef]
9. Pinloche, E.; McEwan, N.; Marden, J.P.; Bayourthe, C.; Auclair, E.; Newbold, C.J. The Effects of a probiotic yeast on the bacterial diversity and population structure in the rumen of cattle. *PLoS ONE* **2013**, *8*, e67824. [CrossRef]
10. Sartori, E.D.; Canozzi, M.E.A.; Zago, D.; Prates, Ê.R.; Velho, J.P.; Barcellos, J.O.J. The Effect of live yeast supplementation on beef cattle performance: A systematic review and meta-analysis. *J. Agric. Sci.* **2017**, *9*, 21. [CrossRef]
11. Phesatcha, K.; Phesatcha, B.; Wanapat, M.; Cherdthong, A. Roughage to concentrate ratio and *Saccharomyces cerevisiae* inclusion could modulate feed digestion and in vitro ruminal fermentation. *Vet. Sci.* **2020**, *7*, 151. [CrossRef]
12. Association of Official Analytical Chemists (AOAC). *Official Methods of Analysis*, 19th ed.; AOAC International: Gaithersburg, MD, USA, 2012.
13. Van Soest, P.J.; Robertson, J.B.; Lewis, B.A. Methods for dietary fiber neutral detergent fiber, and nonstarch polysaccharides in relation to animal nutrition. *J. Dairy Sci.* **1991**, *74*, 3583–3597. [CrossRef]
14. Van Keulen, J.Y.; Young, B.A. Evaluation of acid-insoluble ash as a natural marker in ruminant digestibility studies. *J. Anim. Sci.* **1977**, *44*, 282–287. [CrossRef]
15. Moss, A.R.; Jouany, J.P.; Newbold, J. Methane production by ruminants: Its contribution to global warming. *Anim. Res.* **2000**, *49*, 231–253. [CrossRef]
16. Galyean, M. *Laboratory Procedure in Animal Nutrition Research*; Department of Animal and Range Sciences, New Mexico State University: Las Cruces, NM, USA, 1989; Volume 188.

17. Stevenson, D.M.; Weimer, P.J. Dominance of Prevotella and low abundance of classical ruminal bacterial species in the bovine rumen revealed by relative quantification real-time PCR. *Appl. Microbiol. Biotechnol.* **2007**, *75*, 165–174. [CrossRef]
18. Koike, S.; Kobayashi, Y. Development and use of competitive PCR assays for the rumen cellulolytic bacteria: *Fibrobacter succinogenes, Ruminococcus albus* and *Ruminococcus flavefaciens. FEMS Microbiol. Lett.* **2001**, *204*, 361–366. [CrossRef]
19. Tajima, K.; Aminov, R.I.; Nagamine, T.; Matsui, H.; Nakamura, M.; Benno, Y. Diet-dependent shifts in the bacterial population of the rumen revealed with real-time PCR. *Appl. Environ. Microbiol.* **2001**, *67*, 2766–2774. [CrossRef]
20. Denman, S.E.; Tomkins, N.W.; McSweeney, C.S. Quantitation and diversity analysis of ruminal methanogenic populations in response to the antimethanogenic compound bromochloromethane. *FEMS Microbiol. Ecol.* **2007**, *62*, 313–322. [CrossRef]
21. Crocker, C.L. Rapid determination of urea nitrogen in serum or plasma without deproteinization. *Am. J. Med. Technol.* **1967**, *33*, 361–365.
22. SAS (Statistical Analysis System). *User's Guide: Statistic*, 9.3th ed.; SAS Inst. Inc.: Cary, NC, USA, 2013.
23. Guedes, C.M.; Gonçalves, D.; Rodrigues, M.A.M.; Dias-da-Silva, A. Effects of a *Saccharomyces cerevisiae* yeast on ruminal fermentation and fibre degradation of maize silages in cows. *Anim. Feed Sci. Technol.* **2008**, *145*, 27–40. [CrossRef]
24. Mir, Z.; Mir, P.S. Effect of the addition of live yeast (*Saccharomyces cerevisiae*) on growth and carcass quality of steers fed high-forage or high-grain diets and on feed digestibility and in situ degradability. *J. Anim. Sci.* **1994**, *72*, 537–545. [CrossRef]
25. dos Santos Monnerat, J.P.; Paulino, P.V.R.; Detmann, E.; Valadares Filho, S.C.; Valadares, R.D.F.; Duarte, M.S. Effects of *Saccharomyces cerevisiae* and monensin on digestion, ruminal parameters, and balance of nitrogenous compounds of beef cattle fed diets with different starch concentrations. *Trop. Anim. Health Prod.* **2013**, *45*, 1251–1257. [CrossRef] [PubMed]
26. Ghasemi, E.; Khorvash, M.; Nikkhah, A. Effect of forage sources and *Saccharomyces cerevisiae* (Sc47) on ruminal fermentation parameters. *S. Afr. J. Anim. Sci.* **2012**, *42*, 164–168. [CrossRef]
27. Wanapat, M.; Pimpa, O. Effect of ruminal NH_3-N levels on ruminal fermentation, purine derivatives, digestibility and rice straw intake in swamp buffaloes. *Asian-Australas. J. Anim. Sci.* **1999**, *12*, 904–907. [CrossRef]
28. Li, Y.; Shen, Y.; Niu, J.; Guo, Y.; Pauline, M.; Zhao, X.; Li, Q.; Cao, Y.; Bi, C.; Zhang, X.; et al. Effect of active dry yeast on lactation performance, methane production, and ruminal fermentation patterns in early-lactating Holstein cows. *J. Dairy Sci.* **2021**, *104*, 381–390. [CrossRef]
29. Jiang, Y.; Ogunade, I.M.; Qi, S.; Hackmann, T.J.; Staples, C.R.; Adesogan, A.T. Effects of the dose and viability of *Saccharomyces cerevisiae*. 1. Diversity of ruminal microbes as analyzed by Illumina MiSeq sequencing and quantitative PCR. *J. Dairy Sci.* **2017**, *100*, 325–342. [CrossRef]
30. Xiao, J.X.; Alugongo, G.M.; Chung, R.; Dong, S.Z.; Li, S.L.; Yoon, I.; Wu, Z.H.; Cao, Z.J. Effects of *Saccharomyces cerevisiae* fermentation products on dairy calves: Ruminal fermentation, gastrointestinal morphology, and microbial community. *J. Dairy Sci.* **2016**, *99*, 5401–5412. [CrossRef] [PubMed]
31. McAllister, T.A.; Beauchemin, K.A.; Alazzeh, A.Y.; Baah, J.; Teather, R.M.; Stanford, K. Review: The use of direct fed microbials to mitigate pathogens and enhance production in cattle. *Can. J. Anim. Sci.* **2011**, *91*, 193–211. [CrossRef]
32. Dawson, K.A.; Hopkins, D.M. Differential effects of live yeast on the cellulolytic activities of anaerobic ruminal bacteria. *J. Anim. Sci.* **1991**, *69* (Suppl. 1), 531.
33. Patra, A.K. The use of live yeast products as microbial feed additives in ruminant nutrition. *Asian. J. Anim. Vet. Adv.* **2012**, *7*, 366–375. [CrossRef]
34. Wang, Z.; He, Z.; Beauchemin, K.A.; Tang, S.; Zhou, C.; Han, X.; Wang, M.; Kang, J.; Odongo, N.E.; Tan, Z. Evaluation of different yeast species for improving in vitro fermentation of cereal straws. *Asian-Australas. J. Anim. Sci.* **2016**, *29*, 230–240. [CrossRef]
35. Munoz, C.; Wills, D.A.; Yan, T. Effects of dietary active dried yeast (*Saccharomyces cerevisiae*) supply at two levels of concentrate on energy and nitrogen utilisation and methane emissions of lactating dairy cows. *Anim. Prod. Sci.* **2017**, *57*, 656–664. [CrossRef]
36. Bayat, A.R.; Kairenius, P.; Stefański, T.; Leskinen, H.; Comtet-Marre, S.; Forano, E.; Chaucheyras-Durand, F.; Shingfield, K.J. Effect of camelina oil or live yeasts (*Saccharomyces cerevisiae*) on ruminal methane production, rumen fermentation, and milk fatty acid composition in lactating cows fed grass silage diets. *J. Dairy Sci.* **2015**, *98*, 3166–3181. [CrossRef] [PubMed]
37. Darabighane, B.; Salem, A.Z.M.; Aghjehgheshlagh, F.M.; Mahdavi, A.; Zarei, A.; Elghandour, M.M.Y.; Lopez, S. Environmental efficiency of *Saccharomyces cerevisiae* on methane production in dairy and beef cattle via a meta-analysis. *Environ. Sci. Pollut. Res.* **2018**, *26*, 3651–3658. [CrossRef] [PubMed]
38. Zhu, W.; Wei, Z.; Xu, N.; Yang, F.; Yoon, I.; Chung, Y.; Liu, J.; Wang, J. Effects of *Saccharomyces cerevisiae* fermentation products on performance and rumen fermentation and microbiota in dairy cows fed a diet containing low quality forage. *J. Anim. Sci. Biotechnol.* **2017**, *8*, 36. [CrossRef] [PubMed]
39. Ding, G.; Chang, Y.; Zhao, L.; Zhou, Z.; Ren, L.; Meng, Q. Effect of *Saccharomyces cerevisiae* on alfalfa nutrient degradation characteristics and rumen microbial populations of steers fed diets with different concentrate-to-forage ratios. *J. Anim. Sci. Biotechnol.* **2014**, *5*, 24. [CrossRef]
40. Lu, Q.; Wu, J.; Wang, M.; Zhou, C.; Han, X.; Odongo, E.N.; Tan, Z.; Tang, S. Effects of dietary addition of cellulase and a *Saccharomyces cerevisiae* fermentation product on nutrient digestibility, rumen fermentation and enteric methane emissions in growing goats. *Arch. Anim. Nutr.* **2016**, *70*, 224–238. [CrossRef] [PubMed]

fermentation

MDPI

Article

Survey of Inoculated Commercial *Saccharomyces cerevisiae* in Winery-Based Trials

Filomena L. Duarte [1,2] and M. Margarida Baleiras-Couto [1,2,*]

[1] Instituto Nacional de Investigação Agrária e Veterinária, INIAV-Dois Portos, Quinta da Almoinha, 2565-191 Dois Portos, Portugal; filomena.duarte@iniav.pt
[2] BioISI-Biosystems and Integrative Sciences Institute, Faculty of Sciences, University of Lisboa, 1749-016 Lisboa, Portugal
* Correspondence: margarida.couto@iniav.pt

Abstract: Wine production has developed from spontaneous to controlled fermentations using commercial active dry yeasts (ADY). In this study, *S. cerevisiae* commercial ADY were tested, and yeast community dynamics were monitored at different fermentation stages in three winery-based trials with volumes ranging from 60 L to 250 hL. The differentiation of *S. cerevisiae* strains was achieved using microsatellite markers. In Experiment 1, results showed that both ADY strains revealed similar profiles, despite being described by the producer as having different properties. In Experiment 2, higher genetic diversity was detected when co-inoculation was tested, while in sequential inoculation, the initial ADY seemed to dominate throughout all fermentation. Pilot-scale red wine fermentations were performed in Experiment 3, where one single ADY strain was tested along with different oenological additives. Surprisingly, these trials showed an increase in distinct profiles towards the end of fermentation, indicating that the dominance of the ADY was lower than in the blank modality. The use of ADY is envisaged to promote a controlled and efficient alcoholic fermentation, and their purchase represents an important cost for wineries. Therefore, it is most relevant to survey commercial ADY during wine fermentation to understand if their use is effective.

Keywords: active dry yeast; wine; co-inoculation; sequential inoculation; microsatellite markers; genetic relationships

Citation: Duarte, F.L.; Baleiras-Couto, M.M. Survey of Inoculated Commercial *Saccharomyces cerevisiae* in Winery-Based Trials. *Fermentation* **2021**, 7, 176. https://doi.org/10.3390/fermentation7030176

Academic Editor: Ronnie G. Willaert

Received: 13 August 2021
Accepted: 31 August 2021
Published: 3 September 2021

Publisher's Note: MDPI stays neutral with regard to jurisdictional claims in published maps and institutional affiliations.

Copyright: © 2021 by the authors. Licensee MDPI, Basel, Switzerland. This article is an open access article distributed under the terms and conditions of the Creative Commons Attribution (CC BY) license (https://creativecommons.org/licenses/by/4.0/).

1. Introduction

The use of select commercial *Saccharomyces cerevisiae* strains has been a common practice for wine production mostly since the end of the 20th century. The development of the wine industry has also been followed by an increase in the supply of new species and strains in the active dry yeast (ADY) production industry. Production techniques have undergone many changes and improvements, and strain selection criteria have adapted to the new requirements of winemakers, opinion makers and consumers. The production of ADY has evolved with specificities aimed at its use at the beginning or end of fermentation, suitability for lower temperatures, and reduction in the yield of alcohol or production of specific aromas, among others. In fact, nowadays, the offer of these oenological yeasts is enormous.

To evaluate the use of starter cultures, initial studies have been conducted on the laboratory scale despite the burden that this may introduce in scaling up the results to the winery level. Recent winery-based studies have shown very interesting results on the presence and/or dominance of indigenous versus commercial *S. cerevisiae* strains [1–7]. Scholl et al. investigated the presence of commercial *S. cerevisiae* strains in spontaneous fermentations at wineries that conduct both inoculated and spontaneous fermentations and showed that indigenous *S. cerevisiae* strains were found in relatively low abundance [5]. Differences were also found in the diversity of *S. cerevisiae* strains depending on the cultivar and the winery. Some studies have also reported on the co-inoculation of multiple

S. cerevisiae strains, but overall, studies have mostly been dedicated to the influence on the volatile composition and sensory properties of the produced wines [8–10].

On the other hand, recent studies on the dominance of different inoculated strains have mostly been carried out on mixed fermentations with non-*Saccharomyces* yeasts and *S. cerevisiae* with the aim of understanding the behavior and interactions of strains throughout the fermentation process [11–14]. Fewer studies have focused on the population dynamics of different inoculated strains of *S. cerevisiae*, as reported by Gustafsson et al. [15]. Previous winery-based studies have shown that the frequently used strains Lalvin RC212 and Lalvin ICV-D254 tend to predominate, with over 80% presence even in spontaneous fermentations [16].

In order to follow *S. cerevisiae* population dynamics during fermentation, adequate methods that enable differentiation at the strain level are needed. Simple sequence repeat (SSR) markers are powerful biomolecular tools for the differentiation of yeast strains [17]. The development of these techniques has led to the study of population dynamics during fermentation and has opened the possibility to understand the complex roles of microorganisms in wine fermentation [1,18].

In this paper, wine fermentations carried out in the cellar using *S. cerevisiae* commercial ADY were performed in white and red grapes; the size of the assessed fermentations ranged from microvinifications (60 L) and pilot scale fermentations (500 L) to large-scale fermentations (250 hL). The presence of the added ADY during fermentation was monitored using microsatellite or SSR markers.

2. Materials and Methods

2.1. Yeast Strains

Different commercial ADY were used in the three experiments. ADY were selected by the wine producer depending on the desired characteristics. Table 1 summarizes the information available.

Table 1. ADY used in the experiments, as well as their respective codes and suppliers.

Strain	Code	Supplier	Experiment
S. cerevisiae r.f. *uvarum*	C	AEB Bioquímica, SA Portugal	1
S. cerevisiae r.f. *bayanus*	B	AEB Bioquímica, SA Portugal	1
S. cerevisiae hybrid	V	Anchor, South Africa	2
S. cerevisiae var. *bayanus*	F	DSM, The Netherlands	2
S. cerevisiae, var. cerevisiae	D	Proenol, Vila Nova de Gaia	3

2.2. Winery Procedures

Three winery experimental trials were performed. White wine was produced in Experiment 1 and 2, while Experiment 3 produced red wine. The general setup for Experiment 1 and 2 is schematized in Figure 1.

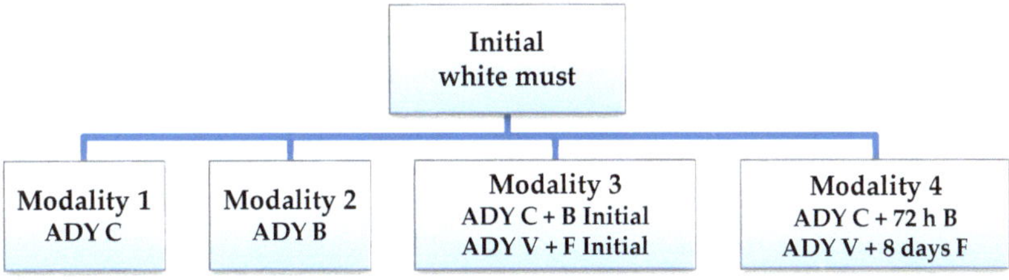

Figure 1. Schematic representation of white must experiments. Experiment 1—ADY C and B; experiment 2—ADY V and F.

In Experiment 3, the influence of oenological additives on the prevalence of ADY in red wine fermentations was evaluated using a single ADY strain. The fermentation activators applied were a blend of yeast nutrients (F1 and F2), and the applied tannin preparations were a blend of condensed and hydrolyzed tannins specific for color stabilization (T1 and T2), as well as a blank (B) without additives. For the five trials carried out, detailed information and procedures used were previously described [1].

2.2.1. Experiment 1

A first set of two commercial ADY was selected by the winery to ferment their wines; namely, strains C and B were selected. The winery planned to use strain C as the starter yeast and to add strain B at mid-fermentation in order to prevent stuck fermentation. Yeast rehydration was performed according to the manufacturer instructions, and 30 g/hL were added.

White must (sugars 188 g/L; total acidity 5.5 g/L tartaric acid; pH 3.29) was clarified by cold settling. SO_2 was then added (50 mg/L), and the mixture was homogenized and distributed in 4 deposits of 60 L, corresponding to the four modalities of yeast addition, as represented in Figure 1: modality 1—single addition of ADY C; modality 2—single addition of ADY B; modality 3—simultaneous addition of ADY C and B; modality 4—sequential addition, wherein ADY C was added to the initial must and ADY B was added after 72 h (must density 1.032). Fermentation took place at room temperature of around 23 °C.

Three samples were collected from each modality at 2 h, 72 h and 7 days after the first ADY addition, corresponding to the initial (I), middle (M) and end (E) time periods of the fermentation process.

2.2.2. Experiment 2

A second set of two ADY was selected; specifically, these were strains V and F (Table 1). The winery planned to use strain V as the starter yeast and to add strain F at mid-fermentation in order to prevent stuck fermentations.

A white must from a different vintage was used (sugars 213 g/L) and similar procedures as Experiment 1 were performed, including cold settling and addition of SO_2 (50 mg/L). Fermentation activators and pectolytic enzymes were also added. The must was proportionally distributed in two stainless steel deposits of 250 hL. Two modalities of yeast addition were performed, corresponding to modalities 3 and 4 (Figure 1): modality 3—simultaneous addition of ADY V and ADY F; modality 4—sequential addition, wherein ADY V was added to the initial must, followed by ADY F after eight days (must density 1.040). Fermentation took place at around 16 °C.

Three samples were collected from each deposit 1, 9 and 17 days after ADY addition, corresponding to the initial (I), middle (M) and end (E) time periods of the fermentation process.

2.2.3. Experiment 3

Homogenized red mush (sugars 179.5 g/L; pH 3.13) with SO_2 (80 mg/L) and ADY (20 g/100 kg) was equally distributed in five tanks of 500 L each. A different fermentation activator (F1 or F2) or tannin (T1 or T2) was added to each of the corresponding four tanks, and the remaining one was set as blank [1].

Must samples were collected at the middle (M) and end (E) of the fermentation process.

2.3. Commercial Yeast Isolation

In order to determine the microsatellite profiles of the commercial ADY used, a new package of each ADY was opened in a laminar flow chamber. The yeast was resuspended in sterilized water following the supplier instructions. After yeast rehydration, serial decimal dilutions using a solution of NaCl and tryptone (8.5 g/L and 1.0 g/L, respectively; autoclaved for 15 min at 121 °C) were performed to obtain a countable number of colonies by spread plating a volume of 100 μL of the appropriate dilutions on YPD agar (yeast extract, 5 g/L, bacto-peptone, 10 g/L, glucose, 20 g/L, agar, 20 g/L). Twenty-five random colonies of each ADY were further purified and used for SSR analysis.

2.4. Fermentation Yeasts Isolation

Three must samples were collected from each fermentation at each time point. Appropriate serial dilutions were performed, and samples were spread plated on a grape must-agar medium (diluted grape must, 50%, *v/v*; pH 5; agar 20 g/L). After 48 h at 25 °C, 10 colonies of each triplicate presenting with the *S. cerevisiae* morphology type (whitish and slightly brilliant, butyrous, smooth, raised and occasionally conical, entire margin and opaque) were randomly collected and further purified. Cultures were maintained on YPD agar at 4 °C.

The yeast isolates were screened on lysine medium agar (lysine hydrochloride, Sigma-Aldrich, added to Yeast Carbon Base, Difco Laboratories) in order to exclude the non-*Saccharomyces* yeasts (lysine+).

2.5. DNA Extraction and Purification

Procedures were performed as described by Duarte et al. [1]. Briefly, cells were suspended in a lysing buffer and disrupted using glass beads. DNA was purified with chloroform:isoamyl alcohol (24:1) and precipitated with 1/10 volume of sodium acetate (3 M, pH 5.2) and two volumes of absolute ethanol followed by incubation with RNase (100 µg/mL in TE: 10 mM tris-HCl and 1 mM EDTA, pH 8) for 30 min at 37 °C. DNA was washed with ethanol (70%, *v/v*), dried and resuspended in TE.

2.6. Microsatellite Analysis

The same six SSR loci mentioned by Duarte et al. [1] were used, namely ScAAT1, ScAAT2, ScAAT3, ScAAT5, SCYOR267C and SC8132X [19,20]. SSR amplifications were conducted in two multiplex reactions as previously described [1]. The first multiplex with primers ScAAT1, ScAAT2 and ScAAT5 was performed following the conditions already described by Pérez et al. [19]. The other multiplex reaction with primers ScAAT3, SCYOR267C and SC8132X was optimized. The amplification conditions consisted of an initial denaturation step at 94 °C for five min, followed by 10 cycles of 15 s at 94 °C, 30 s at 68 °C (decreasing 1 °C per cycle to 57 °C) and 45 s at 72 °C; this was then followed by 25 cycles of 15 s at 94 °C, 30 s at 58 °C, and 45 s at 72 °C, and a final step of 5 min at 72 °C [1]. An aliquot of 1–2 µL of the amplified product, 0.5 µL of Ceq DNA Size Standard kit-600 (Beckman Coulter Inc., Fullerton, California, CA, USA) and 25 µL formamide were sequentially dispensed on a 96-well sample plate. Separation was performed by capillary electrophoresis (6 kV; 50 °C; capillary 30 cm length; 35 min) on a CEQ 8000 Genetic Analysis System (Beckman Coulter Inc., Fullerton, California, CA, USA).

2.7. Data Analysis

To investigate the genetic relationships between isolates, microsatellite profiles were analysed using the poppr package, v2.8.3, under R statistical software, v3.6.1 [21]. Dendrograms were established using Nei distance [22] and UPGMA (unweighted pair group method with arithmetic mean) clustering.

3. Results and Discussion

The microsatellite profiles of the commercial ADY used in the three winery experiments, as well as fermentation isolates, were determined. The genetic distance between the isolates is represented by dendrograms, built based on the Nei distance [22] and the clustering method UPGMA.

3.1. Experiment 1

In this experiment, 60 L fermentation deposits were used and 2 commercial ADY were tested. ADY C is, according to the manufacturer, a *S. cerevisiae r.f. uvarum* which is cryotolerant, with a high production of aromatic compounds with a positive effect on the wine sensory characteristics. ADY B is, according to the manufacturer, a *S. cerevisiae r.f. bayanus* more suitable for the control, regularization and restart of fermentation.

The dendrogram presented in Figure 2 compares the 25 isolates from ADY B with the 25 isolates from ADY C based on the microsatellite profiles. Isolates from ADY B were genetically closer to 24 isolates from ADY C, with a total of six profiles shared by isolates from each of the two ADY, C and B. Only the isolate C15 from ADY C presented a genetically distant profile.

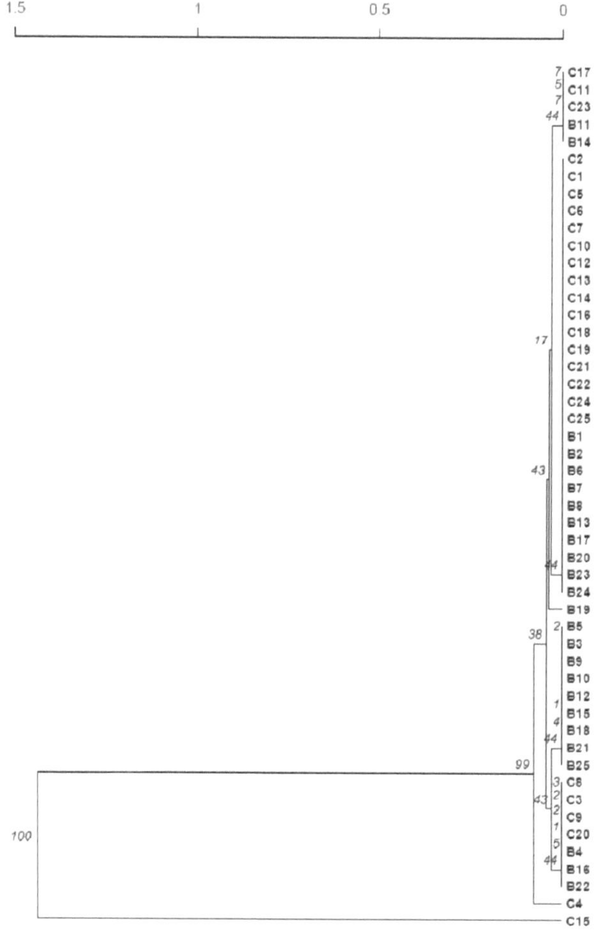

Figure 2. Dendrogram presenting the genetic distance between the ADY used in experiment 1, based on the microsatellite profiles. C—ADY C isolates; B—ADY B isolates; both are followed by the number of the isolate. Scale at top represents genetic distance.

Surprisingly, the SSR profiles of the isolates of commercial ADY C and B showed a high similarity, clearly evidenced by the dendrogram in Figure 2. These two ADY commercial preparations were, most probably, originated from the same yeast strain. This result made it impossible to study the evolution of the added yeast or yeast mixtures during fermentation, as their profiles were similar.

The presence of *S. cerevisiae* isolates presenting profiles genetically distant from those found for the ADY C and B, which might correspond to native yeasts, were only detected at the beginning of fermentation (results not shown).

This result highlights the major importance of using molecular markers for certifying the commercial yeasts present in the market, as is already a current requirement for the grapevine varieties trade.

3.2. Experiment 2

In this experiment, fermentation deposits of 250 hL were used. These deposits were close to the conditions usually used by the winery, and two commercial ADY were also tested. ADY V (Anchor, South Africa) is suited for the production of fresh and fruity white and rosé wines to enhance volatile thiol aromas (passion fruit, grapefruit, gooseberry and guava), to produce esters (tropical fruit salad, floral) and killer positive. ADY F (DSM, The Netherlands) is a fructophilic yeast to prevent and restart stuck fermentation. ADY F does not produce secondary aromas and preserves the specific characteristics of the must when restarting fermentation. This experiment was performed utilizing modalities 3 and 4 (Figure 1).

For the 25 ADY V isolates, 14 different SSR profiles were obtained, although these profiles were genetically very close. This can be observed from the dendrograms of Figures 3 and 4, where those profiles were included together with the profiles of the yeast isolates from the initial (I), middle (M) and end (E) time periods of the fermentation of modalities 3 and 4, respectively. Concerning the 25 isolates of ADY F, 9 different profiles were obtained. However, these profiles are genetically very close, as can be seen in the dendrograms of Figures 3 and 4.

It is important to notice that the profiles for ADY V and ADY F were genetically distant from each other, allowing their differentiation.

Regarding modality 3, where both ADY were added at the beginning of fermentation, the isolates from the initial, middle and end periods of fermentation presented with profiles genetically closer to ADY V and ADY F (Figure 3). Though similar numbers of each ADY were detected at the beginning, towards the end of fermentation, a prevalence of genetically closer ADY V profiles was observed; 20 isolates were identified, and only two ADY F profiles were detected. Isolates genetically distant from these ADY profiles, probably corresponding to native yeasts, were also detected at the initial (one isolate), middle (8) and end (7) periods of fermentation.

For modality 4, where ADY F was added at mid-fermentation, only isolates with profiles genetically closer to ADY V were detected. This occurred at the beginning, where it was expected, but also at the middle and end of fermentation (Figure 4). Moreover, two isolates each from the middle and end of fermentation presented profiles genetically distant from ADY profiles, most likely corresponding to native yeasts.

Higher genetic diversity was detected when the two ADY were added at the beginning of fermentation (modality 3), with both ADY detected throughout the fermentation and a high number of native yeasts detected at the middle and end of fermentation. On the contrary, for modality 4, the ADY V added at the beginning of fermentation seemed to dominate, as no ADY F profile was detected during fermentation and a lower number of native yeasts were also detected. Several works have already shown that the inoculum of *S. cerevisiae* ADY cells, both commercial and indigenous, ensures a very rapid dominance of a single strain and the suppression of natural microbiota [23–25].

Additionally, it was interesting to observe that sequential inoculation (modality 4) resulted in a slower fermentation, a stable number of viable cells and a higher wine quality (Figures S1 and S2 in supplementary data). Co-inoculation (modality 3) resulted in a faster fermentation, resulting in a decline in the number of cells with culture viability and a low-quality wine (Figures S1 and S2).

This study showed that despite ADY sequential addition, only the first inoculated strain was detected, inferring its dominance during fermentation. Other works have reported that the use of starter cultures does not always guarantee the dominance of the inoculated strain [1,26,27]. It is important to further evaluate in future works if single addition of ADY V would result in the same fermentation performance and wine characteristics, thus saving money and work for the winery.

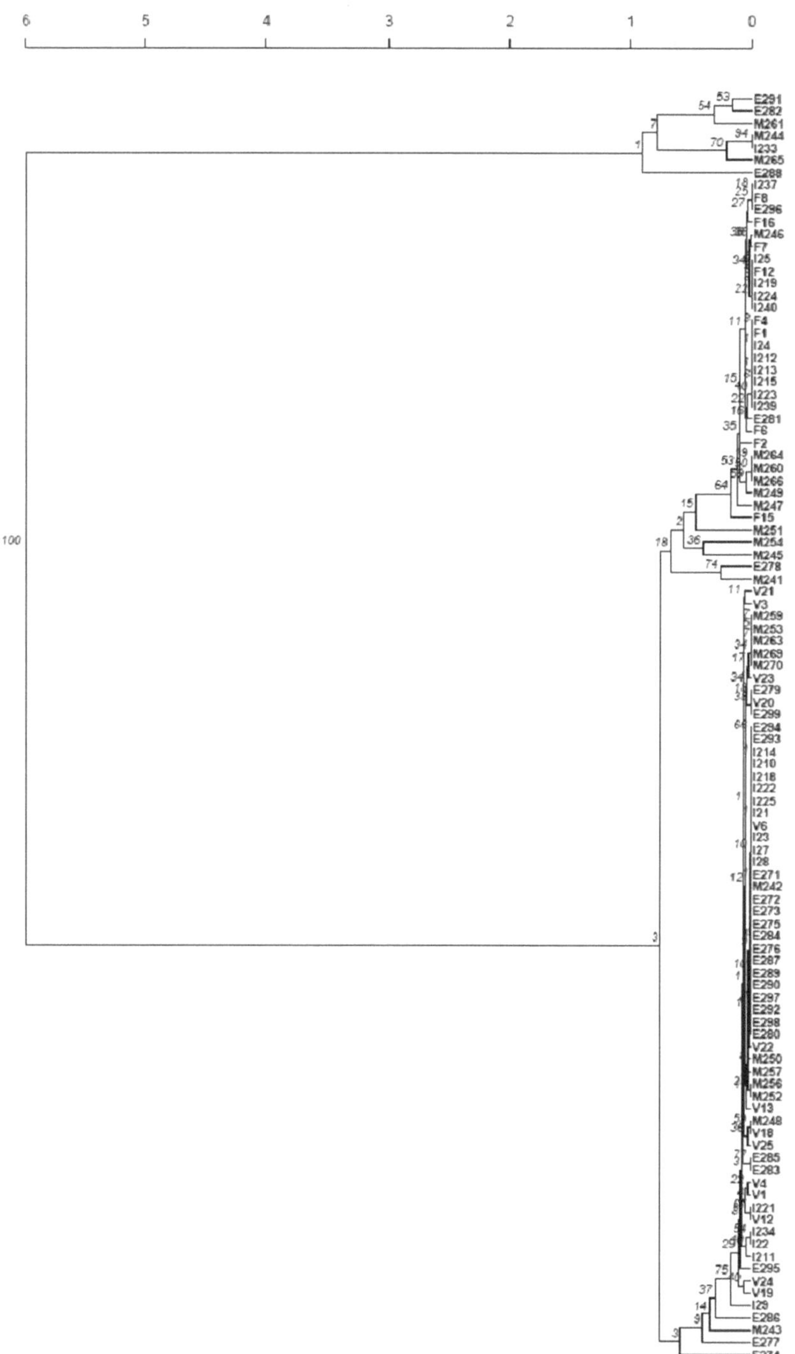

Figure 3. Dendrogram presenting the genetic distance between the isolates from Experiment 2, modality 3 and the ADY isolates, based on the microsatellite profiles. V—ADY V isolates; F— ADY F isolates; isolates from fermentation are represented by a letter; I—initial, M—middle or E—end time periods of fermentation, followed by the number of the isolate. Scale at top represents genetic distance.

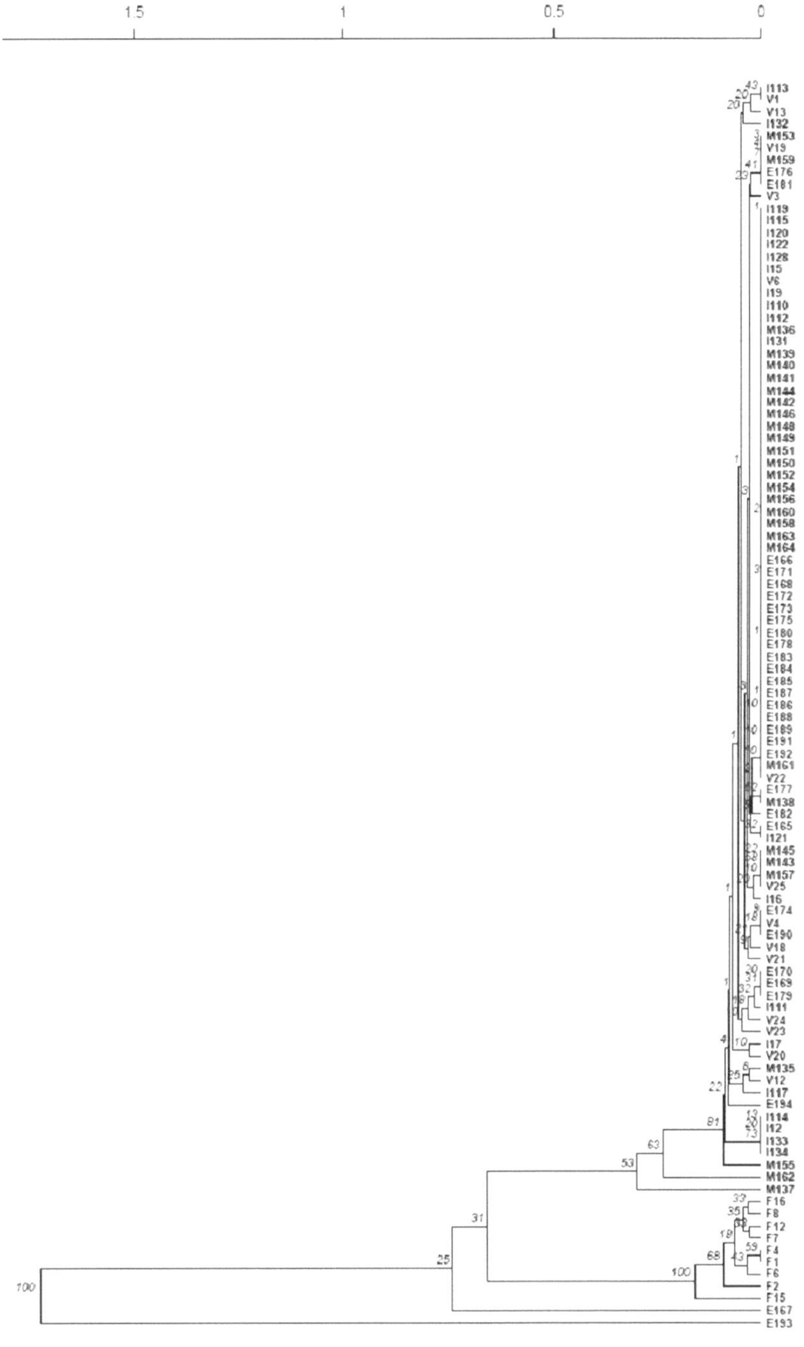

Figure 4. Dendrogram presenting the genetic distance between the isolates from Experiment 2, modality 4 and the ADY isolates, based on the microsatellite profiles. V—ADY V isolates; F—ADY F isolates; isolates from fermentation are represented by a letter; I—initial, M—middle or E—end time periods of fermentation, followed by the number of the isolate. Scale at top represents genetic distance.

3.3. Experiment 3

Pilot-scale red wine fermentations of 500 L volume were performed in the winery, where fermentation modalities included the use of fermentation activators and tannin preparations.

Must samples were collected at the middle (must density between 1.044 and 1.053) (M) and end of fermentation (E) (must density between 0.998 and 1.006; just before the separation of grape solids from the wine).

The isolates from the ADY D used in this experiment presented a single profile with only one allele for each locus. The majority of the isolates of the blank modality, without additives, presented a profile similar to the ADY. Only five isolates during the middle of fermentation presented a profile genetically distant from the ADY profile; these most certainly corresponded to native yeasts (Figure 5). By the end of fermentation, only isolates presenting the ADY profile were detected, indicating a dominance of the starter yeast.

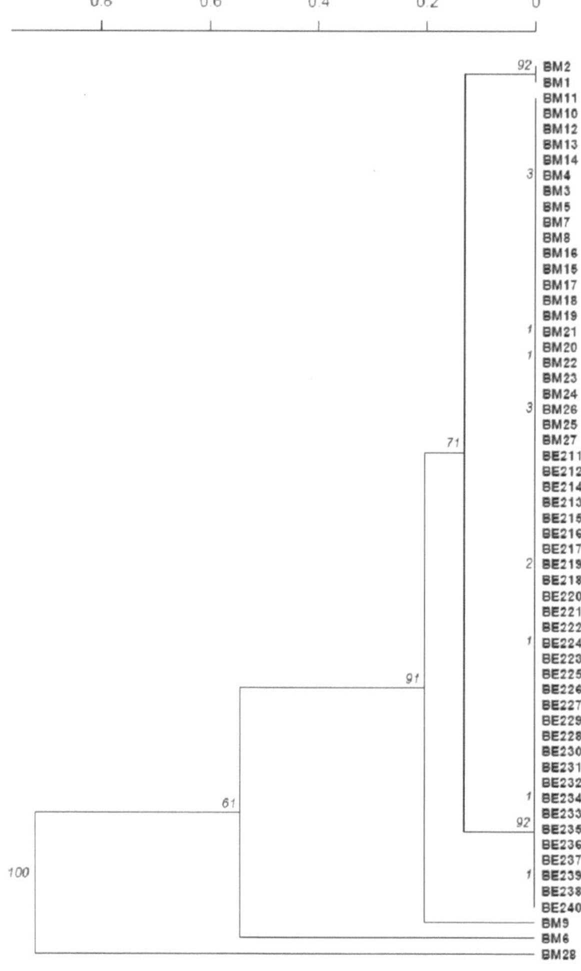

Figure 5. Dendrogram presenting the genetic distance between the isolates from the blank modality based on the microsatellite profiles. Isolates from fermentation are represented by the letters BM or BE, corresponding to middle or end of the blank fermentation, respectively, followed by the number of the isolate. Scale at top represents genetic distance.

Regarding the addition of fermentation activators, two commercial preparations were used; these were F1 and F2. Although the majority of the isolates obtained at the middle and end of the fermentation presented a profile genetically similar to the ADY profile, genetically distant profiles were also detected, most probably corresponding to native yeasts (Figure 6). A higher number of isolates with genetically distant profiles was detected when the F1 fermentation activator was used. Surprisingly, both fermentation activator (F1 and F2) trials showed an increase in distinct profiles towards the end of fermentation, indicating that the dominance of the ADY was lower than in the blank modality. Lower ADY implantation was found with oenological additives, as already observed in a previous work [1].

In relation to tannin preparations T1 and T2, a high number of isolates presenting profiles genetically distant from the ADY profile was observed (Figure 7). For T2, at middle fermentation, almost half of the isolates corresponded to native yeasts. At the end of fermentation, a high number of isolates with profiles genetically distant from the ADY profile was detected. It can therefore be assumed that these additives negatively influenced the ADY implantation.

A putative influence of fermentation activators and tannin preparations on ADY strain implantation during fermentation was observed. In all trials, a negative correlation was observed on the dominance of ADY D when compared with the blank modality. All trials showed an increase in the number of genetically distant profiles from the ADY towards the end of fermentation, except for T2, which, nevertheless, presented a high number of native yeast at the end of fermentation.

The relevance of this study is attributed to the fermentation procedures tested, which were close to industrial winery conditions. The application of SSR markers to *S. cerevisiae* strain characterization in the winery environment, as previously reported [17], was revealed once more to be a powerful tool.

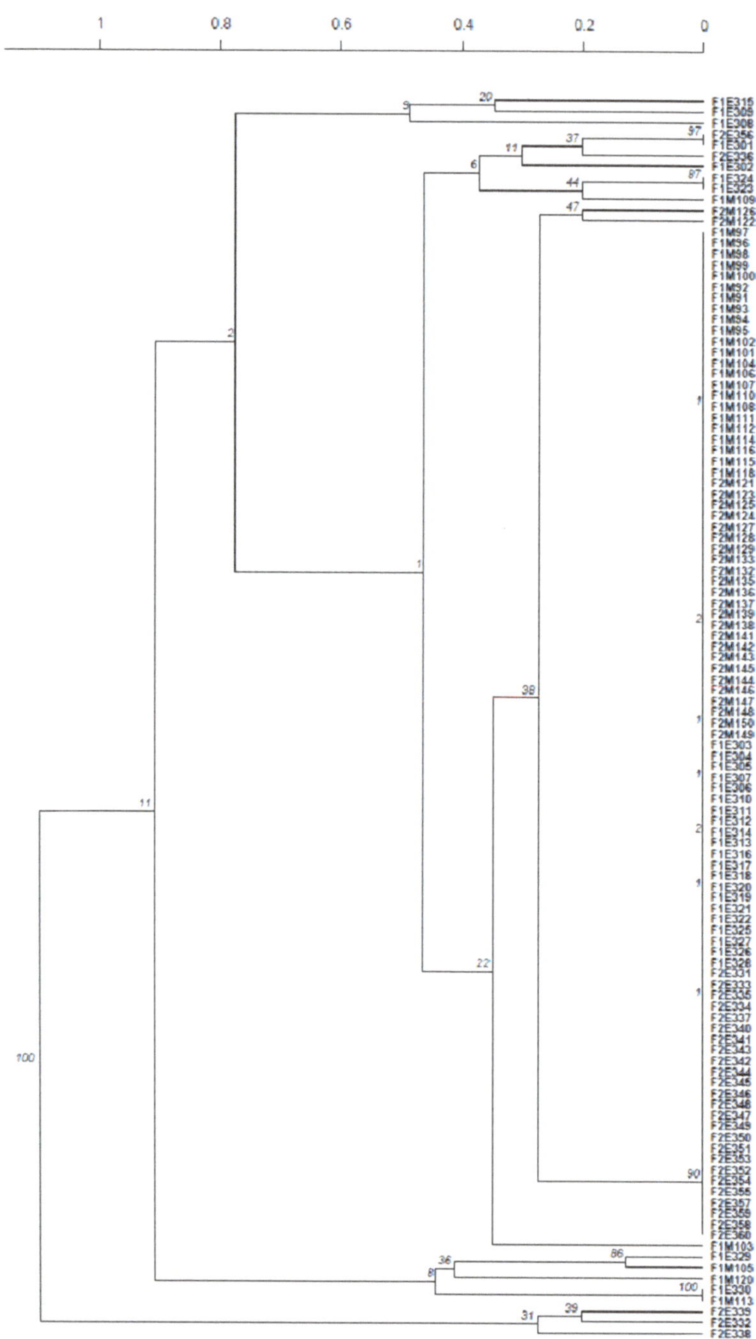

Figure 6. Dendrogram presenting the genetic distance between the isolates from the fermentation activator modality based on microsatellite profiles. Isolates from fermentation are represented by F1 or F2, depending on the fermentation activator used, followed by the letter M—middle or E—end of fermentation, and finally the number of the isolate. Scale at top represents genetic distance.

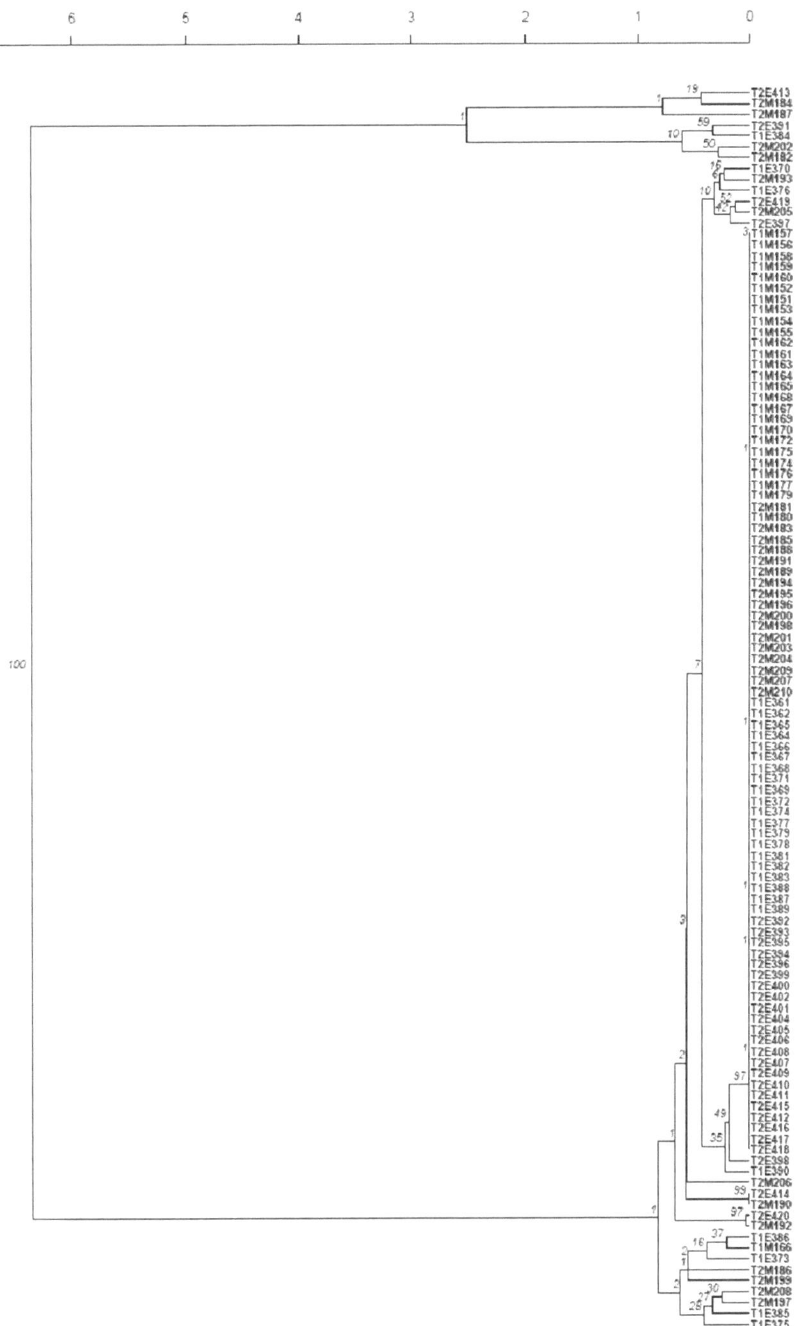

Figure 7. Dendrogram presenting the genetic distance between the isolates from the tannin preparation modality based on the microsatellite profiles. Isolates from fermentation are represented by T1 or T2, depending on the tannin preparation used, followed by the letter M—middle or E—end of fermentation, and finally the number of the isolate. Scale at top represents genetic distance.

4. Conclusions

In this study, *S. cerevisiae* commercial ADY were tested, and yeast dynamics were monitored in white and red wine fermentation winery trials. The purchase of ADY is an important cost for wineries and is mostly relevant to achieve constant and controlled fermentations. The study of ADY implantation could not be achieved in one of the trials, as both commercial ADY strains used revealed similar SSR profiles. These results emphasized the need for rigorous control on the production and commercialization of ADY, ideally through molecular certification.

Higher genetic diversity was detected when co-inoculation was tested. In sequential inoculation, the initial ADY predominated over fermentation, while other SSR profiles, either the ADY added sequentially or native yeasts, were less detected. This result may question the effectiveness of the second inoculation. Therefore, one of the envisaged future works is to evaluate if a single addition of the initial ADY would result in the same fermentation performance and wine characteristics, thus reducing the additional cost and work for the winery inherent to the use of a second ADY. Likewise, a possible negative influence of oenological additives on ADY efficiency was detected, as the ADY was less detected when compared to the blank deposit.

Overall, there are many factors influencing the dynamics of ADY strains in wine fermentation. This type of survey is of utmost importance to ensure their effectiveness during fermentation and to achieve the desired sensorial quality, as the use of ADY is often directed to increase the sensorial complexity of wines using different strains of *S. cerevisiae* and, more recently, even non-*Saccharomyces* species.

Supplementary Materials: The following are available online at https://www.mdpi.com/article/10.3390/fermentation7030176/s1. Figure S1. Fermentation evolution evaluated by must density (D) and yeast cell counts (evaluated by culture) (Y) for modalities 3 and 4 of Experiment 2; Figure S2. Overall quality of the wines obtained for modalities 3 and 4 of experiment 2. The bars represent the average value obtained with 8 experts from the INIAV Dois Portos tasting panel. The standard deviation is indicated by the error bars at the top.

Author Contributions: Conceptualization, F.L.D. and M.M.B.-C.; methodology, F.L.D. and M.M.B.-C.; software, F.L.D. and M.M.B.-C.; formal analysis, F.L.D. and M.M.B.-C.; investigation, F.L.D. and M.M.B.-C.; data curation, F.L.D. and M.M.B.-C.; writing—original draft preparation, F.L.D. and M.M.B.-C.; writing—review and editing, F.L.D. and M.M.B.-C.; visualization, F.L.D. and M.M.B.-C. All authors have read and agreed to the published version of the manuscript.

Funding: This research received no external funding.

Institutional Review Board Statement: Not applicable.

Informed Consent Statement: Not applicable.

Data Availability Statement: The data presented in this study are available on request from the corresponding author. The data are not publicly available due to producers' privacy.

Acknowledgments: The authors wish to acknowledge the wineries involved in the trials.

Conflicts of Interest: The authors declare no conflict of interest.

References

1. Duarte, F.L.; Alves, A.C.; Alemão, M.F.; Baleiras-Couto, M.M. Influence of red wine fermentation oenological additives on inoculated strain implantation. *World J. Microbiol. Biotechnol.* **2013**, *6*, 1139–1144. [CrossRef]
2. Perrone, B.; Giacosa, S.; Rolle, L.; Cocolin, L.; Rantsiou, K. Investigation of the dominance behavior of Saccharomyces *cerevisiae* strains during wine fermentation. *Int. J. Food Microbiol.* **2013**, *2*, 156–162. [CrossRef] [PubMed]
3. Lange, J.N.; Faasse, E.; Tantikachornkiat, M.; Gustafsson, F.S.; Halvorsen, L.C.; Kluftinger, A.; Ledderhof, D.; Durall, D.M. Implantation and persistence of yeast inoculum in Pinot noir fermentations at three Canadian wineries. *Int. J. Food Microbiol.* **2014**, *180*, 56–61. [CrossRef]
4. Martiniuk, J.T.; Pacheco, B.; Russell, G.; Tong, S.; Backstrom, I.; Measday, V. Impact of Commercial Strain Use on *Saccharomyces cerevisiae* Population Structure and Dynamics in Pinot Noir Vineyards and Spontaneous Fermentations of a Canadian Winery. *PLoS ONE* **2016**, *8*, e0160259. [CrossRef] [PubMed]

5.   Scholl, C.M.; Morgan, S.C.; Stone, M.L.; Tantikachornkiat, M.; Neuner, M.; Durall, D.M. Composition of *Saccharomyces cerevisiae* strains in spontaneous fermentations of Pinot Noir and Chardonnay. *Aust. J. Grape Wine Res.* **2016**, *22*, 384–390. [CrossRef]

6.   Aponte, M.; Romano, R.; Villano, C.; Blaiotta, G. Dominance of *S. cerevisiae* Commercial Starter Strains during Greco di Tufo and Aglianico Wine Fermentations and Evaluation of Oenological Performances of Some Indigenous/Residential Strains. *Foods* **2020**, *9*, 1549. [CrossRef]

7.   Tantikachornkiat, M.; Morgan, S.; Lepitre, M.; Cliff, M.; Durall, D. Yeast and bacterial inoculation practices influence the microbial communities of barrel-fermented Chardonnay wines. *Aust. J. Grape Wine Res.* **2020**, *26*, 279–289. [CrossRef]

8.   King, E.S.; Swiegers, J.H.; Travis, B.; Francis, I.L.; Bastian, S.E.P.; Pretorius, I.S. Coinoculated Fermentations Using *Saccharomyces* Yeasts Affect the Volatile Composition and Sensory Properties of *Vitis vinifera* L. cv. Sauvignon Blanc Wines. *J. Agric. Food Chem.* **2008**, *56*, 10829–10837. [CrossRef]

9.   King, E.S.; Kievit, R.L.; Curtin, C.; Swiegers, J.H.; Pretorius, I.S.; Bastian, S.E.P.; Francis, I.L. The Effect of Multiple Yeasts Co-Inoculations on Sauvignon Blanc Wine Aroma Composition, Sensory Properties and Consumer Preference. *Food Chem.* **2010**, *122*, 618–626. [CrossRef]

10. Saberi, S.; Cliff, M.A.; van Vuuren, H.J.J. Impact of Mixed *S. cerevisiae* Strains on the Production of Volatiles and Estimated Sensory Profiles of Chardonnay Wines. *Food Res. Int.* **2012**, *48*, 725–735. [CrossRef]

11. Contreras, A.; Curtin, C.; Varela, C. Yeast population dynamics reveal a potential 'collaboration' between *Metschnikowia pulcherrima* and *Saccharomyces uvarum* for the production of reduced alcohol wines during Shiraz fermentation. *Appl. Microbiol. Biotechnol.* **2015**, *99*, 1885–1895. [CrossRef] [PubMed]

12. Padilla, B.; Zulian, L.; Ferreres, À.; Pastor, R.; Esteve-Zarzoso, B.; Beltran, G.; Mas, A. Sequential Inoculation of Native Non-*Saccharomyces* and *Saccharomyces cerevisiae* Strains for Wine Making. *Front. Microbiol.* **2017**, *8*, 1293–1314. [CrossRef]

13. Dimopoulou, M.; Troianou, V.; Toumpeki, C.; Gosselin, Y.; Dorignac, É.; Kotseridis, Y. Effect of strains from different *Saccharomyces* species used in different inoculation schemes on chemical composition and sensory characteristics of Sauvignon blanc wine. *OENO One* **2020**, *4*, 745–759. [CrossRef]

14. Zhu, X.; Torija, M.-J.; Mas, A.; Beltran, G.; Navarro, Y. Effect of a Multistarter Yeast Inoculum on Ethanol Reduction and Population Dynamics in Wine Fermentation. *Foods* **2021**, *10*, 623. [CrossRef] [PubMed]

15. Gustafsson, F.S.; Jiranek, V.; Neuner, M.; Scholl, C.M.; Morgan, S.C.; Durall, D.M. The Interaction of Two *Saccharomyces cerevisiae* Strains Affects Fermentation-Derived Compounds in Wine. *Fermentation* **2016**, *2*, 9. [CrossRef]

16. Hall, B.; Durall, D.M.; Stanley, G. Population Dynamics of *Saccharomyces cerevisiae* during Spontaneous Fermentation at a British Columbia Winery. *Am. J. Enol. Vitic.* **2011**, *62*, 66–72. [CrossRef]

17. Rex, F.; Hirschler, A.; Scharfenberger-Schmeer, M. SSR-Marker Analysis—A Method for *S. cerevisiae* Strain Characterization and Its Application for Wineries. *Fermentation* **2020**, *6*, 101. [CrossRef]

18. Kioroglou, D.; LLeixá, J.; Mas, A.; Portillo, M.D.C. Massive Sequencing: A New Tool for the Control of Alcoholic Fermentation in Wine? *Fermentation* **2018**, *4*, 7. [CrossRef]

19. Pérez, M.A.; Gallego, F.J.; Martinéz, I.; Hidalgo, P. Detection, distribution and selection of microsatellites (SSRs) in the genome of the yeast *Saccharomyces cerevisiae* as molecular markers. *Lett. Appl. Microbiol.* **2001**, *33*, 461–466. [CrossRef]

20. González-Techera, A.; Jubany, S.; Carrau, F.M.; Gaggero, C. Differentiation of industrial wine yeast strains using microsatellite markers. *Lett. Appl. Microbiol.* **2001**, *33*, 71–75. [CrossRef] [PubMed]

21. R Core Team. *R: A Language and Environment for Statistical Computing*; R Foundation for Statistical Computing: Vienna, Austria, 2019.

22. Nei, M. Estimation of average heterozygosity and genetic distance from a small number of individuals. *Genetics* **1978**, *89*, 583–590. [CrossRef]

23. Ciani, M.; Capece, A.; Comitini, F.; Canonico, L.; Siesto, G.; Romano, P. Yeast Interactions in Inoculated Wine Fermentation. *Front. Microbiol.* **2016**, *7*, 555. [CrossRef]

24. Capece, A.; Pietrafesa, R.; Siesto, G.; Romaniello, R.; Condelli, N.; Romano, P. Selected Indigenous *Saccharomyces cerevisiae* Strains as Profitable Strategy to Preserve Typical Traits of Primitivo Wine. *Fermentation* **2019**, *5*, 87. [CrossRef]

25. Philipp, C.; Bagheri, B.; Horacek, M.; Eder, P.; Bauer, F.F.; Setati, M.E. Inoculation of grape musts with single strains of *Saccharomyces cerevisiae* yeast reduces the diversity of chemical profiles of wines. *PLoS ONE* **2021**, *7*, e0254919. [CrossRef]

26. Barrajón, N.; Arévalo-Villena, M.; Úbeda, J.; Briones, A. Enological properties in wild and commercial *Saccharomyces cerevisiae* yeasts: Relationship with competition during alcoholic fermentation. *World J. Microbiol. Biotechnol.* **2011**, *27*, 2703–2710. [CrossRef]

27. Capece, A.; Romaniello, R.; Poeta, C.; Siesto, G.; Massari, C.; Pietrafesa, R.; Romano, P. Control of inoculated fermentations in wine cellars by mitochondrial DNA analysis of starter yeast. *Ann. Microbiol.* **2011**, *61*, 49–56. [CrossRef]

 *fermentation*

MDPI

Article

Mechanisms of Metabolic Adaptation in Wine Yeasts: Role of Gln3 Transcription Factor

Aroa Ferrer-Pinós, Víctor Garrigós, Emilia Matallana and Agustín Aranda *

Institute for Integrative Systems Biology (I2SysBio), University of Valencia-CSIC, 46980 Paterna, Spain;
afepi2@alumni.uv.es (A.F.-P.); victor.garrigos@uv.es (V.G.); emilia.matallana@uv.es (E.M.)
* Correspondence: agustin.aranda@csic.es; Tel.: +34-96-354-4816

Abstract: Wine strains of *Saccharomyces cerevisiae* have to adapt their metabolism to the changing conditions during their biotechnological use, from the aerobic growth in sucrose-rich molasses for biomass propagation to the anaerobic fermentation of monosaccharides of grape juice during winemaking. Yeast have molecular mechanisms that favor the use of preferred carbon and nitrogen sources to achieve such adaptation. By using specific inhibitors, it was determined that commercial strains offer a wide variety of glucose repression profiles. Transcription factor Gln3 has been involved in glucose and nitrogen repression. Deletion of *GLN3* in two commercial wine strains produced different mutant phenotypes and only one of them displayed higher glucose repression and was unable to grow using a respiratory carbon source. Therefore, the role of this transcription factor contributes to the variety of phenotypic behaviors seen in wine strains. This variability is also reflected in the impact of *GLN3* deletion in fermentation, although the mutants are always more tolerant to inhibition of the nutrient signaling complex TORC1 by rapamycin, both in laboratory medium and in grape juice fermentation. Therefore, most aspects of nitrogen catabolite repression controlled by TORC1 are conserved in winemaking conditions.

Keywords: wine; *Saccharomyces cerevisiae*; glucose repression; Gln3; nitrogen catabolite repression

Citation: Ferrer-Pinós, A.; Garrigós, V.; Matallana, E.; Aranda, A. Mechanisms of Metabolic Adaptation in Wine Yeasts: Role of Gln3 Transcription Factor. *Fermentation* **2021**, *7*, 181. https://doi.org/10.3390/fermentation7030181

Academic Editor: Ronnie G. Willaert

Received: 8 August 2021
Accepted: 2 September 2021
Published: 5 September 2021

Publisher's Note: MDPI stays neutral with regard to jurisdictional claims in published maps and institutional affiliations.

Copyright: © 2021 by the authors. Licensee MDPI, Basel, Switzerland. This article is an open access article distributed under the terms and conditions of the Creative Commons Attribution (CC BY) license (https://creativecommons.org/licenses/by/4.0/).

1. Introduction

Saccharomyces cerevisiae is the yeast with the most biotechnological interest due to its strong fermentative metabolism and the ability to adapt efficiently to harsh and changing environments [1]. In the wine industry, its role is to ferment the high amount of monosaccharides that are present in the grape juice (glucose and fructose) into ethanol, CO_2, and other molecules of enological interest. In grape juice, sugars are plentiful, but nitrogen is usually scarce, resulting in a limiting factor for growth [2–4]. Yeasts have a preference for some nutrients over others, and those favorite ones exert catabolite repression upon the use of less favored ones. For instance, *S. cerevisiae* favors the use of glucose by fermentation, so when glucose is present over a certain threshold, the use of other less favorite monosaccharides (e.g., galactose), disaccharides (e.g., sucrose), or non-fermenting substrates that have to be metabolized by respiration (e.g., glycerol) is repressed. That is made thanks to a complex genetic program that modifies gene expression in order to impose such glucose repression [5]; that is, the molecular cause of the long term Crabtree effect (the fermentative activity even under fully aerobic conditions) that channels the metabolic flux to the ethanol production, but reducing the biomass generation [6]. Short term Crabtree effect is caused by the inability of mitochondria to deal with a strong glycolytic flux. This metabolic adaptation is a good approach to ferment sugars quickly, producing high ethanol that inhibits growth of less tolerant microorganisms present in grape juice [6]. In modern enology, selected yeasts are inoculated as starters in the form of active dry yeasts [7]. Biomass is propagated in molasses, that are a cheap source of sucrose, keeping low the sucrose concentration and high the oxygen supply to circumvent the Crabtree effect and achieve a higher cell density and diminish fermentation. Therefore, commercial wine strains must have a strong but flexible metabolism, allowing transitions in both directions, fermentation and respiration, to perform at an optimal level. Industrial (brewing and wine) strains tend to use faster sucrose than glucose [8], so the glucose repression is no that stringent for this disaccharide. In fact, we have shown that wine yeasts are more tolerant than laboratory strains to an unmetabolizable glucose analog, 2-deoxyglucose, that induces glucose repression in the presence of sucrose [9]. Nitrogen sources also fall into categories, being considered good ones for instance glutamine and ammonia, while proline and allantoin are poor (the variability, in this case, is higher and there are differences in the order each amino acid is consumed according to genetic and environmental factors) [2,4]. In a similar way, good sources (those that are incorporated easily into the metabolic pathways) impose a nitrogen catabolite repression (NCR) to the use of the poor ones (the ones that require more metabolic steps, energy or oxygen to be fully metabolized) [5].

All the processes dealing with nutrients are well known in laboratory conditions [5]. When glucose is plenty, protein kinase A promotes growth and suppresses stress response, leading to the establishment of glucose repression. When assimilable sugars drop, then Snf1 kinase is activated by phosphorylation, increasing the functions involved in gluconeogenesis and respiration, making possible the use of other carbon sources like galactose, glycerol, or the ethanol produced by fermentation, that is consumed by a glucose-repressed isoform of alcohol dehydrogenase, *ADH2* [10]. In a similar fashion, when preferred nitrogen sources are plentiful, the target of rapamycin (TOR) kinase, acting inside the complex TORC1, promotes protein biosynthesis and growth. Its activity imposes the NCR. The activation of genes involved in the metabolism of non-preferred nitrogen sources relies upon the GATA transcription factors Gln3 and Gat1. Those factors remain on the cytosol when TORC1 activity is high [11]. This is achieved through repressor Ure2 that binds them, and this interaction is regulated by a complex balance in phosphorylation. This situation does not apply automatically to winemaking conditions, where Snf1 and Gln3 showed an early activation when sugars and nitrogen are still plenty [12]. Gln3 is also a target of Snf1, as amino acid metabolism has to be balanced with the metabolism of their carbon backbones [13]. We have found that deletion of *GLN3* in a haploid wine yeast has an impact in fermentation quite similar to *SNF1* deletion, so those pathways may be related and indicates the relevance of this transcription factor during fermentation of grape juice. The contribution of *GAT1* was much smaller [9]; it was found that in a haploid wine strain and some laboratory genetic backgrounds, *GLN3* deletion results in blocking growth in respiratory substrates, like glycerol, and in increased glucose repression [9].

In this work, we aimed to a quantitative analysis of glucose repression and respiration in a group of interesting industrial yeast strains to understand the variability and the contribution of such mechanisms in different growth media. Next, we analyzed the relevance of Gln3 transcription factor in two diploid commercial strains by deletion, both in a variety of laboratory media and minivinifications, analyzing some molecular markers and testing inhibitors of the relevant pathways. The results indicate that Gln3 is relevant for processes regulating carbon and nitrogen metabolism, and shows that a genetic background is crucial to understand the contribution of the nutrient signaling mechanisms.

2. Materials and Methods

2.1. Yeast Strains and Genetic Manipulation

S. cerevisiae wine strains (EC1118, T73, 71B, L2056, M2) were from Lallemand Inc., baker's yeast Cinta Roja was from AB/Mauri, and *chicha* strains EYS5 and ERS1 were isolated by our laboratory [14]. *GLN3* deletion mutants were made in M2 and EC1118 diploid strains with the reusable *kan*MX marker, amplified by PCR from the pUG6 plasmid [15]. This marker contains *lox*P sites to be excised it by Cre recombinase from plasmid YEp-cre-cyh [16]. The CRISPR-Cas9 deletion of the *URE2* gene was made using plasmid pRCC-K, a gift from Eckhard Boles (Addgene plasmid # 81191), in accordance with the provided protocol [17]. Yeast transformations were performed by the lithium acetate method [18].

2.2. Growth Media and Conditions

By default, yeasts were grown in a rich YPD medium (1% yeast extract, 2% bactopeptone, 2% glucose). Solid plates contained 2% agar and 20 µg/mL of geneticin for the selection of *kan*MX transformants. Other rich media were derived by changing the carbon source: YPS contained 2% sucrose, YPGal 2% galactose, YPGly 2% glycerol. Minimal medium SD contained 0.17% yeast nitrogen base, 0.5% ammonium sulfate, and 2% glucose [19]. This medium was used to select the transformants with cycloheximide resistance by employing it at 2 µg/mL. Nitrogen was changed from this minimal medium replacing ammonium sulfate by 0.5% proline (SPro), 0.5% glutamine (SGln), 300 mg/L of a mix of amino acids (Saa), or 300 mg/L of ammonium chloride (SNH_4).

Growth curves were performed in a Varioskan Lux plate reader at 30 °C with shaking, inoculating from a stationary culture in YPD at OD_{600} of 0.1. Inhibitors were used at the following concentration: 2-deoxyglucose was added at 200 µg/mL, rapamycin at 200 nM, antimycin A at 3 mg/L. For the spot analysis, serial dilutions from stationary cultures in YPD were carried out and 5 µL drops were placed on selective media containing the right amount of inhibitors (glucosamine 0.05%, 10 mM 3-aminotriazole, 50 mM methylamine, 80 mg/L canavanine). Synthetic grape juice MS300 (containing 300 mg/L of assimilable nitrogen) was made as previously described [20] with some changes [21]. It contains a equimolar amount of glucose and fructose at 10%, malic acid 3 g/L, citric acid 0.3 g/L, tartaric acid 3 g/L, assimilable nitrogen source 300 mg N/L (120 mg as (NH4)Cl and 180 mg as amino acids), mineral salts (KH_2PO_4 750 mg/L, K_2SO_4 500 mg/L, $MgSO_4$ 250 mg/L, $CaCl_2$ 155 mg/L, and NaCl 200 mg/L), oligoelements, vitamins and anaerobic factors (ergosterol 15 mg/L and oleic acid 5 mg/L, Tween 80 0.5 mL/L) at pH 3.3. MS60 was the same, but with a reduction in the amount of amino acids and ammonium proportionally. Cells were inoculated from a stationary culture in YPD at 10^6 cells/mL in 30 mL fill-in tubes and kept at 25 °C with low shaking (50 rpm).

2.3. Biochemical Determinations

Reducing sugars were measured with DNS (dinitro-3,5-salicylic acid) compared with a glucose calibration curve according to Miller's method [22]. α-amino acids were determined by the O-phthaldialdehyde/N-acetyl-L-cysteine method, using a curve of isoleucine as reference [23]. Other metabolites were measured with commercial kits (Megazyme Ltd., Bray, Ireland). Cellular respiration was followed by the 1-(4,5-dimethylthiazol-2-yl)-3,5-diphenyltetrazolium bromide (MTT) assay [24] with modifications [25].

2.4. Western Blot and Zymogram

To analyze Snf1 activation, proteins were extracted by fast cell lysis with trichloroacetic acid (TCA) [26]. To 5 OD_{600} units of cells, 5.5% TCA was added. Cells were incubated on ice for 15 min before centrifuging. The pellet was washed twice with acetone and resuspended in 150 µL of 10 mM Tris-HCl, pH 7.5, 1 mM EDTA, and broken with 150 µL of 0.2 M NaOH. SDS-PAGE was carried out in an Invitrogen mini-gel device, gel was blotted onto PVDF membranes a Novex semy dry blotter (Invitrogen, Waltham, MA, USA). The membrane was probed with anti-AMPKα Thr172, Cell Signalling Technologies, Topsfield, MA, USA). The ECL Western blotting detection system (GE) was used following the manufacturer's instructions.

To perform the zymogram, cells were broken in cold 50 mM phosphate buffer, pH 7.5, with glass beads in a FastPrep 24 (MP-Biomedicals, Irvine, CA, USA) [27]. Electrophoresis were run in cold non-denaturing 6.5% acrylamide PAGE gel in an Invitrogen mini-gel device. Activity was detected by soaking the gel in 2 mg phenazine methosulfonate (PMS), 5 mg nitro-tetrazolium blue (NTB), 25 mg NAD, and 0.05 mL ethanol dissolved in 25 mL of 0.1 M Tris-HC1 HCl buffer, pH 8.5 solution by looking for a dark deposit [28].

3. Results

3.1. Quantitative Analysis of Carbon Metabolism in Food-related S. cerevisiae Strains

First, the behavior under different carbon sources and with the presence of metabolic inhibitors was tested (Figures 1 and 2 and Supplementary Figure S1). Several commercial wine yeasts (T73, EC1118, M2, L2056, and 71B) were compared to baker's yeast Cinta Roja and chicha fermentation yeasts (EYS5, corn, and ERS1, rice [14]) and a laboratory strain that has no auxotrohies for amino acid metabolism, BQS252. Yeast was grown in a rich medium containing, as carbon source, glucose (YPD), sucrose (YPS), and galactose (YPGal). Antimycin A, an electron transport chain inhibitor, was used to test the role of respiration [29]. Glucose analog 2-deoxyglucose was used to study glucose repression on alternative carbon sources other than glucose, such as sucrose and galactose. Growth was carried out in multiwell plates at 30 °C and followed by OD_{600}, obtaining kinetic parameters such as maximum velocity of growth (Vmax), maximum OD600 (OD_{600}max), and lag time (Tables 1 and 2 and Supplemental Table S1). YPD is the standard rich medium with glucose used to propagate all kinds of yeasts. Yeasts in this condition grow by fermentation and mitochondrial respiration is not required. Antimycin A would block such respiration. The effect on growth for laboratory strain BQS252 and most industrial strains (M2 is depicted in Figure 1, for example) is therefore very small, as expected. After consuming glucose, cells enter postdiauxic growth and they consume the resulting ethanol by respiration, so the maximum growth is dependent on mitochondrial activity, and that is reflected in a small reduction in the OD_{600}max reached. There is, however, some phenotypic variation among wine yeasts. The 71B strain is the most sensitive one, which relies mostly on respiratory metabolism, as the ratios of all parameters (Table 1) indicate a growth delay. Surprisingly, the EC1118 strain performs better with antimycin A, starting growing earlier (the lowest ratio in lag time) and the growth speed and saturation OD were the highest. Therefore, inhibition of mitochondria helps this strain to achieve a better fermentation performance. Sucrose is a carbon source commonly used in the food industry, as for instance, for yeast biomass propagation. In this case, the effect of antimycin is also small (Supplementary Figure S1 and Table S1), indicating a fully fermentative metabolism. In this case, strain EC1118 does not have an improved saturation point nor lag phase, although its Vmax is bigger. The lag phase in galactose is a way of stimulating the activation of respiratory genes [29]. Unfortunately, EC1118 and T73 strains are unable to metabolize this monosaccharide, so the spectrum of wine yeast is reduced. The rest of the strains showed an expected delay in the start of growth, similar to the laboratory strain (Supplemental Table S1). Baker's yeast Cinta Roja was insensitive to this chemical.

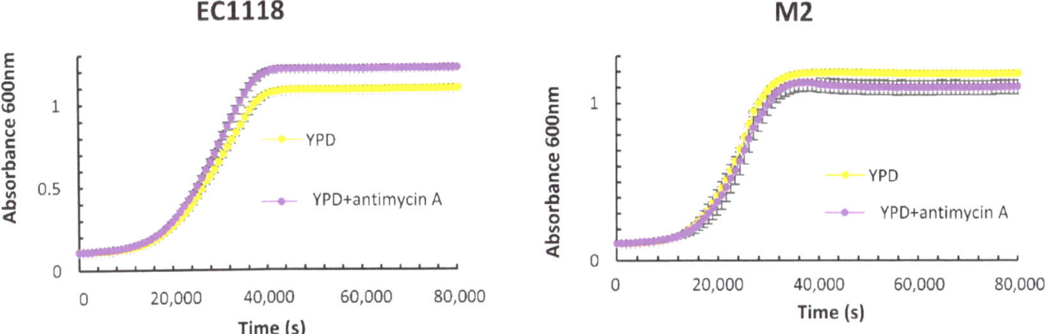

Figure 1. *Cont.*

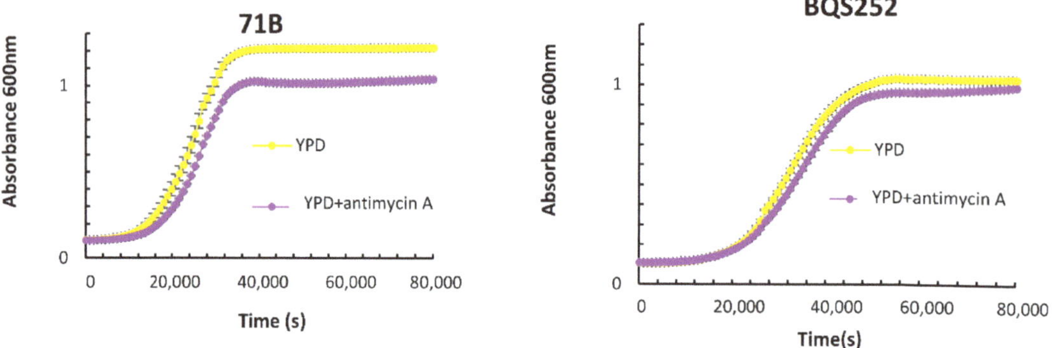

Figure 1. Mitochondrial activity has little effect on fermentative growth. Growth curves by OD_{600} measurement of selected strains grown in rich medium YPD with or without mitochondrial respiration inhibitor antimycin A. The average of three experiments is shown.

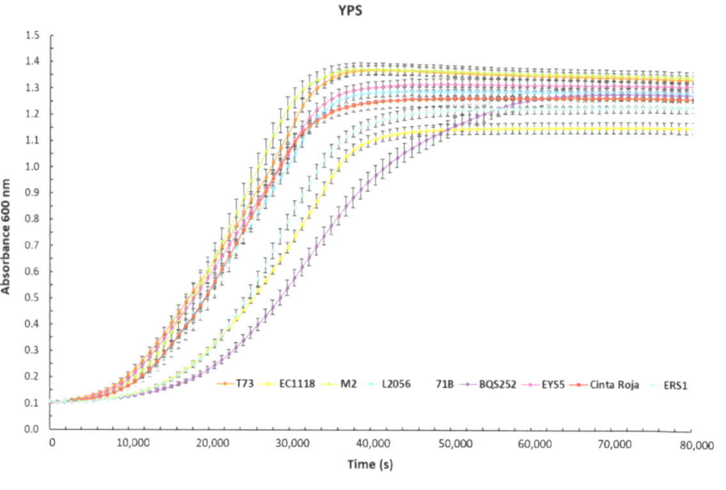

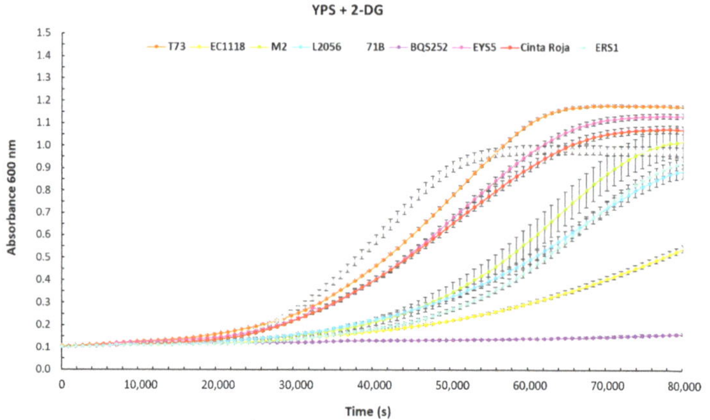

Figure 2. Wine yeasts have a variety of behaviors for glucose repression. Growth curves by OD_{600} measurement of selected strains grown in sucrose-containing medium YPS with or without 2-deoxyglucose. The average of three experiments is shown.

Table 1. Kinetic parameters of industrial strains grown in YPD and YPD+antimycin A. Maximum velocity (Vmax), maximum OD_{600} (ODmax), and lag time are shown. YPD+antimycin A/YPD ratios are also shown. Experiments were done in triplicate, Average (Av) and standard deviation (SD) is shown.

	YPD						YPD + Antimycin A						Ratios		
	Vmax		ODmax		Lag Time		Vmax		ODmax		Lag Time				
	Av	SD	Av	SD	Av	SD	Av	SD	Av	SD	Av	SD	Vmax	MaxOD	Lag
T73	0.219	0.046	1.150	0.024	11,710	269	0.259	0.009	1.060	0.057	15,720	442	1.18	0.82	1.34
EC1118	0.184	0.007	0.989	0.016	17,705	417	0.218	0.007	1.112	0.024	16,570	438	1.18	1.12	0.94
M2	0.330	0.009	1.077	0.008	14,807	247	0.287	0.024	1.017	0.064	15,413	1592	0.87	0.94	1.04
L2056	0.293	0.040	1.084	0.022	14,597	1251	0.239	0.000	1.124	0.011	14,365	375	0.81	1.04	0.98
71B	0.328	0.055	1.114	0.010	14,337	1070	0.254	0.005	0.940	0.005	16,920	57	0.77	0.84	1.18
BQS252	0.175	0.031	0.930	0.018	21,755	375	0.141	0.003	0.902	0.003	22,670	693	0.81	0.97	1.04
EYS5	0.275	0.010	1.092	0.038	14,390	1067	0.240	0.010	0.998	0.038	15,950	622	0.88	0.91	1.11
ERS1	0.236	0.001	1.061	0.029	17,967	724	0.214	0.005	0.972	0.012	19,887	351	0.91	0.92	1.11
Cinta Roja	0.224	0.010	1.014	0.025	14,227	547	0.219	0.005	0.952	0.006	15,185	290	0.98	0.94	1.07

Table 2. Kinetic parameters of industrial strains grown in YPS and YPS+2-deoxyglucose (non-metabolizable glucose analog). Maximum velocity (Vmax), maximum OD_{600} (ODmax), and lag time are shown. YPS+2-deoxyglucose /YPs ratios are also indicated. Experiments were done in triplicate. Average (Av) and standard deviation (SD) are shown.

	YPs						YPS + 2DG						Ratios		
	Vmax		Max OD		Lag Time		Vmax		Max OD		Lag Time				
	Av	SD	Av	SD	Av	SD	Av	SD	Av	SD	Av	SD	Vmax	MaxOD	Lag
T73	0.24	0.0001	1.19	0.0071	9667	400	0.13	0.0001	1.02	0.0218	25,203	410	0.70	0.87	1.92
EC1118	0.19	0.0034	1.04	0.0297	15,840	57	0.06	0.0037	0.48	0.0232	47,137	621	0.32	0.46	2.98
M2	0.26	0.0076	1.25	0.0289	10,993	824	0.12	0.0078	0.91	0.0812	39,020	2365	0.46	0.73	3.55
L2056	0.22	0.0067	1.17	0.0202	11,870	531	0.10	0.0040	0.79	0.0373	37,667	883	0.42	0.67	3.17
71B	0.23	0.0039	1.13	0.0161	12,597	367	0.14	0.0016	0.87	0.0246	26,827	300	0.62	0.77	2.13
BQS252	0.17	0.0026	1.16	0.0297	18,433	717							0.00	0.00	0.00
EYS5	0.22	0.0020	1.20	0.0250	10,342	346	0.16	0.0378	1.02	0.0135	28,543	526	0.76	0.85	2.76
ERS1	0.22	0.0017	1.11	0.0242	16,057	487	0.11	0.0058	0.83	0.0280	43,797	922	0.49	0.75	2.73
Cinta Roja	0.21	0.0022	1.15	0.0038	11,697	175	0.10	0.0030	0.96	0.0205	28,877	437	0.48	0.83	2.47

Glucose analog 2-deoxyglucose can induce a state of glucose repression that inhibits growth in a carbon source subjected to catabolite repression, such as sucrose and galactose. In the case of the strains that grow in galactose, their growth was fully inhibited by the amount of 2DG used (Supplementary Figure S1), so there are no quantitative data. However, surprisingly chicha rice yeast ERS1 was able to grow quite normaly, indicating that its galactose metabolism is insensitive for glucose repression by unknown causes.

The 2DG effect is better known in a medium containing sucrose as the carbon source. In this condition, the growth of laboratory strain is fully inhibited, while all industrial strains were able to cope with glucose repression and growth (Figure 2). That was observed qualitatively before [12], but now using kinetic parameters, we can estimate differences between strains quantitatively (Table 2). There is a wide variety of behaviors among wine strains. The most sensitive strain is EC1118, which does not reach saturation along the time course. The ratio of Vmax is the lowest among strains, reinforcing the idea of higher repression. Interestingly, EC1118 is the strain that reaches the lowest absorbance at saturation point in the control YPS curve, suggesting a different way of dealing with this disaccharide that may influence the degree of glucose repression. The most tolerant wine strain was T73, with the highest ODmax and a high Vmax. 71B strain shows a different profile, with relatively short lag time and growth speed, but with a higher impact in the long term, reaching a low ODmax. That may suggest short-term and long-term effects on glucose repression, or a better quick response but an earlier entry into the stationary phase due to a poor adaptation to sucrose starvation. Wine yeasts do not act differently to other yeasts of biotechnological interest (for instance, wine L2056 is very similar to *chicha* ERS1 at this effect), so the mechanisms of glucose are not particular for this breed of industrial yeast overall.

3.2. Gln3 Has a Complex Role in Carbon Metabolism in Commercial Wine Yeasts

As there are phenotypic differences between commercial wine yeasts regarding respiratory metabolism and glucose repression, the next experiments tried to identify potential molecular players in these processes. Previously, we have shown that transcription factor *GLN3* deletion blocks respiration and increases glucose repression in one haploid wine strain and some laboratory, but not all, genetic backgrounds [9]. In order to investigate the relevance of the Gln3 transcription factor in commercial strains, the two copies of *GLN3* were deleted from diploid strains EC1118 and M2. EC1118 was chosen by its extreme behavior against inhibitors antimycin A and 2-deoxyglucose, and because it is one of the most used strains in industry. As a complement, M2 strain was chosen as a reference due to its average response to those inhibitors. The results of these mutations was tested by spot analysis in selective media (Figure 3). The fastest way to test for respiration is to grow yeasts in a media where the carbon source has to be metabolized by respiration, such as glycerol. Cells were spotted in rich medium containing glucose (YPD) or glycerol (YPglycerol) (Figure 3A). *GLN3* mutation does not cause a deleterious phenotype as both mutants grow fine in YPD. EC1118 *gln3* grows strongly in glycerol, but M2 *gln3* is unable to grow in a non-fermentative carbon source. Therefore, both strains are quite different in terms of respiratory metabolism, and M2 relies more on this transcription factor. To test glucose repression, cells were grown in YPSucrose with 2-deoxyglucose. EC1118 is more sensitive to 2DG, but *GLN3* deletion does not change it. However, deletion of *GLN3* in M2 causes a full inhibition of growth, also indicating a clear role of this factor in this event. Glucosamine is another glucose analog that is known to induce glucose repression. Again M2 *gln3* is unable to grow in the presence of this inhibitor. In the EC1118 *gln3* strain, the effect is smaller but in the same direction. Gln3 is controlled, among other signaling pathways, by the TORC1 complex. Deletion of *GLN3* causes increased rapamycin tolerance in laboratory strains [30], and it is the same for the M2 *gln3* mutant, so the TORC1 branch of Gln3 regulation seems to be fine. The effect on EC1118 is similar but much less intense.

To further characterize the phenotype of *GLN3* deletion, mutants were spotted in minimal medium, where aspects of nitrogen metabolism can be studied (Figure 3B). Canavanine and methylamine are toxic analogs of arginine and ammonia, respectively, that can be used to assess the impact of a mutation in amino acid or ammonia import. In this case, the mutants have no phenotype in either genetic background. 3-aminotriazol is used to measure amino acid biosynthesis, as it inhibits the His3 imidazoleglycerol-phosphate dehydratase I for histidine biosynthesis. M2 *gln3* is more tolerant to this inhibitor, while EC1118 is insensitive to it. Therefore, there is also a phenotypic diversity in some aspects of nitrogen metabolism regarding *GLN3* implication.

Ure2 is a repressor of Gln3 transcription factor that channels signaling from TORC1. We developed a CRISPR-Cas9 based method to delete both copies of such genes in industrial yeasts. We failed to modify EC1118 and M2 strains, but the method was successful in strain L2056. The mutation behaves the expected way [31], as it increased sensitivity to rapamycin (Figure 3A). Ure2 plays no role in respiration, as the mutant grows fine in glycerol, but its mutation increases glucose repression as the deletion mutant is more sensitive to 2DG.

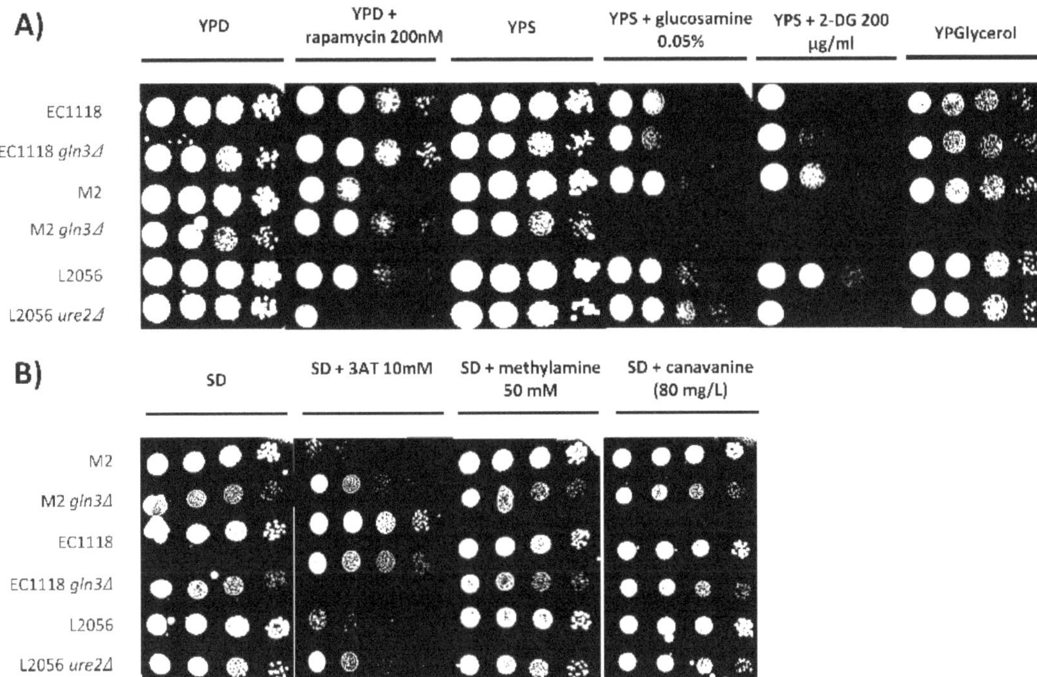

Figure 3. Gln3 has a distinctive role in respiration and glucose repression in some commercial wine strains. Spot analysis in rich (**A**) and minimal (**B**) media in the presence of several metabolic inhibitors (for *GLN3* deletion mutants EC1118 *gln3* and M2 *gln3* and for *URE2* deletion mutant L2056 *ure2* mutants and their parental strains. Serial dilutions were made and 5 µL dropped for each spot.

To further quantitatively analyze the effect of such mutations, growth curves of the aforementioned mutants in the presence of inhibitors, like 2-DG and antimicyn A, were obtained and the kinetic parameters calculated (Figure 4). Cells grown in sucrose confirm the higher sensitivity to 2DG of the M2 *gln3* in all parameters, while showing that deletion of *URE2* extends lag time in the presence on 2-DG and reduces Vmax and ODmax. *GLN3* deletion increases Vmax in the presence of rapamycin in both genetic backgrounds, indicating that the effect on growth is common. Both mutants behave similarly in terms of lag phase, so their mechanisms of adaptation are also similar. Antimycin A was tested for all three mutants in YPS and no major impact was seen. Antimycin was also tested in YPGalactose (in this case, EC1118 was left out, as it does not grow in galactose). In galactose, *GLN3* deletion had a negative impact on M2 growth (increase lag phase, reduced speed, and maximum OD). However, antimycin A relieves those differences, and in fact, the mutant had a reduced lag phase and it reaches an slightly higher final OD. *URE2* deletion has no major impact on the sensitivity to antimycin A during growth in galactose. Inhibition of growth with 2DG in YPgalactose was complete for all strains.

3.3. Molecular Markers of GLN3 Deletion

To further characterize the impact of *GLN3* deletion on wine yeast metabolic regulation, molecular markers were studied (Figure 5). Snf1 kinase activation is required for full derepression of sugar-regulated genes, including respiration ones. That was followed using a specific antibody in our wine strains and in the laboratory strain 1278b, which also was unable to grow by respiration when *GLN3* was deleted [9]. When cells are transferred from high to low glucose, Snf1 was rapidly phosphorylated and activated in all strains (Figure 5A). The phosphorylation peaked at 15 min and then it decreased. The *gln3* mutants in both M2 and

EC11118 strains have the same pattern, so there is no global regulation of the process from the early point of regulation, and Gln3 effect has to be more specific. Interestingly, the activation is reduced in the laboratory strain, so the behavior regarding this aspect is different in industrial strains, reinforcing the necessity to be characterized. A marker of glucose repression is alcohol dehydrogenase. Isoenzyme *ADH2* is glucose repressed, and it is activated only when sugar concentration decreases (as it occurs in stationary phase), while *ADH1* is used in fermentation conditions (as the exponential phase in YPD). A zymogram of ADH activity was done for the two industrial strains using 1278b strain as control (Figure 5B). A distinctive ADHII band appeared in stationary conditions in the wild-type strain M2, but this band failed to show in M2 *gln3* fitting its inability for respiratory growth, while derepression in EC1118 was fine. So the lower part of the derepression pathway is affected by *GLN3* deletion in M2 strain. A similar behavior is shown in laboratory strain, as expected. To try to dissect the degree of repression in M2 *gln3*, cells were grown in different media and subjected to derepression. When the mutant is grown in YPD (glucose) and shifted to YPGlycerol, there is a sign of partial derepression (Figure 5C), so there is not an intrinsic inability to express *ADH2* in this strain. When grown in galactose, the wild type shows both bands, indicating an intrinsic derepression, while the mutant has only ADHI. However, when shifted to glycerol from galactose, induction of ADHII is clear again for both strains. Therefore, the effect of *GLN3* deletion is targeted to specific conditions and genes and does not seem to be a general effect on *ADH2* promoter expression.

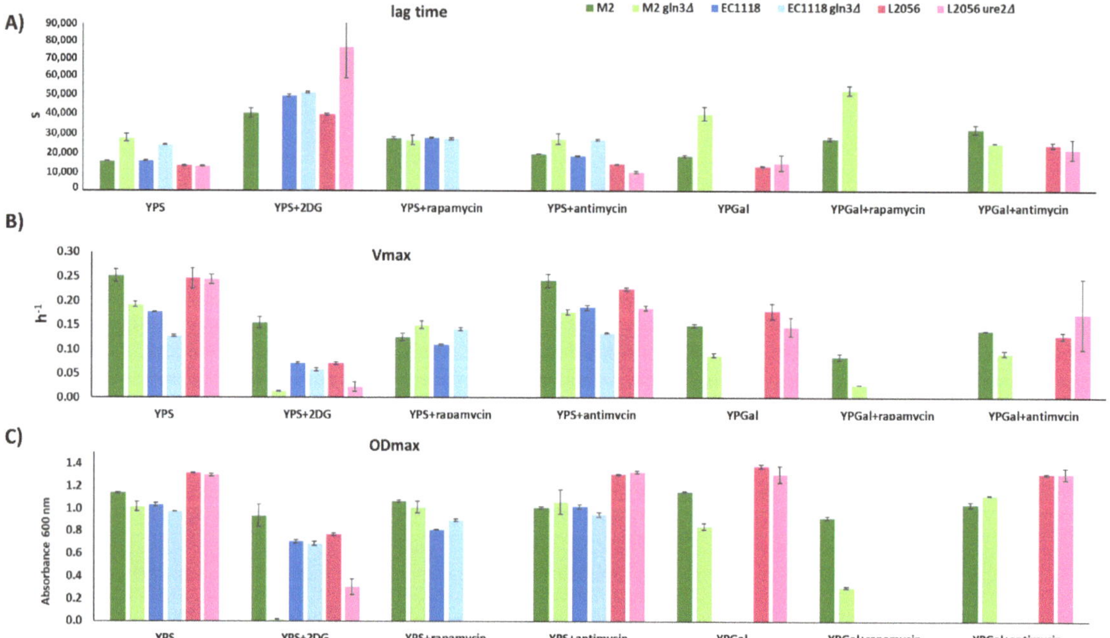

Figure 4. Kinetic parameters of industrial strains carrying *GLN3* and *URE2* mutations grown in sucrose (YPS) galactose (YPGal) with inhibitors. (**A**) lag time, (**B**) maximum speed (Vmax), (**C**) maximum OD$_{600}$ (ODmax).

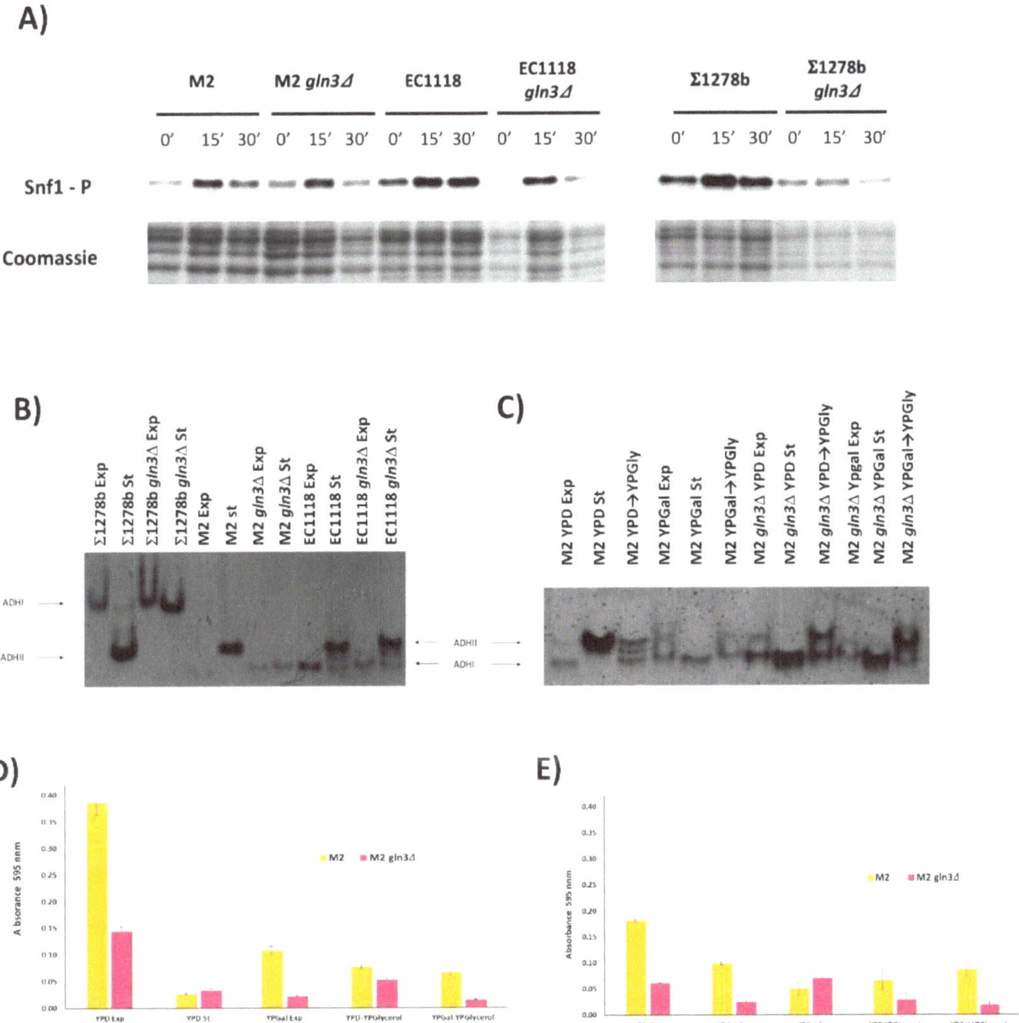

Figure 5. Gln3 impacts at the molecular level. (**A**) Western blot analysis of phosphorylated Snf1 proteins from *GLN3* deletion mutants in M2, EC1118, and 1278b strains (M2 *gln3*, EC1118 *gln3*, and 1278b *gln3*. Cells growing exponentially in YPD were changed to medium with 0.05% glucose and samples were taken at 15 and 30 min. Coomassie staining was used as the loading control. (**B**) Alcohol dehydrogenase zymogram. Samples from the same strains were collected in exponential (Exp) and stationary phases (St) of growth in YPD and stained for ADH activity. (**C**) Same as (**B**) with M2 and M2 *gln3* under different growth conditions and carbon sources. Shifts from one carbon source to another were indicated by an arrow. Exponentially growing cells in the first medium are shifted to the second, and cells were incubated then for two more hours. (**D**) Respiratory activity measured by MTT reaction in the same conditions as panel (**C**) from a preculture in YPD. (**E**) same as panel (**D**) with a preculture in YPGalactose.

To test respiration directly, the activity of the electron transport chain was measured by reduction of the MTT probe (Figure 5D,E). A preculture in YPD was used to inoculate various media (Figure 5D). *GLN3* deletion in M2 causes a decrease of mitochondrial activity in cells growing exponentially in YPD, but not in the stationary phase. A similar reduction is seen when cells are growing in galactose, and when cells are shifted to glycerol from exponentially growing cells in glucose and galactose (more evident in the latter).

When the preculture is made in YPGalactose (Figure 5E), again, *GLN3* deletion reduces mitochondrial activity in exponentially grown cells in YPD and YPGalactose. When the culture in YPGalactose reaches the stationary phase, such difference disappears again. Change to glycerol again retains the difference. So, Gln3 has an impact on mitochondrial activity in many growth conditions.

3.4. Role of GLN3 in Winemaking Conditions

Next, the impact of *GLN3* deletion was tested in winemaking conditions by conducting minivinifications of synthetic grape juice (Figure 6), which allows the change in media composition regarding nitrogen. The standard synthetic grape juice had 300 mg/L of assimilable nitrogen (MS300) while the limiting one had 60 mg/L (MS60). In addition, rapamycin and antimycin A was added to MS300 to follow the impact of such inhibitors. First, strain EC1118 was tested. Fermentation was followed by measuring sugar consumption (Figure 6A). *GLN3* deletion causes a minor delay at short times in MS300, but the fermentation finished rapidly. In the wild type, antimycin A does not have a significant impact on fermentation, indicating that in those fully fermentative conditions, respiration may not be very relevant. However, there is a significant delay in fermentation when antimycin A is added to the EC1118 *gln3* mutant, indicating an alteration in mitochondrial activity caused by *GLN3* deletion. Rapamycin has a massive impact on the advance of fermentation, indicating that TORC1 is key for cellular growth and proliferation, as expected. EC1118 *gln3* is more tolerant to rapamycin, as seen in laboratory media, so the relevance of Gln3 in controlling metabolism in a TORC1-mediated way is conserved during vinification. As expected, fermentation at limiting nitrogen is delayed in the wild type and in a similar way in the mutant. The production of ethanol, glycerol and acetic acid after finishing fermentation was not significantly altered (data not shown). To get some information about nitrogen metabolism, amino acid concentration in the media was measured at the beginning of fermentation (Supplementary Figure S2). In the rich medium, the MS300 wild-type strain consumes more than half amino acids at day 3 and then the consumption speed slowed down, suggesting an excess of amino acids. EC1118 *gln3* strain showed a similar pattern, so there is no need for this transcription factor to activate the bulk of amino acid transport. Limiting grape juice MS60 showed a rapid depletion of the scarce amino acids to a residual level. Again, *GLN3* deletion does not alter this pattern. Antimycin A does not alter this particular metabolic aspect, but rapamycin does it, as expected. Again EC1118 *gln3* mutant performs better, indicating a better amino acid assimilation.

The behavior of M2 and M2 *gln3* strains was followed in the same conditions (Figure 6B). In this case, *GLN3* deletion has a more prominent impact on the advance on fermentation, delaying its completion. Again, the mutation is more relevant in this genetic background. The impact in this case of antimycin A is negligible for both strains. Again, *GLN3* mutation delays fermentation in nitrogen limiting grape must MS60. As it happened with the EC1118 strain, rapamycin causes a big upset that is relieved by *GLN3* deletion (Supplementary Figure S3), indicating that the functions related to TORC1 are conserved among strains. Regarding the amino acid profile (Supplementary Figure S2), the pattern under MS60 growth and rapamycin and antimycin A is similar to EC1118 background, but the impact of *GLN3* deletion in MS300 is more apparent. Wild type M2 strain consumes most amino acids very fast, although those increase later (maybe due to export or partial cell lysis). M2 *gln3* is clearly delayed in this aspect, so again, Gln3 is more relied upon in this genetic background to achieve full amino acid import.

Regarding *URE2* deletion (Figure 6C), it slows down fermentation in MS300. Despite being a repressor of Gln3 function, it does not have the opposite role, probably due to the high number of targets involved that have to be regulated in a balanced way. Antimycin A, in this case, improves the function of the mutant, so it alleviates some of the negative aspects of such mutations that are be related to mitochondrial function. Rapamycin impact on fermentation is worse in the *ure2* strain, so in this regard, the mutation is indeed opposite to the *GLN3* deletion. Finally, in poor MS60 *URE2* deletion improves fermentation as expected. In terms of amino acid transport, surprisingly, *URE2* deletion mutant behaves in a similar way to its parental strain in all conditions tested (Supplementary Figure S2).

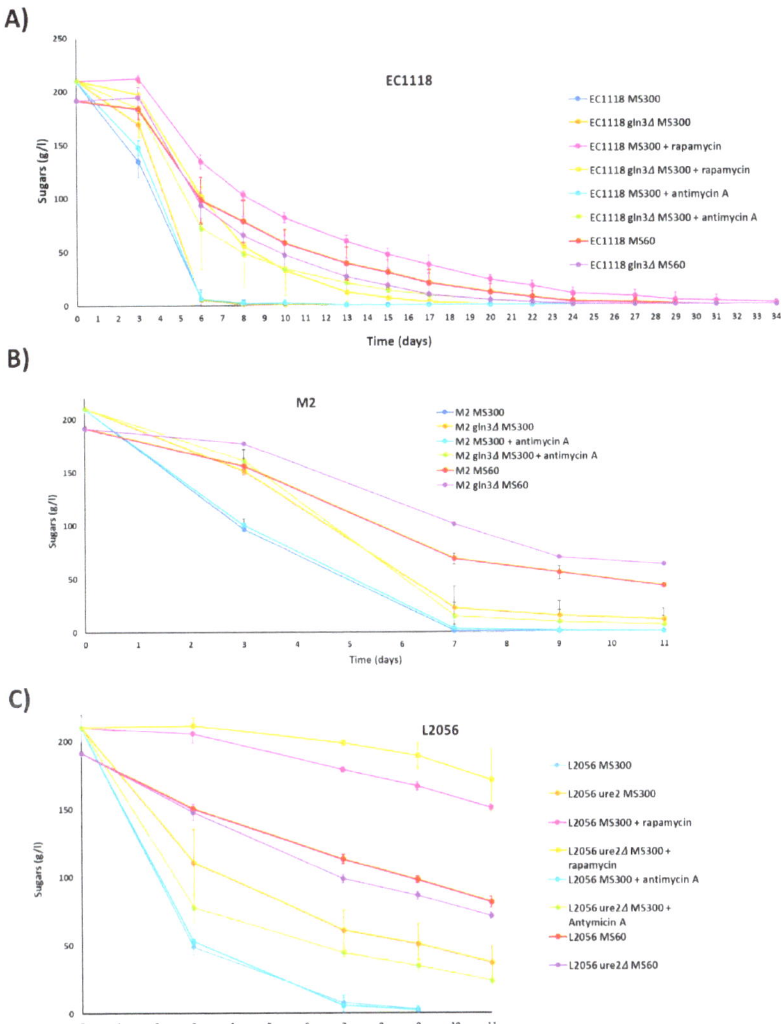

Figure 6. Gln3 role in wine fermentation. Progress of fermentation is followed by measuring reducing sugars. High nitrogen must (MS300) and low nitrogen must (MS60) were used. In the MS300, rapamycin (200 nM) and antimycin A (3 mg/L) were added. (**A**) EC1118 and *GLN3* deletion mutant EC1118 *gln3* M2 and *GLN3* deletion mutant M2 *gln3* (**C**) L2046 and *URE2* deletion mutant L2056 *ure2*.

3.5. GLN3 Relevance in Nitrogen Metabolism in Industrial Wine Yeasts

GLN3 functions regarding TORC1 activity in wine yeast and in winemaking conditions are consistent with their known role in nitrogen metabolism. Growth curves with different nitrogen sources were made in minimal medium and kinetic parameters were obtained to better understand its role regarding specific nitrogen sources (Figure 7). Reference minimal medium SD contains 0.5% ammonium sulfate. In this medium, the lag phase was delayed and the maximum OD_{600} was decreased in the mutant strains for both genetic backgrounds,

M2 and EC1118, while Vmax was basically unaffected. So in ammonium, Gln3 is relevant to resume growth and reach stationary phase, but not so relevant for steady growth.

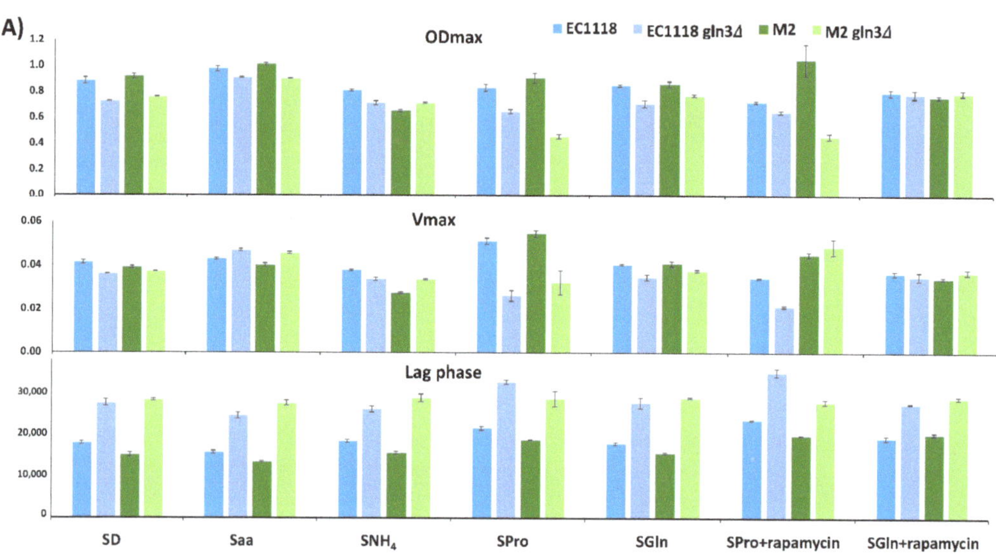

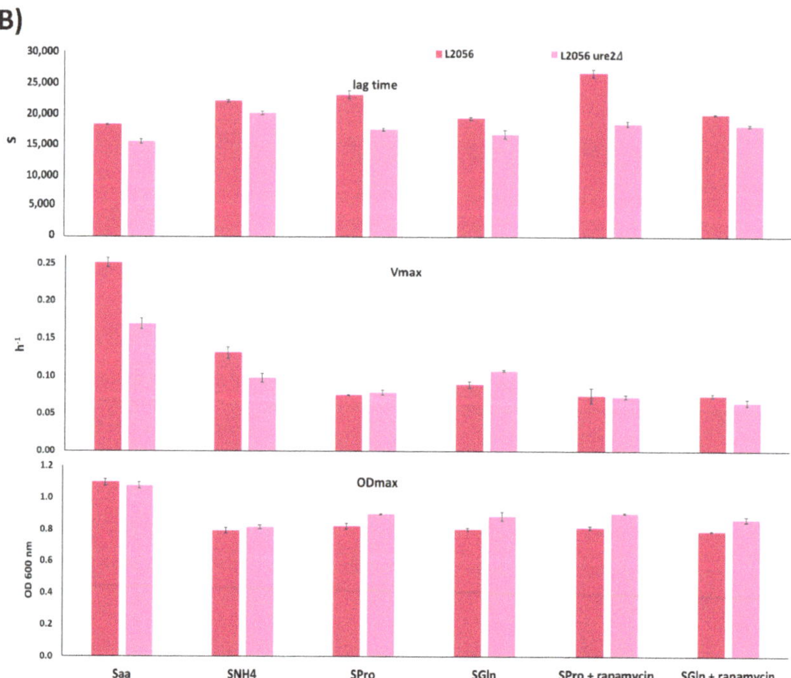

Figure 7. Gln3 impacts nitrogen assimilation in wine yeasts. (**A**) Kinetic parameters (ODmax, Vmax, and lag phase) obtained from *GLN3* deletion mutants in EC1118 and M2 strains grown in minimal medium with a different nitrogen source. SD: standard minimal medium with = 0.5% ammonium sulfate, Saa (300 mg/L of assimilable nitrogen as amino acid mix), SNH₄ (300 mg/L of assimilable nitrogen as ammonium chloride), SPro (0.5% proline as sole nitrogen source), SGln (0.5% glutamine as sole nitrogen source). Rapamycin was used at 200 nM. (**B**) Same for L2056 *URE2* deletions strain.

Next, another two unique nitrogen sources were tested. Glutamine is considered a rich, nitrogen repression inducing source, while proline is a poor nitrogen source that triggers Gln3 activity. Results in glutamine are similar to SD, the expected for a rich nitrogen source. However, proline has a bigger impact for both mutant strains, particularly at the Vmax. Therefore, Gln3 is required to fully assimilate proline as a nitrogen source. Rapamycin shuts off TORC1 complex and its effects were followed in proline and glutamine containing media. Rapamycin eliminates the differences caused by *GLN3* deletion in ODmax and Vmax in glutamine, so part of the effect of the mutation is TORC1-dependent. Proline-grown EC1118/EC1118 *gln3* do not change when ratios when exposed to rapamicyn, suggesting that proline is not fully activating TORC1. However, Vmax was increased in M2 *gln3*, as if TORC1 was causing growth problems in a Gln3 mediated way. Finally, synthetic media with 300 mg of assimilable nitrogen (like synthetic must MS300) in the form of just amino acids (Saa) or just ammonia (S_{NH4}) were tested to check the potential contributions of the nitrogen source of musts. *GLN3* mutations increase Vmax in both backgrounds, suggesting that biosynthesis is more deleterious to the cell than importing all the amino acids from the medium, a situation that may be relevant for a rich environment like grape juice. The reduced amount of nitrogen in S_{NH4} is not relevant for the EC1118/EC1118 *gln3* pair compared to SD, but M2 *gln3* has a relative increase of Vmax and ODmax, so both strains show different sensitivities to ammonia that may be at the core of nitrogen requirements for both strains.

Regarding Ure2 (Figure 7B), their influence is lower on average. Surprisingly its deletion improves not only growth in proline, as expected, but also a slight increase in Vmax and ODmax in glutamine, indicating that alleviating NCR is good whatever the input. The same pattern is seen in ammonium. However, when grown in a mix of amino acids, *URE2* deletion improves lag time, but it has no impact on the final ODmax and reduces Vmax. So, in a complex situation (as the one seen on industrial growth media), the presence of a regulator like Ure2 may be required for optimal growth. Rapamycin presence reduces the difference in Vmax in glutamine, suggesting that the described positive effect was TORC1 mediated, but Ure2-independent.

4. Discussion

The mechanisms of yeast metabolic adaptations are complex and poorly characterized in industrial yeasts during biotechnological conditions, but it is an issue that has to be tackled to improve yeast industrial performance. A limitation for such a study is that there is not a reference industrial strain, not even a reference wine yeast. In the studies of respiration and glucose repression, wine yeasts offer a broad variability, with extreme behaviors that are more distant than yeasts used for other purposes, like bread or *chicha* making (Figures 1 and 2, and Supplementary Figure S1). A positive aspect is that what we learn from wine strains may be useful to improve the performance of yeasts in other food-making processes. The mentioned phenotypic variability reinforces the idea that the right yeast has to be chosen for each specific process and that conditions have to be adapted for each strain. The latter is more difficult to be carried out in an industry with increasingly standardized protocols. The fact that wine yeast starters have to be propagated in sucrose-rich molasses with high respiration, to be later used in a low oxygen environment, such as grape juice fermentation, just increases the complexity of the overall performance of a selected yeast. Inhibition of respiration by antimycin A does not impact greatly on wine fermentation (Figure 5) or initial growth on many fermentable sugars (Supplementary Table S1) but it is required to achieve maximum biomass production. However, some strains are more sensitive than others, and stopping respiration improves EC1118 strain growth (Figure 1). EC1118 showed an extreme behavior regarding other carbon-related phenotypes, like glucose repression (Figure 2), suggesting a particular genetic background regarding these events. EC1118 is a widely used strain that belongs to a subset of wine yeasts known as the Prise de Mousse clade (PdM, for being strain suited to perform sparkling wine fermentations) [32]. Those are genetically very similar strains that share the same recent common origin, although there is not a single common feature that could be considered exclusive of that clade. Industrial strains contain genes or chromosomal regions that are exclusive of some strains [32]. EC1118 was the first wine yeast to be completely sequenced [33] and three unique large chromosomal regions were found (caused in part by horizontal transfer). However, those are not exclusive of the clade and are found scattered along many other wine strains [32]. According to Novo et al. (2009), none of those regions were found, for instance, in strains L2056 and 71B, but all of them are in T73, so there is not an easy correlation with the phenotype showed by these strains under different conditions and those specific regions. M2 has no any of these regions either [32]. None of the regions contains a gene easily linked to metabolic control. Therefore, there is a long way to be able to link the increasing number of sequences obtained from industrial yeasts to specific phenotypes, and further work has to be done to clarify those phenotypes, like the one to study nitrogen requirements by studying quantitative trait loci (QTL) [34].

The bulk of this study is based on the analysis of the Gln3 transcription factor. This is a key factor in metabolic regulation, as it is involved both in nitrogen and carbon metabolism, being regulated by both nitrogen-induced TORC1 complex and carbon catabolite derepressing kinase Snf1. Many previous works have focused in the Gln3 repressor Ure2 to improve wine yeasts [35–38], mainly by improving the assimilation of poor nitrogen sources, such as proline. Here we aimed to analyze systematically *GLN3* function by deleting it in two commercial wine strains with different behavior, M2 and EC1118. The *GLN3* deletion mutants had the expected role in some aspects, mainly the ones related to nitrogen and TORC1 regulation. For instance, they are more tolerant to TORC1 inhibition by rapamycin during winemaking conditions (Figure 6), and they grow worse with a poor nitrogen source like proline (Figure 7). However, the effect of the mutation is background-dependent when it comes to carbon metabolism. M2 *gln3* is unable to grow by respiration and shows complete glucose repression by 2-deoxyglucose that the EC1118 counterpart does not suffer (Figure 3), and its impact in wine fermentation is bigger too (Figure 6). We previously found that *GLN3* deletion impaired respiration in laboratory strain 1278b but not in reference to S288c genetic background [9], so this elusive phenotype is not linked to industrial strains per se. Comparison of sequences of NCR-related genes in the databases did not reveal a distinctive pattern linked to this phenotype (data not shown). In the absence of Gln3, differences in Gat1 function and regulation can become more apparent, and although previous works indicated that Gat1 role in winemaking is smaller [9], that has to be carefully studied in the future. The real molecular causes behind the impact of Gln3 in carbon metabolism have to be clarified. In this paper, we showed that there is not a global alteration in glucose repression via Snf1 due to lack of activation (Figure 5A), so probably a reduced subset of molecular targets are affected. Glucose repressible Adh2 expression is blocked in the mutant in the stationary phase, but induction is possible when transferred to glycerol (Figure 5C), although cells are unable to grow in it. Therefore, the lack of regulation of glucose repression is partial. Transcriptomic global analysis indicates that Gln3 is a positive transcriptional activator of *ADH2* gene [39], although detailed analysis of *ADH2* promoters suggest the bulk of its regulation is made by other transcription factors, such as Adr1 [40]. An indirect effect of *GLN3* deletion acting on more specific activators cannot be ruled out. That would explain the fact that we have detected decreased mitochondrial activity that may contribute to the phenotype (Figure 5D and 5E). A detailed global analysis at different levels is required in the future to pinpoint the molecular targets of *GLN3* in those specific backgrounds. *URE2* deletion has no phenotype in respiration, but it does increase glucose repression (Figure 5). Therefore, both functions of Gln3 could be differently regulated. Deletion of *URE2* and *GLN3* give opposite phenotypes regarding TORC1 inhibition by rapamycin, so those carbon-related events probably are not regulated by TORC1. It may seem surprising that deletion of *GLN3* and *URE2* have similar, not opposite, phenotypes regarding glucose repression, but this is shared by other events, like pseudohyphal growth [41].

5. Conclusions

The mechanisms involved in glucose repression in industrial yeasts, particularly wine yeasts, are complex and may influence their performance under biotechnological conditions. Transcription factor Gln3 is at the crossroads between respiration, glucose repression, and nitrogen repression, so it is a protein worth studying to better understand the fine-tuning adaptation to environmental changes.

Supplementary Materials: The following are available online at https://www.mdpi.com/article/10.3390/fermentation7030181/s1, Table S1. Kinetic parameters of *S. cerevisiae* strains under different growing conditions. Figure S1. Growth curves of industrial wine yeasts in YPD, YPS, and YPGal, plus with the indicated inhibitor. Figure S2. Quantification of alpha-amino acids of the vinifications described in Figure 6. Figure S3. MS300 synthetic grape juices fermentation of M2 and M2 *gln3* strains in the presence of 200 nM rapamycin. Conditions as in Figure 6.

Author Contributions: Conceptualization, A.A. and E.M.; investigation, A.F.-P., V.G. and A.A.; resources, E.M. and A.A.; writing—original draft preparation, A.A.; writing—review and editing, A.F.-P., V.G., E.M., and A.A.; project administration, E.M.; funding acquisition, E.M. and A.A. All authors have read and agreed to the published version of the manuscript.

Funding: This work was funded by a grant from the Spanish Ministry of Science (AGL2017-83254-R) to E.M. and A.A. V.G. has Generalitat Valenciana fellowship ACIF/2020/122.

Institutional Review Board Statement: Not applicable.

Informed Consent Statement: Not applicable.

Data Availability Statement: Not aplicable.

Conflicts of Interest: The authors declare no conflict of interest.

References

1. Matallana, E.; Aranda, A. Biotechnological impact of stress response on wine yeast. *Lett. Appl. Microbiol.* **2016**. [CrossRef] [PubMed]
2. Kessi-Perez, E.I.; Molinet, J.; Martinez, C. Disentangling the genetic bases of *Saccharomyces cerevisiae* nitrogen consumption and adaptation to low nitrogen environments in wine fermentation. *Biol. Res.* **2020**, *53*, 2. [CrossRef] [PubMed]
3. García-Ríos, E.; Guillamón, J.M. Mechanisms of yeast adaptation to wine fermentations. *Prog. Mol. Subcell. Biol.* **2019**, *58*, 37–59. [CrossRef] [PubMed]
4. Gobert, A.; Tourdot-Maréchal, R.; Sparrow, C.; Morge, C.; Alexandre, H. Influence of nitrogen status in wine alcoholic fermentation. *Food Microbiol.* **2019**, *83*, 71–85. [CrossRef] [PubMed]
5. Conrad, M.; Schothorst, J.; Kankipati, H.N.; Van Zeebroeck, G.; Rubio-Texeira, M.; Thevelein, J.M. Nutrient sensing and signaling in the yeast *Saccharomyces cerevisiae*. *FEMS Microbiol. Rev.* **2014**, *38*, 254–299. [CrossRef]
6. Piskur, J.; Rozpedowska, E.; Polakova, S.; Merico, A.; Compagno, C. How did Saccharomyces evolve to become a good brewer? *Trends Genet.* **2006**, *22*, 183–186. [CrossRef] [PubMed]
7. Pérez-Torrado, R.; Gamero, E.; Gómez-Pastor, R.; Garre, E.; Aranda, A.; Matallana, E. Yeast biomass, an optimised product with myriad applications in the food industry. *Trends Food Sci. Technol.* **2015**, *46*, 167–175. [CrossRef]
8. Meneses, F.J.; Henschke, P.A.; Jiranek, V. A Survey of industrial strains of *Saccharomyces cerevisiae* reveals numerous altered patterns of maltose and sucrose utilisation. *J. Inst. Brew.* **2002**, *108*, 310–321. [CrossRef]
9. Vallejo, B.; Peltier, E.; Garrigós, V.; Matallana, E.; Marullo, P.; Aranda, A. Role of *Saccharomyces cerevisiae* nutrient signaling pathways during winemaking: A phenomics approach. *Front. Bioeng. Biotechnol.* **2020**. [CrossRef]
10. Beier, D.R.; Sledziewski, A.; Young, E.T. Deletion analysis identifies a region, upstream of the ADH2 gene of *Saccharomyces cerevisiae*, which is required for ADR1-mediated derepression. *Mol. Cell Biol* **1985**, *5*, 1743–1749. [CrossRef]
11. De Virgilio, C. The essence of yeast quiescence. *FEMS Microbiol. Rev.* **2012**, *36*, 306–339. [CrossRef] [PubMed]
12. Vallejo, B.; Matallana, E.; Aranda, A. *Saccharomyces cerevisiae* nutrient signaling pathways show an unexpected early activation pattern during winemaking. *Microb. Cell Fact.* **2020**. [CrossRef]
13. Bertram, P.G.; Choi, J.H.; Carvalho, J.; Chan, T.F.; Ai, W.; Zheng, X.F. Convergence of TOR-nitrogen and Snf1-glucose signaling pathways onto Gln3. *Mol. Cell Biol.* **2002**, *22*, 1246–1252. [CrossRef]
14. Grijalva-Vallejos, N.; Aranda, A.; Matallana, E. Evaluation of yeasts from Ecuadorian chicha by their performance as starters for alcoholic fermentations in the food industry. *Int. J. Food Microbiol.* **2020**, *317*, 108462. [CrossRef]
15. Guldener, U.; Heck, S.; Fielder, T.; Beinhauer, J.; Hegemann, J.H. A new efficient gene disruption cassette for repeated use in budding yeast. *Nucleic Acids Res.* **1996**, *24*, 2519–2524. [CrossRef]

16. Delneri, D.; Tomlin, G.C.; Wixon, J.L.; Hutter, A.; Sefton, M.; Louis, E.J.; Oliver, S.G. Exploring redundancy in the yeast genome: An improved strategy for use of the cre-loxP system. *Gene* **2000**, *252*, 127–135. [CrossRef]
17. Generoso, W.C.; Gottardi, M.; Oreb, M.; Boles, E. Simplified CRISPR-Cas genome editing for *Saccharomyces cerevisiae*. *J. Microbiol. Methods* **2016**, *127*, 203–205. [CrossRef] [PubMed]
18. Gietz, R.D.; Woods, R.A. Transformation of yeast by lithium acetate/single-stranded carrier DNA/polyethylene glycol method. *Methods Enzym.* **2002**, *350*, 87–96.
19. Adams, A.; Kaiser, C.; Cold Spring Harbor Laboratory. *Methods in Yeast Genetics: A Cold Spring Harbor Laboratory Course Manual*, 1997th ed.; Cold Spring Harbor Laboratory Press: Plainview, NY, USA, 1998; ISBN 0879695080.
20. Riou, C.; Nicaud, J.M.; Barre, P.; Gaillardin, C. Stationary-phase gene expression in *Saccharomyces cerevisiae* during wine fermentation. *Yeast* **1997**, *13*, 903–915. [CrossRef]
21. Viana, T.; Loureiro-Dias, M.C.; Prista, C. Efficient fermentation of an improved synthetic grape must by enological and laboratory strains of *Saccharomyces cerevisiae*. *AMB Express* **2014**, *4*, 1–9. [CrossRef] [PubMed]
22. Robyt, J.F.; Whelan, W.J. Reducing value methods for maltodextrins. I. Chain-length dependence of alkaline 3,5-dinitrosalicylate and chain-length independence of alkaline copper. *Anal. Biochem.* **1972**, *45*, 510–516. [CrossRef]
23. Dukes, B.C.; Butzke, C.E. Rapid determination of primary amino acids in grape juice using an *o*-phthaldialdehyde/N-acetyl-L-cysteine spectrophotometric Assay. *Am. J. Enol. Vitic.* **1998**, *49*, 125–134.
24. Sánchez, N.S.; Königsberg, M. Using yeast to easily determine mitochondrial functionality with 1-(4,5-dimethylthiazol-2-yl)-3,5-diphenyltetrazolium bromide (MTT) assay. *Biochem. Mol. Biol. Educ. Bimon. Publ. Int. Union Biochem. Mol. Biol.* **2006**, *34*, 209–212. [CrossRef] [PubMed]
25. Márquez, I.G.; Ghiyasvand, M.; Massarsky, A.; Babu, M.; Samanfar, B.; Omidi, K.; Moon, T.W.; Smith, M.L.; Golshani, A. Zinc oxide and silver nanoparticles toxicity in the baker's yeast, *Saccharomyces cerevisiae*. *PLoS ONE* **2018**. [CrossRef]
26. Orlova, M.; Barrett, L.; Kuchin, S. Detection of endogenous Snf1 and its activation state: Application to Saccharomyces and *Candida* species. *Yeast* **2008**, *25*, 745–754. [CrossRef] [PubMed]
27. Williamson, V.M.; Bennetzen, J.; Young, E.T.; Nasmyth, K.; Hall, B.D. Isolation of the structural gene for alcohol dehydrogenase by genetic complementation in yeast. *Nature* **1980**, *283*, 214–216. [CrossRef]
28. Fowler, P.W.; Ball, A.J.; Griffiths, D.E. The control of alcohol dehydrogenase isozyme synthesis in *Saccharomyces cerevisiae*. *Can. J. Biochem.* **1972**, *50*, 35–43. [CrossRef]
29. Perez-Samper, G.; Cerulus, B.; Jariani, A.; Vermeersch, L.; Barrajón Simancas, N.; Bisschops, M.M.M.; van den Brink, J.; Solis-Escalante, D.; Gallone, B.; De Maeyer, D.; et al. The crabtree effect shapes the *Saccharomyces cerevisiae* lag phase during the switch between different carbon sources. *MBio* **2018**. [CrossRef]
30. Huber, A.; Bodenmiller, B.; Uotila, A.; Stahl, M.; Wanka, S.; Gerrits, B.; Aebersold, R.; Loewith, R. Characterization of the rapamycin-sensitive phosphoproteome reveals that Sch9 is a central coordinator of protein synthesis. *Genes Dev.* **2009**, *23*, 1929–1943. [CrossRef]
31. Xie, M.W.; Jin, F.; Hwang, H.; Hwang, S.; Anand, V.; Duncan, M.C.; Huang, J. Insights into TOR function and rapamycin response: Chemical genomic profiling by using a high-density cell array method. *Proc. Natl. Acad. Sci. USA* **2005**, *102*, 7215–7220. [CrossRef]
32. Borneman, A.R.; Forgan, A.H.; Kolouchova, R.; Fraser, J.A.; Schmidt, S.A. Whole genome comparison reveals high levels of inbreeding and strain redundancy across the spectrum of commercial wine strains of *Saccharomyces cerevisiae*. *G3* **2016**. [CrossRef]
33. Novo, M.; Bigey, F.; Beyne, E.; Galeote, V.; Gavory, F.; Mallet, S.; Cambon, B.; Legras, J.L.; Wincker, P.; Casaregola, S.; et al. Eukaryote-to-eukaryote gene transfer events revealed by the genome sequence of the wine yeast *Saccharomyces cerevisiae* EC1118. *Proc. Natl. Acad. Sci. USA* **2009**, *106*, 16333–16338. [CrossRef] [PubMed]
34. Brice, C.; Sanchez, I.; Bigey, F.; Legras, J.L.; Blondin, B. A genetic approach of wine yeast fermentation capacity in nitrogen-starvation reveals the key role of nitrogen signaling. *BMC Genom.* **2014**, *15*, 495. [CrossRef] [PubMed]
35. Salmon, J.M.; Barre, P. Improvement of nitrogen assimilation and fermentation kinetics under enological conditions by derepression of alternative nitrogen-assimilatory pathways in an industrial *Saccharomyces cerevisiae* strain. *Appl. Environ. Microbiol.* **1998**, *64*, 3831–3837. [CrossRef] [PubMed]
36. Deed, N.K.; van Vuuren, H.J.; Gardner, R.C. Effects of nitrogen catabolite repression and di-ammonium phosphate addition during wine fermentation by a commercial strain of S. cerevisiae. *Appl. Microbiol. Biotechnol.* **2011**, *89*, 1537–1549. [CrossRef]
37. Tate, J.J.; Tolley, E.A.; Cooper, T.G. Sit4 and PP2A dephosphorylate nitrogen catabolite repression-sensitive Gln3 when TorC1 is up- as well as down-regulated. *Genetics* **2019**. [CrossRef]
38. Dufour, M.; Zimmer, A.; Thibon, C.; Marullo, P. Enhancement of volatile thiol release of *Saccharomyces cerevisiae* strains using molecular breeding. *Appl. Microbiol. Biotechnol.* **2013**, *97*, 5893–5905. [CrossRef]
39. Reimand, J.; Vaquerizas, J.M.; Todd, A.E.; Vilo, J.; Luscombe, N.M. Comprehensive reanalysis of transcription factor knockout expression data in *Saccharomyces cerevisiae* reveals many new targets. *Nucleic Acids Res.* **2010**, *38*, 4768–4777. [CrossRef]
40. Young, E.T.; Yen, K.; Dombek, K.M.; Law, G.L.; Chang, E.; Arms, E. Snf1-independent, glucose-resistant transcription of Adr1-dependent genes in a mediator mutant of *Saccharomyces cerevisiae*. *Mol. Microbiol.* **2009**, *74*, 364–383. [CrossRef]
41. Lorenz, M.C.; Heitman, J. The MEP2 ammonium permease regulates pseudohyphal differentiation in *Saccharomyces cerevisiae*. *EMBO J.* **1998**, *17*, 1236–1247. [CrossRef]

 *fermentation*

Article

Manipulation of In Vitro Ruminal Fermentation and Feed Digestibility as Influenced by Yeast Waste-Treated Cassava Pulp Substitute Soybean Meal and Different Roughage to Concentrate Ratio

Gamonmas Dagaew, Anusorn Cherdthong *, Sawitree Wongtangtintharn, Metha Wanapat and Chanon Suntara

Tropical Feed Resources Research and Development Center (TROFREC), Department of Animal Science, Faculty of Agriculture, Khon Kaen University, Khon Kaen 40002, Thailand; gamonmasdagaew@gmail.com (G.D.); sawiwo@kku.ac.th (S.W.); metha@kku.ac.th (M.W.); chanon_su@kkumail.com (C.S.)

* Correspondence: anusornc@kku.ac.th; Tel./Fax: +66-4320-2362

Citation: Dagaew, G.; Cherdthong, A.; Wongtangtintharn, S.; Wanapat, M.; Suntara, C. Manipulation of In Vitro Ruminal Fermentation and Feed Digestibility as Influenced by Yeast Waste-Treated Cassava Pulp Substitute Soybean Meal and Different Roughage to Concentrate Ratio. *Fermentation* **2021**, *7*, 196. https://doi.org/10.3390/fermentation7030196

Academic Editor: Ronnie G. Willaert

Received: 1 September 2021
Accepted: 15 September 2021
Published: 17 September 2021

Publisher's Note: MDPI stays neutral with regard to jurisdictional claims in published maps and institutional affiliations.

Copyright: © 2021 by the authors. Licensee MDPI, Basel, Switzerland. This article is an open access article distributed under the terms and conditions of the Creative Commons Attribution (CC BY) license (https://creativecommons.org/licenses/by/4.0/).

Abstract: Cassava pulp (CS) is high in fiber and low in protein; hence, improving the nutritive value of CS is required to increase its contribution to enhancing ruminant production. The present work hypothesized that CS quality could be enhanced by fermentation with yeast waste (YW), which can be used to replace soybean meal (SBM), as well as lead to improved feed utilization in ruminants. Thus, evaluation of in vitro ruminal fermentation and feed digestibility, as influenced by YW-treated CS and different roughage (R) to concentrate (C) ratios, was elucidated. The design of the experiment was a 5 × 3 factorial arrangement in a completely randomized design. Each treatment contained three replications and three runs. The first factor was replacing SBM with CS fermented with YW (CSYW) in a concentrate ratio at 100:0, 75:25, 50:50, 25:75, and 0:100, respectively. The second factor was R:C ratios at 70:30, 50:50, and 30:70. The level of CSYW showed significantly higher ($p < 0.01$) gas production from the insoluble fraction (b), potential extent of gas production (a + b), and cumulative gas production at 96 h than the control group ($p < 0.05$). There were no interactions among the CSYW and R:C ratio on the in vitro digestibility ($p > 0.05$). Furthermore, increasing the amount of CSYW to replace SBM up to 75% had no negative effect on in vitro neutral detergent fiber degradability (IVNDFD) ($p > 0.05$) while replacing CSWY at 100% could reduce IVNDFD ($p > 0.05$). The bacterial population in the rumen was reduced by 25.05% when CSYW completely replaced SBM ($p < 0.05$); however, 75% of CSWY in the diet did not change the bacterial population ($p > 0.05$). The concentration of propionate (C3) decreased upon an increase in the CSYW level, which was lowest with the replacement of SBM by CSYW up to 75%. However, various R:C ratios did not influence total volatile fatty acids (VFAs), and the proportion of VFAs ($p > 0.05$), except the concentration of C3, increased when the proportion of a concentrate diet increased ($p < 0.05$). In conclusion, CSYW could be utilized as a partial replacement for SBM in concentrate diets up to 75% without affecting gas kinetics, ruminal parameters, or in vitro digestibility.

Keywords: yeast waste; cassava pulp; rumen fermentation; in vitro gas production technique

1. Introduction

Cassava (*Manihot esculenta*) is a root crop planted mostly in the tropical and subtropical regions of the world. Per hectare, 25 to 60 tons is produced, and cassava is resistant to poor soils, diseases, and drought [1]. The world's cassava production is expected to be around 230 million metric tons per year, produced predominantly in Nigeria, Brazil, Thailand, Vietnam, Indonesia, and the Democratic Republic of Congo [2,3]. Cassava pulp (CS) is a byproduct of the extraction of starch from cassava roots, and its disposal can have negative consequences for the environment. As a result, the starch industry has attempted to phase it out or find alternative uses for it. The use of CS as an animal feed is an alternative to solve this problem [4]. However, CS is high in fiber and low in protein; hence, there have been various elucidated methods to improve its nutritional value [5].

Yeast (*Saccharomyces cerevisiae*) is a source of probiotics that have a beneficial effect on rumen fermentation. Crude protein (CP) in CS increased by nearly 7% in microbial mixed culture of *S. cerevisiae* and fermentation procedures using solid media [6]. In ruminant feeding, the utilization of microorganisms, including *S. cerevisiae*, has become common [7]. Boonnop et al. [8] found that *S. cerevisiae* fermented cassava chip-enhanced CP levels from 2% to 30.4%. In addition, Polyorach et al. [9,10] reported that yeast fermented cassava chip protein (YEFECAP) might well be created to promote a CP level of up to 47%. However, *S. cerevisiae* products tend to be expensive; thus, alternate yeast sources should be considered.

Since the concentrated amounts of active yeast can be obtained from the local industry, the process of employing yeast for animal feed is exciting. In ethanol production processes, the initial substrates are molasses and inoculants of the yeast *S. cerevisiae*. Yeast waste is the byproduct of *S. cerevisiae*, fermenting sugarcane juice and molasses to produce bioethanol (YW). YW is generated throughout the year and contains 60–70% of yeast live cells and a CP content of about 30–35% [11,12]. Cherdthong et al. [13] found that using YW as a replacement for soybean meal (SBM) had no negative impact on feed intake or rumen fermentation in ruminant diets up to 100%. An earlier study demonstrated that the quality of citric waste can improve by being treated with YW, which could be a potential replacement for SBM up to 75% [14].

We hypothesized not only the significance of the costs suffered by SBM, but also the impact on the environment. The usage of cassava starch plant waste (cassava pulps, CS) and ethanol industry byproduct (yeast waste, YW) has never been reported. As a consequence, optimization of industrial use and zero waste of raw materials throughout every operation of the plant is a challenging idea, and our main goal is to enhance feed utilization in ruminants. Thus, evaluation of in vitro ruminal fermentation and feed digestibility, as influenced by CS fermented with YW (CSYW) and different roughage to concentrate ratios, was elucidated.

2. Materials and Methods

2.1. Preparation of Cassava Pulp Fermented with Yeast Waste (CSYW)

YW was supported by Khon Kaen Sugar Industry Public Co., Ltd. Cassava pulp (CS), commercial grade urea, and molasses were purchased from the local shop. The media and solution were prepared as follows: (1) CSYW was obtained by the combination of 100 mL of YW and was weighed equally into a flask containing 100 mL distilled water, then was mixed and incubated at room temperature for 2 h; (2) the medium solution was prepared by mixing 24 g of molasses and 50 g of urea into 100 mL distillation water; then the pH of the medium solution was adjusted using H_2SO_4 until the final pH was obtained at 3.5–5; (3) the solution of (1) and (2) was mixed at the ratio 1:1 and then flushed with oxygen for 18 h; (4) after 18 h, CS was mixed well with the yeast medium solution (3) at the ratio 100 g to 50 mL; (5) then the product was allowed to ferment for 14 days, followed by sun drying for 72 h to keep the moisture lower than 10%, and used for a substrate test in the in vitro gas production.

2.2. Experimental Design and Dietary Treatments

The design of the experiment was a 5 × 3 factorial arrangement in a completely randomized design (CRD). Each treatment contained three replications and three runs. The first factor replaced SBM with CSYW in a concentrate ratio at 100:0, 75:25, 50:50, 25:75, and 0:100, respectively. The second factor was roughage (R) to concentrate (C) ratios at 70:30, 50:50, and 30:70. All samples of substrates were dried at 60 °C for 48 h. Before chemical analysis, samples were dried in an oven at lower temperatures (60 °C) for 48 h and then ground by forcing them through a 1 mm steel screen (Wiley mill, Arthur H. Thomas Co., Philadelphia, PA, USA). All samples were analyzed for dry matter (DM; ID 967.03), ash (ID 492.05), ether extract (EE; ID 455.08), and CP (CP; ID 984.13) by using the procedures of

Association of Official Analytical Chemists (AOAC) [15]. Neutral detergent fiber (NDF) in substrates was measured following the work of Van Soest et al. [16], with supplementation of alpha-amylase but no sodium sulphite, and results are demonstrated with residual ash.

2.3. Ruminal Fluid Donors and Substrates of Inoculum

Two rumen fluid donors were obtained from 2-year-old dairy steers (400 ± 15.0 kg body weight; BW) and collected via fistulae rumens. The animals were fed concentrate containing CP 180 g/kg DM, OM 920 g/kg DM, NDF 220 g/kg DM, ADF 108 g/kg DM, and 806 g/kg total digestible nutrient (TDN) at 0.5% of BW (07:00 and 16:00); rice straw was provided to the animals on an ad libitum basis. The steers were housed separately and supplied with water ad libitum. Before morning feeding, 1500 mL of rumen fluid was obtained from the animals via cannula. The samples were filtered through four layers of cheesecloth and placed in a container with thermal insulation (39 °C) before being delivered to the lab in 15 min. According to Menke and Steingass [17], artificial saliva preparations contained distilled water (1095 mL), a micro mineral mixture (0.23 mL; $MnCl_2 \cdot 4H_2O$ 10.0 g/100 mL, $CaCl_2 \cdot 2H_2O$ 13.2 g/100 mL, $CoCl_2 \cdot 6H_2O$ 1.0 g/100 mL, and $FeCl_3 \cdot 6H_2O$ 8.0 g/100 mL), a macro mineral mixture (365 mL; KH_2PO_4 6.2 g/L, Na_2HPO_4 5.7 g/L, NaCl 2.22 g/L, and $MgSO_4 \cdot 7H_2O$ 0.6 g/L), a resazurin mixture 0.1% (1 mL), a reduction mixture (60 mL; $Na_2S \cdot 9H_2O$ 80.0 mg/ 60 mL of NaOH), and a buffer mixture (730 mL; $NaHCO_3$ 35.0 g/L and NH_4HCO_3 4.0 g/L). The artificial saliva was then combined with rumen fluid (660 mL) in a non-oxygen atmosphere. Dietary treatments were weighed at 0.5 g in the 50 mL bottles; a total of 40 mL of rumen liquor medium was added to each treatment bottle using an 18 gauge × 1.5-inch needle. Finally, all experimental bottles were sealed with butyl rubber stoppers and metal caps before being incubated in a hot-air oven at 39 °C for further measurement.

Three groups of experimental bottles were established: Group 1 had gas kinetics and gas production measurement, and 3 bottles per treatment (15 treatments + 3 bottles of blank) were used. The bottles were gently shaken every 3 h throughout the incubation time, and each run included three treated bottles and three blank bottles. The blank bottles contained only rumen liquor, and net gas yield was calculated by subtracting the average value of the gas yields from experimental bottles. A 20 mL glass aloe precision hypodermic syringe (U4520, Becton, Dickinson and Company, Franklin Lakes, NJ, USA) was used to measure gas production. The bottles in the heating chamber were punctured using an 18-gauge injection needle. Group 2 had the pH, ruminal NH_3-N, volatile fatty acids, and microbial count all examined in the same bottle. The samples were taken at 4 h of incubation time from three replicates of a bottle. The last nutrient degradability was measured in Group 3; samples were obtained at 12 h after incubation from three bottle replicates.

2.4. In Vitro Gas Production and Fermentation Characteristics

The amount of gas produced was measured at 0, 0.5, 1, 2, 4, 6, 8, 12, 18, 24, 48, 72, and 96 h of incubation. The bottles were divided into 3 sets. The first set was used for gas kinetics and gas production measurement. The second set was used for measurement of ruminal parameters at 4 and 8 h post-incubation, including pH (Hanna Instruments Pte Ltd., Kallang Way, Singapore), ruminal ammonia-nitrogen (NH_3-N) (Kjeldahl methods [15]), volatile fatty acids (VFA) [18], and ruminal microorganism direct counts (Boeco, Hamburg, Germany). The last set was used for the determination of in vitro degradability (IVDMD), in vitro NDF degradability (IVNDFD), and in vitro ADF degradability (IVADFD) [16,19]. The fermented residues were filtered into an Ankom filter bag (ANKOM 200, ANKOM Technology, New York, NY, USA), dried at 60 °C in an oven for 72 h, and assessed for IVDMD according to Galyean [19] IVDMD% = 100 × [(initial dry sample wt-(residue-blank))/initial dry sample wt] [19]. Dried residues were added with an NDF and ADF solution to measure IVNDFD and IVADFD [16].

2.5. Statistical Analysis

The model of Sommart et al. [20] was used for determining the kinetics of gas production.

$$Y = a + b\,(1 - e^{-ct}) \tag{1}$$

where "a" is the intercept, which ideally reflects the fermentation of the soluble fraction, "b" is the fermentation of the insoluble fraction (which is with the time fermentable), "c" is the rate of gas production, "| a | + b" is the potential extent of gas production, and "Y" is the gas produced at time "t". All data were analyzed as a 5×3 factorial arrangement in a completely randomized design (CRD) using the PROC GLM of SAS program [21]. Multiple comparisons among treatment means were performed by Duncan's New Multiple Range Test (DMRT) [22]. Differences among means with $p < 0.05$ were accepted as being statistically significant.

3. Results

3.1. Nutritional Composition of Feed

The nutritional composition and formulation of the experimental diet are presented in Table 1. The control diet contained a high level of SBM at 180 g/kg and urea 10 g/kg as the main nitrogen source. The CP content in CS was enhanced by fermented YW, and 537 g/kg CP was obtained when the CSYW product was generated. The CSYW contained 349 g/kg DM, 243 g/kg NDF, and 113 g/kg ADF, and the CP content was high at 537 g/kg DM. SBM was replaced by CSYW as a protein source in concentrate diets from 25–100%, resulting in a reduction in the usage of urea in formulations. The CP content of the concentrate diets was similar among the formulas and ranged from 140 to 143 g/kg DM, while the ash, NDF, and ADF content increased as the quantity of CSWY was added.

Table 1. Feed ingredients and chemical composition used in the experimental ration.

Ingredients	Levels of CSYW (g/kg Dry Matter)					CSYW [1]	Rice Straw
	0	25	50	75	100		
Cassava chip	580	580	550	555	550		
Rice bran	120	150	147	122	120		
Palm kernel meal, solvent	80	80	113	135	143		
Soybean meal	180	113	75	37	0		
CSYW [1]	0	37	75	113	150		
Mineral premix	5	5	5	5	5		
Urea	10	10	10	8	7		
Molasses	10	10	10	10	10		
Pure sulfur	10	10	10	10	10		
Salt	5	5	5	5	5		
Chemical composition							
Dry matter (g/kg)	906	901	903	904	912	349	924
—g/kg of dry matter—							
Organic matter	958	930	915	901	902	845	86.5
Ash	42	70	85	99	98	103	125
Crude protein	143	141	141	140	140	537	23
Neutral detergent fiber	150	207	236	258	272	243	755
Acid detergent fiber	92	126	151	174	183	113	553

[1] CSYW = cassava pulp fermented with yeast waste.

3.2. Gas Kinetics and Cumulative Gas Production

In terms of gas production kinetics, no interactions between CSYW levels and the R:C ratio were detected ($p > 0.05$; Table 2). It was found that gas produced immediately from a soluble fraction (a) and gas rate constant for the insoluble fraction (c) did not change among treatments ($p > 0.05$). The level of CSYW showed significantly higher ($p < 0.01$) gas production from the insoluble fraction (b), potential extent of gas production (a + b), and cumulative gas production at 96 h than the control group ($p < 0.05$). The highest b value and a + b value were 126.72 and 126.71 mL/g DM, respectively, when SBM was replaced by CSYW at 100% ($p < 0.05$). However, the R:C ratios did not alter the kinetics of gas (b or a + b) or cumulative gas ($p > 0.05$).

Table 2. Effect of cassava pulp fermented with yeast waste (CSYW) replaced soybean meal (SBM) and various roughage to concentrate ratio (R:C) on gas kinetics and cumulative gas at 96 h of incubation.

Item	SBM:CSYW	R:C	Gas Kinetics [1]				Cumulative Gas (96 h) mL/g DM Substrate
			a	b	c	$\lvert a \rvert + b$	
T1	100:0	70:30	−3.05	70.12	0.03	67.07	144.82
T2	100:0	50:50	−2.55	72.19	0.04	69.64	148.96
T3	100:0	30:70	−0.36	69.54	0.04	69.18	143.66
T4	75:25	70:30	−0.46	88.56	0.03	88.11	181.71
T5	75:25	50:50	−1.73	80.80	0.04	79.06	166.18
T6	75:25	30:70	−0.74	90.14	0.03	89.40	184.86
T7	50:50	70:30	−0.85	93.54	0.02	92.69	191.66
T8	50:50	50:50	−0.08	97.75	0.03	97.67	200.08
T9	50:50	30:70	−0.95	88.48	0.04	87.53	181.54
T10	25:75	70:30	0.10	113.81	0.02	113.91	232.20
T11	25:75	50:50	0.00	105.68	0.03	105.68	215.94
T12	25:75	30:70	−0.85	115.14	0.02	114.29	234.86
T13	0:100	70:30	−0.11	123.06	0.03	122.95	250.70
T14	0:100	50:50	0.29	137.82	0.02	138.11	280.22
T15	0:100	30:70	−0.22	119.29	0.02	119.07	243.16
SEM			1.01	25.66	0.01	21.55	18.25
p-value							
SBM:CSYW			1.75	<0.01	0.17	<0.01	0.05
R:C			0.92	0.33	0.15	0.29	0.45
SBM:CSYW × R:C			0.67	0.25	0.40	0.22	0.43
Average							
SBM:CSYW	100:0		−1.99	70.62 [f]	0.04	68.63 [f]	145.81 [f]
	75:25		−0.98	86.50 [f]	0.03	85.52 [ef]	177.58 [ef]
	50:50		−0.63	93.26 [e]	0.03	92.63 [de]	191.09 [e]
	25:75		−0.25	115.54 [d]	0.03	111.29 [d]	227.67 [d]
	0:100		−0.02	126.72 [d]	0.02	126.71 [d]	258.03 [d]
R:C ratio	70:30		−0.87	97.82	0.03	96.94	200.22
	50:50		−0.82	98.85	0.03	98.03	202.28
	30:70		−0.63	96.52	0.03	95.98	197.61

[1] a = the gas production from the immediately soluble fraction, b = the gas production from the insoluble fraction, c = the gas production rate constant for the insoluble fraction (b), $\lvert a \rvert + b$ = the gas potential extent of gas production. [d–f] Values on the same column with different superscripts differ ($p < 0.05$); SEM = standard error of mean.

3.3. In Vitro Digestibility

The influences of the CSYW level and R:C ratio on in vitro digestibility are illustrated in Table 3. There were no interactions among CSYW and R:C ratio on the in vitro digestibility ($p > 0.05$). When a high level of concentrate diet was supplied, the IVDMD and IVADFD improved ($p < 0.05$). Furthermore, increasing the amount of CSYW to replace SBM up to 75% had no negative effect on IVNDFD ($p > 0.05$), while replacing CSWY at 100% could reduce IVNDFD ($p > 0.05$).

Table 3. Effect of cassava pulp fermented with yeast waste (CSYW) replaced soybean meal (SBM) and various roughage to concentrate ratio (R:C) on the in vitro dry matter degradability (IVDMD), in vitro neutral detergent fiber degradability (IVNDFD), and in vitro acid detergent fiber degradability (IVADFD) at 12 h of incubation.

Item	SBM:CSYW	R:C	IVDMD (g/kg)	IVNDFD (g/kg)	IVADFD (g/kg)
T1	100:0	70:30	440	608	234
T2	100:0	50:50	549	613	210
T3	100:0	30:70	634	637	283
T4	75:25	70:30	509	557	155
T5	75:25	50:50	529	566	239
T6	75:25	30:70	545	633	293
T7	50:50	70:30	502	575	174
T8	50:50	50:50	484	580	227
T9	50:50	30:70	601	624	276
T10	25:75	70:30	427	570	202
T11	25:75	50:50	502	573	201
T12	25:75	30:70	647	615	185
T13	0:100	70:30	456	430	200
T14	0:100	50:50	512	435	187
T15	0:100	30:70	576	512	231
SEM			5.39	3.33	5.31
_p_-value					
SBM:CSYW			0.26	<0.01	0.92
R:C			<0.01	0.50	<0.01
SBM:CSYW × R:C			0.06	0.14	0.06
Average					
SBM:CSYW	100:0		541	619 [a]	222
	75:25		528	585 [a]	197
	50:50		529	593 [a]	201
	25:75		522	586 [a]	202
	0:100		515	459 [b]	194
R:C ratio	70:30		467 [b]	548 [b]	193 [b]
	50:50		515 [b]	553 [b]	213 [b]
	30:70		600 [a]	604 [a]	253 [a]

[a,b] Values on the same column with different superscripts differ ($p < 0.05$); SEM = standard error of mean.

3.4. Ruminal pH, Ammonia-Nitrogen (NH_3-N) Concentration, and Microorganisms

There were no interactions on the ruminal pH, NH_3-N, and microbial population between the CSYW level and R:C ratio ($p > 0.05$; Table 4). The pH and NH_3-N levels in the rumen were measured and ranged from 6.81 to 7.04 and 15.79 to 17.89 mg/dL, respectively. Except for fungal zoospore, the quantity of bacteria and protozoa changed significantly when the concentrate diet was high (bacteria, 36.58×10^7 and protozoa, 3.93×10^5; $p < 0.05$). The bacterial population in the rumen was reduced by 25.05% when CSYW completely replaced SBM ($p < 0.05$); however, 75% of CSWY in the diet did not change the bacterial population ($p > 0.05$).

Table 4. Effect of cassava pulp fermented with yeast waste (CSYW) replaced soybean meal (SBM) and various roughage to concentrate ratio (R:C) on pH, ammonia-nitrogen (NH_3-N), and ruminal microorganisms at 4 h of incubation.

Item	SBM:CSYW	R:C	pH	NH_3-N (mg/dL)	Bacteria ($\times 10^7$ cells/mL)	Protozoa ($\times 10^5$ cells/mL)	Fungal Zoospore ($\times 10^4$ cells/mL)
T1	100:0	70:30	7.04	15.79	25.50	2.05	7.00
T2	100:0	50:50	6.90	15.38	29.75	2.70	10.00
T3	100:0	30:70	6.87	16.24	44.50	4.30	11.25
T4	75:25	70:30	6.96	15.63	27.75	2.25	10.25
T5	75:25	50:50	6.92	17.04	27.25	2.20	11.50
T6	75:25	30:70	6.86	16.43	38.50	4.00	13.50
T7	50:50	70:30	6.95	17.43	31.20	1.55	7.50
T8	50:50	50:50	6.94	16.12	29.00	2.85	10.50
T9	50:50	30:70	6.88	15.90	33.00	4.10	11.25
T10	25:75	70:30	7.10	16.27	26.00	1.90	10.50
T11	25:75	50:50	6.94	17.11	29.75	2.25	12.00
T12	25:75	30:70	6.80	16.09	34.75	4.50	15.00
T13	0:100	70:30	7.03	16.29	18.75	2.15	8.75
T14	0:100	50:50	6.91	15.95	19.75	2.50	9.00
T15	0:100	30:70	6.81	17.89	32.15	2.75	9.50
SEM			0.07	0.66	3.06	0.70	2.07
***p*-value** SBM:CSYW			0.41	0.49	<0.05	0.70	0.22
R:C			0.06	0.86	<001	<001	0.07
SBM:CSYW$\times$R:C			0.09	0.45	0.34	0.42	0.99
Average							
SBM:CSYW	100:0		6.94	15.80	33.25 [a]	3.02	9.99
	75:25		6.91	16.37	31.17 [a]	2.82	9.42
	50:50		6.92	16.48	31.07 [a]	2.83	11.75
	25:75		6.94	16.49	30.17 [a]	2.88	9.75
	0:100		6.92	16.71	23.55 [b]	2.47	12.50
R:C ratio	70:30		7.02	16.28	25.84 [b]	1.98 [b]	9.08
	50:50		6.92	16.32	27.10 [b]	2.50 [b]	8.80
	30:70		6.84	16.51	36.58 [a]	3.93 [a]	10.60

[a,b] Values on the same column with different superscripts differ ($p < 0.05$); SEM = standard error of mean.

3.5. Volatile Fatty Acid

The in vitro total VFAs and proportion of VFAs are shown in Table 5. No interaction occurred between the CSYW level or R:C ratio ($p > 0.05$). The total VFAs and VFA profiles in the rumen did not change when CSYW was replaced by SBM ($p > 0.05$), except the concentration of C3 was changed ($p < 0.05$). The concentration of C3 decreased upon an increase in the CSYW level, which was lowest with the replacement of SBM by CSYW up to 75%. However, various R:C ratios did not influence total VFAs or the proportion of VFAs ($p > 0.05$), except the concentration of C3 increased when the proportion of a concentrate diet increased ($p < 0.05$).

Table 5. Effect of cassava pulp fermented with yeast waste (CSYW) replaced soybean meal (SBM) and various roughage to concentrate ratio (R:C) on concentrations of volatile fatty acid (VFA), acetate (C2), propionate (C3), and butyrate (C4) at 4 h of incubation.

Item	SBM:CSYW	R:C	Total VFA (mmol/L)	Molar Proportions of VFA (mmol/L)			C2:C3 Ratio
				C2	C3	C4	
T1	100:0	70:30	64.36	71.03	15.63	10.56	4.55
T2	100:0	50:50	73.81	67.20	25.54	12.26	2.63
T3	100:0	30:70	79.11	61.46	27.56	14.86	2.23
T4	75:25	70:30	68.10	72.67	15.91	12.57	4.57
T5	75:25	50:50	64.95	74.16	26.95	11.08	2.75
T6	75:25	30:70	77.50	74.23	24.28	9.95	3.06
T7	50:50	70:30	68.56	71.00	20.41	10.50	3.48
T8	50:50	50:50	69.97	69.40	20.64	11.53	3.36
T9	50:50	30:70	69.54	64.94	25.68	10.94	2.53
T10	25:75	70:30	67.17	65.89	20.75	10.55	3.17
T11	25:75	50:50	74.02	64.52	18.76	8.85	3.44
T12	25:75	30:70	66.02	70.78	26.82	12.27	2.64
T13	0:100	70:30	66.27	71.68	18.07	12.55	3.97
T14	0:100	50:50	66.97	66.35	21.00	9.74	3.16
T15	0:100	30:70	64.69	62.52	20.12	8.84	3.11
SEM			3.19	3.82	1.41	1.99	0.67
p-value							
SBM:CSYW			0.23	0.18	0.05	0.69	0.07
R:C			0.11	0.34	0.01	0.83	0.70
SBM:CSYW×R:C			0.08	0.58	0.09	0.61	1.00
Average							
SBM:CSYW	100:0		72.43	66.56	22.91 [a]	12.56	3.14
	75:25		70.19	73.69	22.38 [a]	11.20	3.46
	50:50		69.36	68.45	22.24 [a]	10.99	3.12
	25:75		69.07	67.06	22.11 [a]	10.56	3.08
	0:100		65.98	66.85	19.73 [b]	10.37	3.41
R:C ratio	70:30		66.89	70.45	18.74 [c]	11.35	3.79
	50:50		69.94	68.33	22.58 [b]	10.69	3.07
	30:70		71.37	66.78	25.05 [a]	11.37	2.63

[a–c] Values on the same column with different superscripts differ ($p < 0.05$); SEM = standard error of mean.

4. Discussion

4.1. Chemical Composition

The chemical composition of CSYW in this experiment had lower OM, NDF, and ADF content than the compositions within the study conducted by Sommai et al. [5]. These variations may be a result of different materials, growing locations, and plant factory processing [23]. However, the OM, NDF, and ADF contents in CSYW were similar to those of the report of Chuelong et al. [24], with 845, 243, and 113 g/kg DM, respectively. Furthermore, the use of CSYW instead of SBM resulted in increased ash and fiber content in the concentrate diet, while the CP in the formula was regulated at the same level to investigate the probable use of CSYW replacement.

Combining CS and YW could deliver a product with a high CP content of 537 g/kg DM. The apparent increase in CP could be explained by an increase in microorganisms contained in YW and proliferation in the form of single-cell proteins occurring throughout the fermentation process [6]. Before YW was fermented with CS, the quantity of carbon and nitrogen sources in the medium solution was the key factor that differentiated the amount of yeast and CP contained in the product. Polyorach et al. [9] discovered that protein and lysine levels in cassava chips increased from 3.4% to 32.5% and 3.8% to 8.5%, respectively, when the *S. cerevisiae* grew in media solution containing 9.6% molasses and 19.2% urea. Similarly, Khampa et al. [25] found that *S. cerevisiae* grown in a media solution containing

10% molasses and 24% urea could increase the amount of protein in cassava chips by up to 36.1%. In addition, 23.3% of CP was obtained from CS fermented *S. cerevisiae* with a media solution containing 12% molasses and 25% urea [5].

The greater CP in this study could be related to the product of yeast that was used as a starter. YW obtained from bioethanol production contains a high content of 60% to 70% live cells of *S. cerevisiae* [12]. In the preliminary investigation, the amount of *S. cerevisiae* in YW was found to be around 3.1×10^{13} cells/mL. This indicates that higher protein levels can be obtained than in previous studies utilizing baker's yeast, which had a low yeast cell count (around 10^6 to 10^8 cells/mL, [26]). As a consequence, this experiment implies that utilizing YW to ferment CS has a higher protein productivity potential than previous CS improvement approaches. Its properties could be used as a protein source in animal diets, provided it is economically viable.

4.2. Effects on Gas Kinetics and In Vitro Digestibility

The level of CSYW substitution for SBM in the concentrate diet altered the in vitro rumen gas kinetics. The volume of gas produced from the insoluble fraction (b) increased as CSYW was raised, which is the main reason why CSYW has a significant impact on the prospective scope of the potential extent of gas production (a + b). The use of CSYW could increase the cumulative gas production because the product containing yeast may promote the growth of some cellulolytic bacteria in the rumen. According to Sommai et al. [5], yeast-fermented CS can activate the cellulolytic bacterial population from 2.0×10^9 to 5.6×10^9 cfu/mL, meaning that the more products supplied, the more cellulolytic bacteria there are. These findings were in accordance with Chuelong et al. [24], who confirmed that *S. cerevisiae* fermented with CS increased bacterial populations by 32.2%.

As the proportion of concentrate diet increased from 30% to 70%, the in vitro digestibility of DM and ADF improved by 449 and 308 g/kg DM, respectively. This was in agreement with the statement of Polyorach et al. [7], who found that increasing the concentrate diet from 20% to 80% raised IVDMD by about 13%. In this study, a 30:70 R:C ratio diet provided more readily available energy, which resulted in improved bacterial growth and digestibility [27]. Furthermore, this study confirmed that a concentrate diet increased the ruminal microbiota, particularly bacteria, by approximately 55%. Hungate [28] suggested a more significant effect in the rumen when carbohydrate, rather than forages, is used. These findings corroborated previous research by Sommai et al. [5].

The IVNDFD was maintained when CSYW replaced SBM up to 75%. This relates to the volume of gas produced from the previously mentioned insoluble fraction (b-value). In addition, cellulolytic bacterial colonization of plant cell walls is supported by probiotic yeasts. This effect has many mechanisms of action, one of which is the distribution of thiamin, a vitamin that rumen microorganisms need [14]. Chuelong et al. [24] stated that, when yeast was added to the diet, the activity of most polysaccharidase and glycosidehydrolase enzymes increased and rumen digestion fiber was improved. In addition, the ability of yeast cells to scavenge oxygen is one of the key factors that may justify the beneficial effect of live yeasts on fiber-degrading bacteria [29]. Furthermore, the media solution containing urea might act as an alkaline substance (ammonium hydroxide) and result in a breakdown of the fiber structure in CS [14]. The alkaline substance may then support enzyme activity from yeast to degrade the NDF content contained in CS.

Cherdthong and Supapong [27] found that supplementing yeast would increase the bacterial population by 3.6 times, resulting in enhanced NDF digestibility, as shown in Tables 3 and 4. This agrees with Boonnop et al. [8], who indicated that feeding yeast-fermented cassava chip (YEFFECAP) to dairy steers could increase feed consumption and nutritional digestibility. In addition, yeast efficacy was frequently established when used in combination with low-quality roughage. Tang et al. [30] observed that feeding *S. cerevisiae* to low-quality roughage enhanced in vitro digestibility. These results indicated that CSYW at 75% could be incorporated with no influence on rumen digestibility when added to concentrate diets. However, replacement of SBM by CSYW at 100% could reduce IVNDFD, which might be due to limited high fiber content, resulting in reduced fiber digestibility.

4.3. Ruminal Fermentation and Quantity of Rumen Microorganisms

The ruminal fermentation parameters did not change when SBM was replaced by CSYW, and the value remained stable between 6.82 and 7.05. Wanapat and Cherdthong [31] suggested that the optimum level of pH in the rumen for microbial digestion of fiber and protein is 6.5 to 7.0. Furthermore, raising the CSYW and R:C ratio did not affect the concentrations of NH_3-N and ranged from 15.88 to 17.99 mg/dL in the rumen fluid, which is considered acceptable. The NH_3-N was under the optimum concentrations for bacterial growth and microbial activity in the range of 5–25 mg/dL [31]. This range would be improved by voluntary feed consumption and microbial protein synthesis [32].

Several factors influence the organization of the ruminal microbial community, and diet is a key to rumen community composition [33]. Our investigation showed that CSYW could replace SBM up to 75% without any negative effects on microbial activity. In particular, the bacterial population was comparable to the use of 100% SBM in the concentrate diet. Sommai et al. [5] revealed that yeast fermentation of CS has no negative impact on the bacteria population in Thai native beef cattle. This is in agreement with Cherdthong and Supapong [27], who found that using *S. cerevisiae*-fermented cassava bioethanol waste (YECAW) seems to have no adverse influence on bacterial, protozoa, or fungal populations or values in dairy calf rations. Our study observed that increasing the level of concentrate diet enhanced bacterial and protozoa populations (with a 70% concentration diet, the bacteria and protozoa increase was 41.6% and 98.5%, respectively) in the rumen fluid. The increased fermentable substrate (sugar and starch) in the concentrate diet may have favored the growth of bacteria and protozoa, resulting in a change in the structure and diversity of microbial populations [34]. Accordingly, Phesatcha et al. [35] also noticed an enhancement in the total amount of ruminal bacteria, with a comparable shift in rumen microbial population numbers in animals that were fed a high-concentrate diet versus animals that were fed a low-concentrate diet vs. those fed a low level of concentrate diet. This demonstrated that, when a rapid fermentation carbohydrate is supplied, microbial bacteria in the rumen tend to be enhanced. Cherdthong and Wanapat [36] found that ruminal microbial bacteria's synthesis depends on an appropriate carbohydrate supply of NH_3-N, which is used to synthesize peptide bonds and as an energy source. In accordance with Anantasook and Wanapat [37], when a high-concentrate diet is included in the formula, the bacterial population in the rumen increases dramatically. However, the replacement of CSWY did not affect bacterial populations until 100% SBM was replaced. SBM replacement with CSYW at 100% resulted in a decrease in the bacteria population, possibly because of the high fiber content of CS, which inhibited bacterium digestion and utilization. Therefore, the advice for those using CSYW is that the user should carefully assess the replacement quantity to SBM and that further experiments should be carried out on animals.

4.4. Ruminal Volatile Fatty Acid (VFA)

The TVFA did not change when CSWY was utilized instead of SBM at any level. In this experiment, the concentration of rumen VFA ranged from 64.36 to 79.11 mmol/L, which was close to the previous study (68.8 to 89.7 mmol/L [38]; 50.1 to 68.5 mmol/L [7]).

The use of CSYW instead of SBM in the concentrate diet can be employed up to 75% without altering the VFA profile, which demonstrates that the product has a potential use as animal feed as compared to SBM. According to Cherdthong et al. [13], the highest amounts of SBM were replaced by YECAW, with no alterations in TVFA concentration or VFA profiles. In addition, Polyorach et al. [7] found no difference in TVFA and VFA profiles when YEFECAP was used at the 80% SBM substitution level. However, the concentration of C3 was slightly lowered after CSWY was completely substituted with SBM in the concentrate diet. It could be that the high fiber content and low fermentation fraction in CSWY lead to a low substrate supply to generate C3 in the rumen [39,40]. In addition, C3 is a product of the rumen's bacterial fermentation activity; a change in the number of bacteria in the rumen can impact C3. Our studies have revealed that using 100% CSYW reduces the amount of bacterial population. The decrease in bacteria could be related to the increase in fiber; when the quantity of fiber in the composition is higher, the amount of digestible nutrients is significantly lower and results in a reduction of the fermentation yield [41,42]. This incidence was similar to that reported by Polyorach et al. [7], who found that replacing the SBM with 100% yeast-fermented cassava chip protein lowered the amount of C3.

The concentrated diet ratio, C3, increased by 22.6%. Normally, C3 is generated from the rumen fermentable starch, which is caused by bacteria in the rumen. Thus, high-fermentable starch in the concentrate diet resulted in a high concentration of C3 production [43,44]. This agrees with Cherdthong et al. [40], who revealed that the addition of a high-concentrated diet supplied an enhanced proportion of C3. Furthermore, the C3 content could significantly increase when a substrate containing 80% of the concentrate was tested [35].

5. Conclusions

The quality of CS could be improved by using YW and the optimum media solution. CSYW could be utilized as a partial replacement for SBM in concentrate diets up to 75% without affecting gas kinetics, ruminal parameters, or in vitro digestibility. Furthermore, a 30:70 R:C ratio may be useful for gas kinetics, ruminal ecology, digestibility, volatile fatty acids, and propionic acid. However, more in vivo investigations are needed to determine the success of animal production.

Author Contributions: Planning and design of the study, G.D. and A.C.; conducting and sampling, G.D., A.C. and C.S.; sample analysis, G.D., A.C. and C.S.; statistical analysis, G.D. and A.C.; manuscript drafting, G.D., A.C. and C.S.; manuscript editing and finalizing, G.D., A.C., C.S., S.W. and M.W. All authors have read and agreed to the published version of the manuscript.

Funding: The authors express their most sincere gratitude to the Research Program on the Research and Development of Winged Bean Root Utilization as Ruminant Feed, Increase Production Efficiency and Meat Quality of Native Beef and Buffalo Research Group, and Research and Graduate Studies, Khon Kaen University (KKU). Gamonmas Dagaew was granted by Animal Feed Inter Trade Co., Ltd. and Thailand Science Research and Innovation (TSRI) through the RRi program (contract grant PHD62I0020).

Institutional Review Board Statement: The study was conducted under approval procedure no. IACUC-KKU-108/63 of Animal Ethics and Care issued by Khon Kaen University.

Informed Consent Statement: Not applicable.

Data Availability Statement: Not applicable.

Acknowledgments: The authors would like to express our sincere thanks to the KSL Green Innovation Public Co. Limited, Thailand, and the Tropical Feed Resources Research and Development Center (TROFREC), Department of Animal Science, Faculty of Agriculture, Khon Kaen University (KKU), for the use of the research facilities.

Conflicts of Interest: The authors declare no conflict of interest.

References

1. Chauynarong, N.; Elangovan, A.V.; Iji, P.A. The potential of cassava products in diets for poultry. *World's Poult. Sci. J.* **2009**, *65*, 23–36. [CrossRef]
2. FAOSTAT. Food and Agricultural Commodities Production. Food and Agriculture Organization of the United Nations Statistics Database. Available online: http://faostat.fao.org/ (accessed on 1 April 2021).
3. Sugiharto, S. A review on fungal fermented cassava pulp as a cheap alternative feedstuff in poultry ration. *J. World's Poult. Res.* **2019**, *9*, 1–6. [CrossRef]
4. Rukboon, P.; Prasanpanich, S.; Kongmun, P. Effects of cassava pulp mixed with monosodium glutamate by-product (CPMSG) as a protein source in goat concentrate diet. *Indian J. Anim. Res.* **2019**, *53*, 774–779. [CrossRef]

5. Sommai, S.; Ampapon, T.; Mapato, C.; Totakul, P.; Viennasay, B.; Matra, M.; Wanapat, M. Replacing soybean meal with yeast-fermented cassava pulp (YFCP) on feed intake, nutrient digestibilities, rumen microorganism, fermentation, and N-balance in Thai native beef cattle. *Trop. Anim. Health Prod.* **2020**, *52*, 2035–2041. [CrossRef]

6. Pilajun, R.; Wanapat, M. Chemical composition and in vitro gas production of fermented cassava pulp with different types of supplements. *J. Appl. Anim. Res.* **2018**, *46*, 81–86. [CrossRef]

7. Polyorach, S.; Wanapat, M.; Cherdthong, A. Influence of yeast fermented cassava chip protein (YEFECAP) and roughage to concentrate ratio on ruminal fermentation and microorganisms using in vitro gas production technique. *Asian-Australas. J. Anim. Sci.* **2014**, *27*, 36–45. [CrossRef]

8. Boonnop, K.; Wanapat, M.; Nontaso, N.; Wanapat, S. Enriching nutritive value of cassava root by yeast fermentation. *Sci. Agric.* **2009**, *66*, 629–633. [CrossRef]

9. Polyorach, P.; Wanapat, M.; Wanapat, S. Increasing protein content of cassava (Manihot esculenta, Crantz) using yeast in fermentation. *Khon Kaen Agr. J.* **2012**, *40* (Suppl. S2), 178–182.

10. Polyorach, S.; Wanapat, M.; Wanapat, S. Enrichment of protein content in cassava (Manihot esculenta Crantz) by supplementing with yeast for use as animal feed. *Emirat. J. Food Agric.* **2013**, *25*, 142–149. [CrossRef]

11. Laluce, C.; Leite, G.R.; Zavitoski, B.Z.; Zamai, T.T.; Ventura, R. *Fermentation of Sugarcane Juice and Molasses for Ethanol Production. Sugarcane-Based Biofuels and Bioproducts*; John Willey & Sons: Hoboken, NY, USA, 2016. [CrossRef]

12. Díaz, A.; Ranilla, M.J.; Saro, C.; Tejido, M.L.; Pérez-Quintana, M.; Carro, M. Influence of increasing doses of a yeast hydrolyzate obtained from sugarcane processing on in vitro rumen fermentation of two different diets and bacterial diversity in batch cultures and Rusitec fermenters. *Anim. Feed Sci. Technol.* **2017**, *232*, 129–138. [CrossRef]

13. Cherdthong, A.; Prachumchai, R.; Supapong, C.; Khonkhaeng, B.; Wanapat, M.; Foiklang, S.; Polyorach, S. Inclusion of yeast waste as a protein source to replace soybean meal in concentrate mixture on ruminal fermentation and gas kinetics using in vitro gas production technique. *Anim. Prod. Sci.* **2019**, *59*, 1682–1688. [CrossRef]

14. Suriyapha, C.; Cherdthong, A.; Suntara, C.; Polyorach, S. Utilization of yeast waste fermented citric waste as a protein source to replace soybean meal and various roughage to concentrate ratios on in vitro rumen fermentation, gas kinetic, and feed digestion. *Fermentation* **2021**, *7*, 120. [CrossRef]

15. Association of Official Analytical Chemists (AOAC). *Official Methods of Analysis*, 16th ed.; Association of Official Analytical Chemists: Arlington, VA, USA, 1995.

16. Van Soest, P.J.; Robertson, J.B.; Lewis, B.A. Methods for dietary fiber, neutral detergent fiber, and non-starch polysaccharides in relation to animal nutrition. *J. Dairy Sci.* **1991**, *74*, 3583–3597. [CrossRef]

17. Menke, K.H.; Steingass, H. Estimation of the energetic feed value obtained from chemical analysis and gas production using rumen fluid. *Anim. Res. Develop.* **1988**, *28*, 7–55.

18. Osaki, T.Y.; Kamiya, S.; Sawamura, S.; Kai, M.; Ozawa, A. Growth inhibition of *Clostridium difficile* by intestinal flora of infant faeces in continuous flow culture. *J. Med. Microbiol.* **1994**, *40*, 179–187. [CrossRef]

19. Galyean, M. *Laboratory Procedure in Animal Nutrition Research*; Department of Animal and Life Science, New Mexico State University: Las Cruces, NM, USA, 1989; p. 188.

20. Sommart, K.; Parker, D.S.; Rowlinson, P.; Wanapat, M. Fermentation characteristics and microbial protein synthesis in an in vitro system using cassava, rice straw and dried ruzi grass as substrates. *Asian-Australas. J. Anim. Sci.* **2000**, *13*, 1084–1093. [CrossRef]

21. SAS (Statistical Analysis System). *SAS/STAT User's Guide*, 4th ed.; Statistical Analysis Systems Institute, Version 9; SAS Institute Inc.: Cary, NC, USA, 2013.

22. Steel, R.G.D.; Torrie, J.H. *Principles and Procedures of Statistics: A Biometerial Approach*, 2nd ed.; McGraw-Hill: New York, NY, USA, 1980.

23. Kamphaya, S.; Kumagai, H.; Angthong, W.; Narmseelee, R.; Bureenok, S. Effects of different ratios and storage periods of liquid brewer's yeast mixed with cassava pulp on chemical composition, fermentation quality and in vitro ruminal fermentation. *Asian-Australas. J. Anim. Sci.* **2017**, *30*, 470–478. [CrossRef]

24. Chuelong, S.; Siriuthane, T.; Polsit, K.; Ittharat, S.; Koatdoke, U.; Cherdthong, A.; Khampa, S. Supplementation levels of palm oil in yeast (*Saccharomyces cerevisiae*) culture fermented cassava pulp on rumen fermentation and average daily gain in crossbred native cattle. *Pak. J. Nutr.* **2011**, *10*, 1115–1120. [CrossRef]

25. Khampa, S.; Chaowarat, P.; Singhalert, R.; Wanapat, M. Supplementation of yeast fermented cassava chip as a replacement concentrate on rumen fermentation efficiency and digestibility on nutrients in cattle. *Asian-Australas. J. Anim. Sci.* **2009**, *3*, 18–24. [CrossRef]

26. Promkot, C.; Wanapat, M.; Mansathit, J. Effects of yeast fermented-cassava chip protein (YEFECAP) on dietary intake and milk production of Holstein crossbred heifers and cows during pre-and post-partum period. *Livest. Sci.* **2013**, *154*, 112–116. [CrossRef]

27. Cherdthong, A.; Supapong, C. Improving the nutritive value of cassava bioethanol waste using fermented yeast as a partial replacement of protein source in dairy calf ration. *Trop. Anim. Health Prod.* **2019**, *51*, 2139–2144. [CrossRef]

28. Hungate, R.E. *The Rumen and Its Microbes*; Academic Press: New York, NY, USA, 1966; p. 53.

29. Habeeb, A.A.M. Importance of yeast in ruminants feeding on production and reproduction. *Ecol. Evol.* **2017**, *2*, 49.

30. Tang, L.; Wang, W.; Zhou, W.; Cheng, K.; Yang, Y.; Liu, M.; Cheng, K.; Wang, W. Three-pathway combination for glutathione biosynthesis in *Saccharomyces cerevisiae*. *Microb. Cell Factories.* **2015**, *14*, 1–12. [CrossRef]

31. Wanapat, M.; Cherdthong, A. Use of real-time PCR technique in studying rumen cellulolytic bacteria population as affected by level of roughage in swamp buffaloes. *Cur. Microbiol.* **2009**, *58*, 294–299. [CrossRef] [PubMed]
32. Wanapat, M.; Pimpa, O. Effect of ruminal NH$_3$-N levels on ruminal fermentation, purine derivatives, digestibility and rice straw intake in swamp buffaloes. *Asian-Australas. J. Anim. Sci.* **1999**, *12*, 904–907. [CrossRef]
33. Bi, Y.; Zeng, S.; Zhang, R.; Diao, Q.; Tu, Y. Effects of dietary energy levels on rumen bacterial community composition in Holstein heifers under the same forage to concentrate ratio condition. *BMC Microbiol.* **2018**, *18*, 69. [CrossRef]
34. Fernando, S.C.; Purvis, H.; Najar, F.; Sukharnikov, L.; Krehbiel, C.; Nagaraja, T.; Roe, B.; Desilva, U. Rumen microbial population dynamics during adaptation to a high-grain diet. *Appl. Environ. Microbiol.* **2010**, *76*, 7482–7490. [CrossRef]
35. Phesatcha, K.; Phesatcha, B.; Wanapat, M.; Cherdthong, A. Roughage to concentrate ratio and *Saccharomyces cerevisiae* inclusion could modulate feed digestion and in vitro ruminal fermentation. *Vet. Sci.* **2020**, *7*, 151. [CrossRef]
36. Cherdthong, A.; Wanapat, M. Rumen microbes and microbial protein synthesis in Thai native beef cattle fed with feed blocks supplemented with a urea–calcium sulphate mixture. *Arch. Anim. Nutr.* **2013**, *67*, 448–460. [CrossRef]
37. Anantasook, N.; Wanapat, M. Influence of rain tree pod meal supplementation on rice straw based diets using in vitro gas fermentation technique. *Asian-Australas. J. Anim. Sci.* **2012**, *5*, 325–334. [CrossRef] [PubMed]
38. Suntara, C.; Cherdthong, A.; Uriyapongson, S.; Wanapat, M.; Chanjula, P. Comparison effects of ruminal Crabtree-negative yeasts and Crabtree-positive yeasts for improving ensiled rice straw quality and ruminal digestion using in vitro gas production. *J. Fungi* **2020**, *6*, 109. [CrossRef]
39. Terry, S.A.; Badhan, A.; Wang, Y.; Chaves, A.V.; McAllister, T.A. Fibre digestion by rumen microbiota—A review of recent metagenomic and metatranscriptomic studies. *Can. J. Anim. Sci.* **2019**, *99*, 678–692. [CrossRef]
40. Cherdthong, A.; Wanapat, M.; Kongmun, P.; Pilajun, R.; Khejornsart, P. Rumen fermentation, microbial protein synthesis and cellulolytic bacterial population of swamp buffaloes as affected by roughage to concentrate ratio. *J. Anim. Vet. Adv.* **2010**, *9*, 1667–1675. [CrossRef]
41. Gunun, P.; Gunun, N.; Wanapat, M.; Cherdthong, A.; Polyorach, S.; Kang, S. In vitro rumen fermentation and methane production as affected by rambutan peel powder. *J. Appl. Anim. Res.* **2018**, *46*, 626–631. [CrossRef]
42. Kang, S.; Wanapat, M.; Cherdthong, A. Effect of banana flower powder supplementation as a rumen buffer on rumen fermentation efficiency and nutrient digestibility in dairy steers fed on high concentrate diet. *Anim. Feed Sci. Technol.* **2014**, *196*, 32–41. [CrossRef]
43. Anantasook, N.; Wanapat, M.; Cherdthong, A.; Gunun, P. Changes of microbial population in the rumen of dairy steers as influenced by plant containing tannins and saponins and roughage to concentrate ratio. *Asian-Australas. J. Anim. Sci.* **2013**, *26*, 1583–1591. [CrossRef]
44. Anantasook, N.; Wanapat, M.; Cherdthong, A.; Gunun, P. Effect of plants containing secondary compounds with palm oil on feed Intake, digestibility, microbial protein synthesis and microbial population in dairy cows. *Asian-Australas. J. Anim. Sci.* **2013**, *26*, 820–826. [CrossRef]

fermentation

Article

Influence of Environmental Microbiota on the Activity and Metabolism of Starter Cultures Used in Coffee Beans Fermentation

Vanessa Bassi Pregolini [1], Gilberto Vinícius de Melo Pereira [2,*], Alexander da Silva Vale [2], Dão Pedro de Carvalho Neto [3] and Carlos Ricardo Soccol [2,*]

[1] Departament of Bioprocess Engineering and Biotechnology, Federal Technological University of Paraná (UTFPR), Ponta Grossa 84017-220, PR, Brazil; vanepregolini@gmail.com
[2] Department of Bioprocess Engineering and Biotechnology, Federal University of Paraná (UFPR), Curitiba 81531-970, PR, Brazil; alexander.biotec@gmail.com
[3] Federal Institute of Education, Science and Technology of Paraná (IFPR), Londrina 86060-370, PR, Brazil; daopcn@gmail.com
* Correspondence: gilbertovinicius@gmail.com (G.V.d.M.P.); soccol@ufpr.br (C.R.S.)

Citation: Pregolini, V.B.; de Melo Pereira, G.V.; da Silva Vale, A.; de Carvalho Neto, D.P.; Soccol, C.R. Influence of Environmental Microbiota on the Activity and Metabolism of Starter Cultures Used in Coffee Beans Fermentation. *Fermentation* 2021, 7, 278. https://doi.org/10.3390/fermentation7040278

Academic Editor: Ronnie G. Willaert

Received: 28 October 2021
Accepted: 23 November 2021
Published: 25 November 2021

Publisher's Note: MDPI stays neutral with regard to jurisdictional claims in published maps and institutional affiliations.

Copyright: © 2021 by the authors. Licensee MDPI, Basel, Switzerland. This article is an open access article distributed under the terms and conditions of the Creative Commons Attribution (CC BY) license (https://creativecommons.org/licenses/by/4.0/).

Abstract: Microbial activity is an integral part of agricultural ecosystems and can influence the quality of food commodities. During on-farm processing, coffee growers use a traditional method of fermentation to remove the cherry pulp surrounding the beans. Here, we investigated the influence of the coffee farm microbiome and the resulting fermentation process conducted with selected starter cultures (*Pichia fermentans* YC5.2 and *Pediococcus acidilactici* LPBC161). The microbiota of the coffee farm (coffee fruits and leaves, over-ripe fruits, cherries before de-pulping, depulped beans, and water used for de-pulping beans) was dominated by Enterobacteriaceae and Saccharomycetales, as determined by llumina-based amplicon sequencing. In addition, 299 prokaryotes and 189 eukaryotes were identified. Following the fermentation process, *Pichia* and the family Lactobacillaceae (which includes *P. acidilactici*) represented more than 70% of the total microbial community. The positive interaction between the starters resulted in the formation of primary metabolites (such as ethanol and lactic acid) and important aroma-impacting compounds (ethyl acetate, isoamyl acetate, and ethyl isobutyrate). The success competitiveness of the starters towards the wild microbiota indicated that coffee farm microbiota has little influence on starter culture-added coffee fermentation. However, hygiene requirements in the fermentation process should be indicated to prevent the high microbial loads present in coffee farm soil, leaves, fruits collected on the ground, and over-ripe fruits from having access to the fermentation tank and transferring undesirable aromas to coffee beans.

Keywords: fermentation hygiene; coffee fermentation; *Pichia fermentans*; *Pediococcus acidilactici*

1. Introduction

Coffee is a tropical crop grown in more than 50 countries, with the largest producers being Brazil, Vietnam, Colombia, Indonesia, Ethiopia, and Honduras [1]. To produce coffee beans suitable for transportation and roasting, it is necessary to separate the seeds from the outer layers (skin, pulp, mucilage, and parchment). This process can be carried out by three different methods, namely dry, semi-dry, or wet processing, which reduces the moisture of the coffee beans from 65% to 12% [2]. In the dry processing, intact coffee fruits are exposed to the sun for approximately 30 days until they reach 12% moisture. Then the fruits are mechanically crushed to separate the beans from the outer layers. Wet processing is relatively more complex, as the fruits are mechanically depulped and placed in tanks containing large volumes of water for a natural fermentation to occur for 24 to 48 h. The fermentation process removes the mucilage layer adhered to the seeds, which are finally sun-dried to reach the desired moisture content. Finally, semi-dry processing presents stages of both dry and wet methods, where the coffee fruits are mechanically depulped and then submitted to sun-drying [3].

Recent studies using next-generation sequencing (NGS) have shown that more than 100 microbial genera are involved in coffee beans fermentation from Brazil, Ecuador, Colombia, Honduras, and Australia [4–10]. However, temporal analyses of the fermentations revealed that the core microbiome is distinguished in three stages; (i) Enterobacteriaceae, acetic acid bacteria (AAB), and lactic acid bacteria (LAB) are present in higher frequencies at the beginning of the fermentation, (ii) LAB (mainly *Leuconostoc*, *Lactococcus* or *Lactobacillus*) dominate the fermentation after 6 h and (iii) acid-tolerant LAB remains until the end of the process. Among the eukaryotic community, generally, yeasts belonging to the genus *Pichia* show a high prevalence throughout fermentation [6,7,11,12].

The success of coffee fermentation depends on the indigenous microbiota associated with coffee farm microbiome. Recently, the use of starter cultures has been indicted to replace this empirical process, making coffee fermentation more predictable and controlled [11,12]. One limitation in the use of starter cultures is that fermentations are usually carried out in open tanks, which can favor contamination by the natural microbiota. Due to this lack of control, the starters must establish dominance over the high load of indigenous microorganisms. In recent years, several studies have shown that yeast and LAB species (e.g., *Lactobacillus plantarum*, *Leuconostoc mesenteroides*, *Pichia kluyveri*, *Pichia anomala*, *Pichia fermentans*, *Hanseniaspora uvarum*, *Saccharomyces cerevisiae*, *Debaryomyces hansenii*, and *Torulaspora delbrueckii*, *Candida railenensis*) were able to suppress the indigenous microbiota during coffee fermentation, as well as produce coffee beverages with desired characteristics [13–17]. However, the existence of a connection between coffee farm microbiome and the resulting starter culture-added, coffee fermentation has not yet been investigated.

It was recently demonstrated that co-inoculation of *Pichia fermentans* YC5.2 and *Pediococcus acidilactici* LPBC161 increases the production of metabolites (lactic acid, ethanol, and ethyl acetate) during coffee fermentation, evidencing a positive interaction between these two microbial groups [18]. In this sense, this study aimed to use a next-generation sequencing (NGS) to evaluate the influence of the microbial communities from coffee farm processing and coffee fermentation using *P. fermentans* YC5.2 and *P. acidilactici* LPBC161. In addition, the metabolite changes that occurred during the fermentation process were studied by chromatographic techniques.

2. Material and Methods

2.1. Cultivation of Lactic Acid Bacteria and Yeast

The starter cultures used in this study include the lactic acid bacterium *Pediococcus acidilactici* LPBC161 and the yeast *Pichia fermentans* YC5.2. These microbial cultures were previously isolated, identified and selected from spontaneous coffee fermentations, as reported by Pereira et al. [13] and Muynarsk [19], and are deposited in the Microbial Cultures Collection of the Department of Bioprocess and Biotechnology Engineering of the Federal University of Paraná, UFPR, Curitiba, PR, Brazil. *P. acidilactici* LPBC161 and *P. fermentans* YC5.2 were reactivated in MRS (Merck Millipore, Burlington, MA) and YEPG broth (Himedia, Marg, India), respectively, at 30 °C for 24 h. To achieve a concentration of approximately 10^9 CFU/mL, 0.4 L of a culture of *P. acidilactici* LPBC161 was inoculated in Erlenmeyer containing 3.6 L of a culture medium composed of glucose (5 g/L), yeast extract (5 g/L), ammonium citrate (5 g/L), ammonium phosphate (5 g/L), sodium acetate (2 g/L), manganese sulfate (0.05 g/L) and Tween 80 (0.1%) [20]. *P. fermentans* YC5.2 was also cultivated and inoculated in the same way, but a culture medium containing only glucose (5 g/L) yeast extract (5 g/L) was used. Yeasts and bacteria were incubated at 30 °C for 24 h. After this period, the cells were centrifuged at $5000 \times g$ for 10 min and resuspended in 250 mL of sterile saline solution (0.9% NaCl) and stored at 4 °C until their proper use.

2.2. Area of Study and Sampling Procedure

The experiments were conducted at Shalom farm situated in Patrocínio, Minas Gerais state, Brazil (18°56′38″ S 46°59′34″ O). The environmental samples were collected in triplicate and are composed of: (i) 100 g of soil collected at 10 cm depth in an area of 1 m² in the treetops; (ii) 10 g of coffee leaves collected from the soil surface; (iii) 20 fruits collected from the soil surface; (iv) 10 g of leaves collected from the branches 30 cm from the apical region; (v) 20 cherries collected from the coffee tree; (vi) 20 cherries before de-pulping; (vii) 20 over-ripe fruits collected from the coffee tree; (viii) 100 g depulped beans; (ix) 50 mL water used for de-pulping beans. The samples were stored in falcon sterile tubes (50 mL) and transported to the Center of Agroindustrial Biotechnology of Paraná (CENBAPAR, Curitiba, Brazil) under refrigeration and maintained at $-20\,°C$ until further analysis.

Later, coffee cherries (*Coffea arabica* var. Catuaí Amarelo) were harvested manually and mechanically pulped. Approximately 100 kg of pulped coffee beans were deposited in a 1 m³ concrete tank containing about 50 L of water, in accordance with the local wet processing method. Before inoculation, the cell viability of the starter cultures was determined by spread plate and the results were expressed in CFU/mL. Thus, the appropriate inoculum concentration of *P. fermentans* YC5.2 and *P. acidilactici* LPBC161 were adjusted to achieve an initial concentration of 10^7 CFU/mL and inoculated simultaneously in the fermentation tanks. About 100 g of coffee beans plus the liquid fraction of the coffee pulp were collected at 0, 8, 19, and 24 h. The liquid fraction was frozen in Falcon tubes sterilized at $-20\,°C$ for further analysis. Finally, the coffee beans sampled from each point were sun-dried to 12% moisture. Fermentation and sample collection were performed in triplicates.

2.3. Microbial Community Analysis by High-Throughput Sequencing

Before performing the total DNA extraction of the environmental samples, the microbial cells present in the samples were taken out. About 2.5 g of each sample was added to a tube containing 10 mL of saline solution (0.9% NaCl) and vigorously agitated by vortex (two treated for 2 min with an interval of 15 min). To eliminate coarse impurities, the solution was filtered through a 0.45 μm filter. Fermentation samples and water used for de-pulping beans were not submitted to this process. The total DNA extraction of the samples was performed by the Phenol-Chloroform method described by Carvalho Neto et al. [21]. Group-specific loci of both bacterial and fungal DNA were amplified through PCR. 20 ng of DNA containing Illumina platform adapters [22], were used to amplify the hypervariable region V3/V4 of the bacterial 16S rRNA gene using the 515F-806R primers. The ITS fungal region was amplified by ITS1-ITS2 primers. Barcoded amplicons were generated by PCR under the following conditions: 95 °C for 3 min, and 18 cycles at 95 °C for 30 s, 50 °C for 30 s and 68 °C for 60 s, followed by a final extension at 68 °C for 10 min. Samples were sequenced in the MiSeq platform using the 500 V2 kits, following standard Illumina protocols. Bioinformatics analyses were performed according to Vale et al. [10].

2.4. HPLC Analysis of Fermenting Coffee Pulp

The concentration of reducing sugars (glucose and fructose) and organic acids (lactic, citric, succinic, acetic and propionic acids) in the liquid fraction of the coffee pulp mass in fermentation was determined by High-Performance Liquid Chromatography (HPLC). The samples were analyzed in an HPLC® Agilent Technologies coupled to a refraction index (RID) and diode matrix (DAD) detector. The separation of the compounds was obtained using a Hiplex-H column (300 × 7.7 mm) (Bio-Rad, Richmond, CA, USA) with an isocratic mobile phase composed of 4.0 mM H_2SO_4, with a flow rate of 0.5 mL min^{-1} for 30 min. The temperatures of the sample, column, and RID detector used during the entire race were 25, 70 and 50 °C, respectively. The quantification of organic acids was performed in DAD at 210 nm while reducing sugars were determined in RID.

2.5. GC/MS Analysis of Coffee Pulp and Beans

The determination of volatile compounds presents in the fermentation liquid fraction and the chemical constitution of the coffee beans collected during the fermentation process were determined by Gas Chromatography Coupled to Mass Spectrophotometry (GC/MS). Briefly, the compounds were analyzed by Solid Phase Microextraction (SPME), using a DVB/CAR/PDMS Fibre (Supelco Co., Bellefonte, PA, USA) and injected into GC/MS connected to an autosampler (GCMS2010 Plus, TQ8040, AO 5000; Shimadzu, Tokyo, Japan). The injection parameters used were according to the procedures described by Junqueira et al. [6].

2.6. Statistical Analysis

Statistical significance was calculated using a post-hoc comparison of means using Tukey's test. Analyses were performed using the Statistica program, version 10.0 (Statsoft Inc., Tulsa, OK, USA). The level of significance was established using a two-sided p-value < 0.05. A principal component analysis (PCoA) based on Weighted UniFrac Distances was constructed using the microbial relative prevalence data (at both family and genus levels). Analyses were performed using the Statistica program, version 10.0 (Statsoft Inc., Tulsa, OK, USA).

3. Results and Discussion

3.1. Farm Microbiome

To elucidate the influences of the environmental microbiota on the fermentation process, coffee fruits and leaves (collected from the coffee farm and on the ground), over-ripe fruits, cherries before de-pulping, depulped beans and water used for de-pulping beans, were collected to analyze the bacterial and fungal communities. A total of 282.288 sequences from the hypervariable region V3-V4 of the 16S rRNA gene and 158.316 from the Internal Transcribed Spacer (ITS) rDNA gene were obtained from all samples. Sequences that presented identity above 97% were considered the same Operational Taxonomic Unit (OTUs). Thus, 299 prokaryotes and 189 eukaryotes were identified (Table S1). The rarefaction curves were satisfactory, suggesting that most bacterial and fungal communities were covered (Figure 1).

The presence and abundance of the major bacteria and fungi, defined as taxa with proportional abundance $\geq$2%, are reported in Figure 2. In general, Enterobacteriaceae and Saccharomycetales were the most abundant microbial groups. To facilitate visualization of the results, PCA analysis was constructed to group the coffee farm samples according to microbial abundance and diversity (Figure 3). Among the eukaryotes, the samples were divided into three clusters; (i) cherries before de-pulping, depulped beans and water, characterized by the marked presence of the Saccharomycetales family; (ii) soil, leaves and fruits collected on the ground, and fruits collected from the coffee tree, characterized by a low frequency of Saccharomycetales and dominance of specific groups such as Pleosporales, Mycosphaerellaceae and *Colletotrichum*; and (iii) fruits collected on the ground, leaves collected from the coffee tree and over-ripe fruits, characterized by the dominance of *Candida*, Lecanoromycetes, and *Fusarium*, respectively. These results suggest that the first grouping (mainly cherries before de-pulping and depulped beans) may favor the fermentative process due to the high incidence of Saccharomycetales. This order comprises about 1000 known species, including the yeasts *Pichia, Saccharomyces, Meyerozyma, Candida* and *Hanseniaspora* that have been widely selected as starter cultures for coffee fermentation [8,18,23]. Moreover, it is likely that most of the yeasts found in these samples are endophytic, since the coffee cherries collected from the coffee trees had a low incidence of Saccharomycetales (1.33%), while the depulped fruits contained a high frequency (69.15%) (Figure 2B). On the other hand, the marked presence of filamentous fungi, such as *Cladosporium, Colletotrichum, Fusarium* and Mycosphaerellaceae are undesirable in the fermentation process and control measures should be adopted to reduce the contact of soil, leaves, fruits collected from the ground and over-ripe fruits with the fermentation tank.

Interestingly, the genera *Cladosporium* and *Fusarium* were also observed at high frequency in soil from a Honduran coffee farm [10].

The bacterial group was also divided into three clusters; (i) leaves and fruits collected from the ground, cherries collected from the coffee tree and over-ripe fruits, by the high incidence of Enterobacteriaceae and *Erwinia*; (ii) cherries before de-pulping, depulped beans and water, with high frequencies of Enterobacteriaceae and *Leuconostoc*; and (iii) soil and leaves collected from the coffee tree, with high dominated of *Amycolatopsis* and *Methylobacterium*, respectively, and low frequency of Enterobacteriaceae. *Lactococcus* and *Lactobacillus* were also identified in high populations in the coffee leaves (Figure 2A).

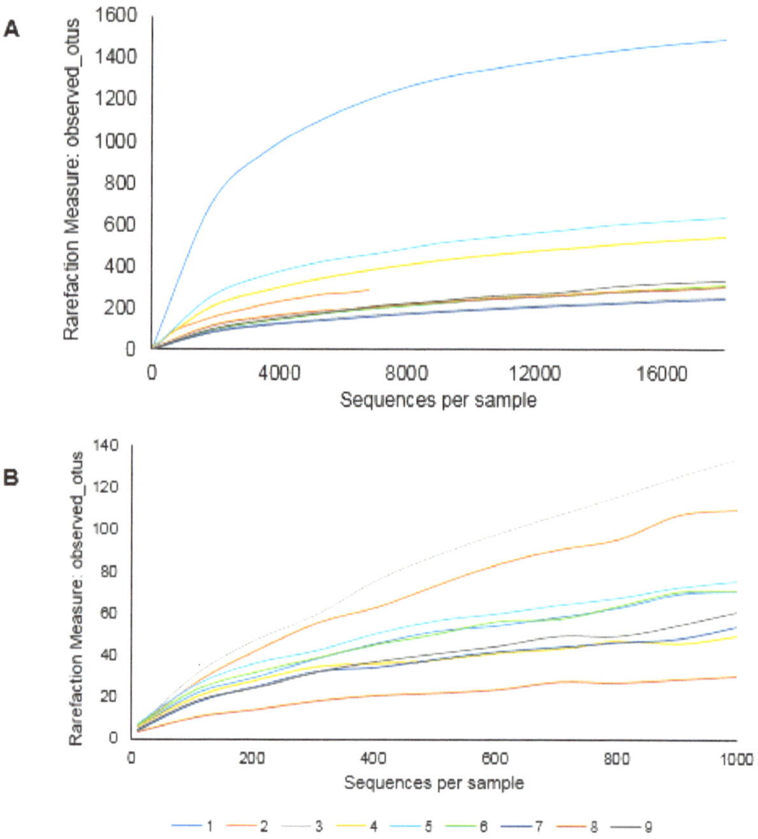

Figure 1. Alpha rarefaction curves of observed OTUs (operational taxonomic units) from the environmental samples. (**A**) Bacterial analysis, (**B**) Fungal analysis. 1 = Soil; 2 = Leaves collected from the ground; 3 = Fruits collected from the ground; 4 = Leaves collected from the coffee tree; 5 = Cherries collected from the coffee tree; 6 = Cherries before de-pulping; 7 = Over-ripe fruits; 8 = Depulped beans; 9 = Water used for de-pulping beans.

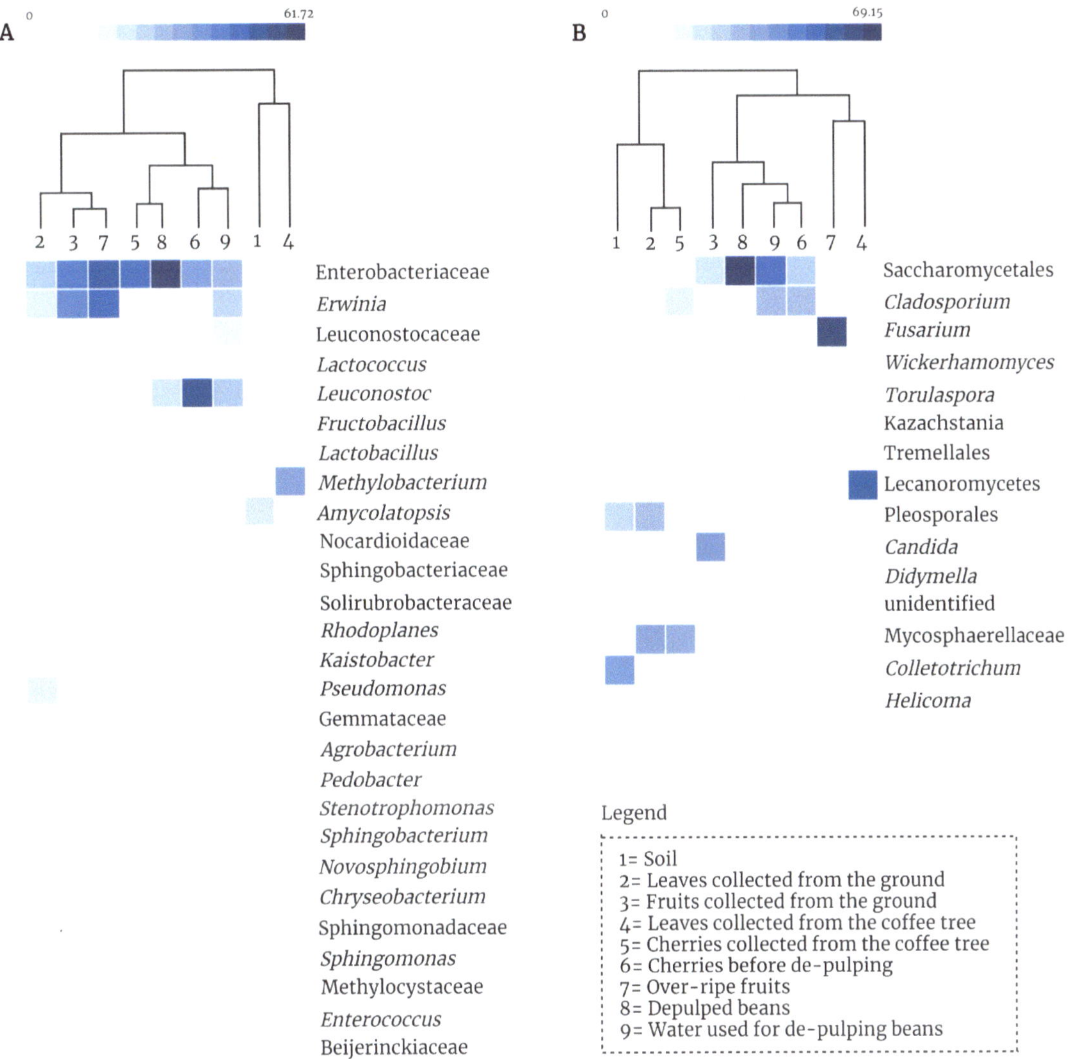

Figure 2. Composition of bacteria (**A**) and fungi (**B**) from coffee farm samples. Only microorganisms with prevalence superior to 2% are showed. The complete list of minor microbial groups is reported in the Supporting Information (Table S1).

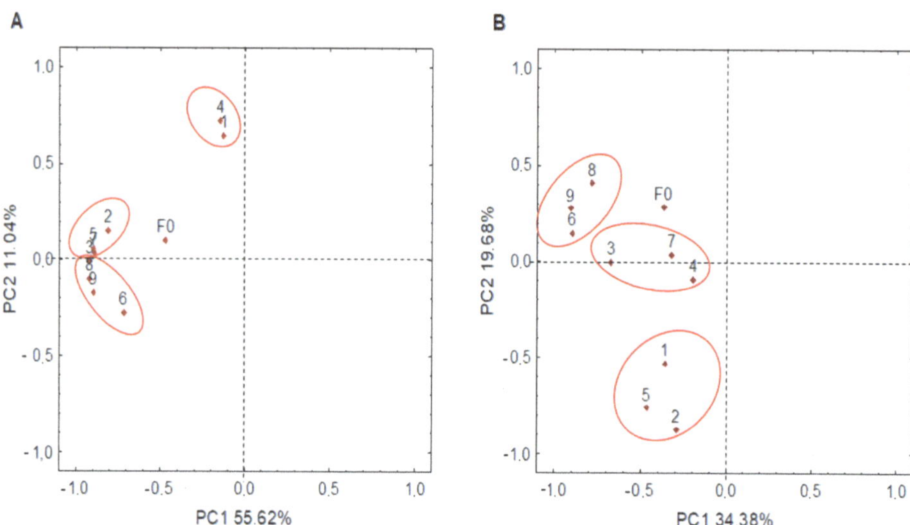

Figure 3. Principal component analysis, based on Weighted UniFrac Distances, according to microbial diversity and abundance. (**A**) Bacterial analysis, (**B**) Fungal analysis. 1 = Soil; 2 = Leaves collected from the ground; 3 = Fruits collected from the ground; 4 = Leaves collected from the coffee tree; 5 = Cherries collected from the coffee tree; 6 = Cherries before de-pulping; 7 = Over-ripe fruits; 8 = Depulped beans; 9 = Water used for de-pulping beans; F0 = Fermentation (0 h).

High incidence of Enterobacteriaceae has already been reported in grape and coffee plantations [9,10,24]. In vineyards, this family is considered beneficial because they produce proteases, chitinases and glucanases, which make them excellent antagonistic microorganisms [25]. In addition, over 80% of the bacterial community of coffee fruits collected on the ground and over-ripe fruits was composed of Enterobacteriaceae and *Erwinia*, showing the ability of this group to secrete hydrolytic enzymes to support their growth on decaying organic matter [26]. However, these microorganisms are not undesirable in coffee fermentation, as a correlation analysis performed by Zhang et al. [9] showed that the main alcohols and esters produced from the course of the fermentation process were not associated with enterobacteria. In addition, Enterobacteriaceae are mainly associated formation of off-flavor metabolites, such as 3-isopropyl-2-methoxy-5-methylpyrazine, 2,3-butanediol and butyric acid [27]. *Leuconostoc* was observed in the coffee cherries before de-pulping (53.48%), depulped beans (20.24%) and in the water used for de-pulping beans (25.60%) (Figure 2A). After harvesting, coffee fruits are usually placed in storage bags for a few hours until they are processed. During this period, micro-cracks can be generated in the fruit skin and several amino acids and phosphorylated carbohydrates present in the coffee pulp become accessible to the epiphytic microbiota of the cherries. The abundance of *Leuconostoc* in these samples may be associated with its ability to produce a wide range of saccharolytic enzymes, as well as having an elaborate carbohydrate uptake system [7]. Shotgun metagenomic analysis of a coffee fermentation identified a gene that encodes hexose 6 phosphate: phosphate antiporter (*uhpT*). This transporter may favor *Leuconostoc* proliferation since phosphorylated hexoses can be transported into bacterial cells without ATP consumption [7].

The coffee farm soil showed a rich and complex microbial diversity, with 214 bacterial groups (Table S1). The high diversity found in the soil was also observed in other coffee-producing regions, such as China, Ecuador, Mexico, Brazil, and Honduras [9,10,28–30]. However, most of the bacterial groups that have been identified in coffee farm soils (e.g., *Amycolatopsis, Bacillus, Bradyrhizobium, Pseudolabrys, Rhodoplanes*, and *Sphingomonas*) are not associated with the fermentation process, so it should be avoided from having access to the fermentation tank.

Interestingly, coffee leaves showed a relatively high population of *Lactococcus* and *Lactobacillus*, which had not been reported so far. The ability of these strains to grow on leaves suggests that these microorganisms possess a metabolic versatility not yet explored, as LAB are auxotrophic for some amino acids and vitamins that are not bioavailable in coffee leaves [9,10,28–31]. In addition, these LAB produce lactic acid as a primary fermentation product, so they are useful for coffee processing [32].

3.2. Microbiota Dynamics during Inoculated Coffee Fermentation

To improve the quality and complexity of fermented beverages, co-inoculation of LAB and yeast has been widely employed in the wine industry and more recently in coffee fermentations [10,33,34]. However, the starter cultures used in these processes must be able to suppress the growth of wild microbiota. The results reported in this study showed that co-inoculation favored the growth of the starter cultures (Figure 4). *Pichia* showed a population over 70% throughout the fermentative process. The family Lactobacillaceae, which includes *Pediococcus*, was also dominant at the beginning of the fermentative process (45.13%) followed by Enterobacteriaceae (23.85%) and *Lactococcus* (10.11%). After 8 h, the OTU readings attributed to Lactobacillaceae increased to 59.38%, suppressing the growth of Enterobacteriaceae (0.45%) and other species. At the end of the process, the reading of Lactobacillaceae increased to 77.40% (Figure 4).

The predominance of *P. fermentans* YC5.2 over the wild microbiota is mainly associated with the metabolic versatility of this yeast in tolerating osmotic pressure (growth detected with up to 50% glucose and fructose), ability to grow over a wide pH range (pH 2.0–8.0), growth at temperatures ranging from 30 to 43 °C, and tolerance to the main metabolites (ethanol, lactic acid and acetic acid) produced in the course of the fermentation [13]. Recently, analysis of the *P. acidilactici* LPBC161 genome performed by Muynarsk et al. [19] showed that its high fermentative capacity can be explained by the presence of several genes involved in the metabolism of sugars present in the coffee pulp, in addition to genes encoding proteins related to oxidative and alkaline stress. Vale et al. [18] also demonstrated that there is a positive synergistic interaction between LAB and yeast. The possible hypotheses to explain these interactions are (i) yeast autolysis releases nutrients, such as polysaccharides, riboflavin, and amino acids, favorable for bacterial growth and (ii) acidification of the fermentation medium by LAB creates a favorable environment for yeast development. Enabling the development and co-dominance of these two species.

Although LAB and yeast are the main microbial groups involved in coffee fermentations, the dominance of both these microbial groups may not be observed during spontaneous process. For instance, codominance of AAB (*Gluconobacter* and *Acetobacter*), LAB (*Leuconostoc* and *Lactococcus*), yeasts (*Hanseniaspora*, *Candida*, and *Pichia*) and the filamentous fungus *Fusarium* was recently observed in a spontaneous fermentation conducted in Honduras [10]. The microbial dynamics of another spontaneous fermentation conducted in Australia also showed some peculiarities. Although LAB increased during the fermentation process, it was noted that the population of Enterobacteriaceae and other subdominant microbial groups remained high after 36 h of fermentation [14]. Thus, the use of starter cultures is seen as essential to ensure the growth of beneficial microorganisms (i.e., yeast and LAB) during coffee fermentation.

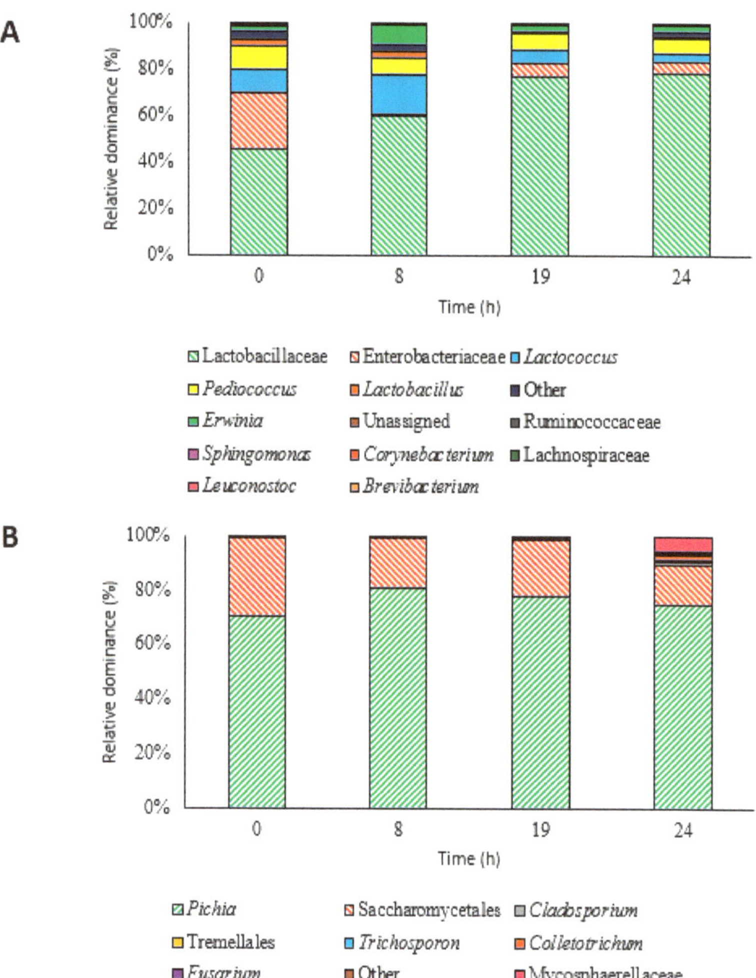

Figure 4. Prevalence and detection threshold of the highly persistent bacteria (**A**) and fungi (**B**) detected during coffee beans fermentation process conducted with selected starter cultures (*Pichia fermentans* YC5.2 and *Pediococcus acidilactici* LPBC161). The complete list of minor microbial groups is reported in the Supporting Information (Table S2).

Interestingly, during the entire fermentation process, a total of 135 and 115 prokaryotic and eukaryotic groups were identified, respectively (Table S2). The fermentation started with only 86 prokaryotes and 45 eukaryotes. It was noted that some species (e.g., *Trichococcus, Terriglobus, Fusobacterium, Passalora* and *Lecanosticta*) were not associated with the farm microbiome, suggesting that these microorganisms may have been introduced into the fermentation by human contact. In addition, several microbial genera (e.g., *Penicillium, Malassezia, Debaryomyces, Enterobacter, Pseudomonas* and *Kaistobacter*), associated with farm microbiome, were detected only 8 h after the start of the fermentation process. Besides human contact, one of the contamination routes is likely to air since the coffee fermentations are carried out in open tanks. However, it is important to note that all these contaminants were present at frequencies $\leq 0.01\%$, showing that the starter cultures were able to suppress their growth. Thus, the use of starter cultures proves to be extremely important for maintaining desirable microbial groups during coffee beans fermentation.

3.3. Chemical Analysis of the Fermentation Liquid Fraction and Coffee Beans

The sugar consumption profile and metabolic formation showed the activity of the initial cultures used during the fermentation process. Glucose and fructose were consumed during fermentation and transformed mainly into lactic acid (associated with the metabolism of *P. acidilactici* LPBC161) and ethanol (associated with the metabolism of *P. fermentans* YC5.2) as shown in Table 1. It is possible to observe a decrease in sugars (glucose and fructose) during the 24 h of fermentation (Table 1). Both glucose and fructose were partially consumed until the end of the process. During the fermentation process, sugars were used for microbial growth and a significant amount of ethanol and lactic acid were produced, causing a drop in pH (5.16 to 4.13). The reduction of pH levels below 4.5 is a method widely used by coffee producers to determine the end of the fermentation of coffee beans during the wet processing method [11,35]. In addition, in conjecture with organic acid production, ethanol generation inhibits the growth of undesirable microorganisms during the fermentation process.

Table 1. Proportion of volatile compounds (area * 10^5), organic acids, ethanol, and reducing sugars (g/L) during coffee beans fermentation process conducted with selected starter cultures (*Pichia fermentans* YC5.2 and *Pediococcus acidilactici* LPBC161). Different letters indicate significant differences over fermentation time.

Compound	0 h	8 h	19 h	24 h
GC-MS (area)				
Higher alcohols (**2**)				
1-Butanol, 3-methyl	0.27 ± 0.02 [a]	0.39 ± 0.02 [b]	0.63 ± 0.00 [c]	0.60 ± 0.00 [c]
2-Heptanol	0.17 ± 0.01 [a]	0.13 ± 0.00 [a]	ND	ND
Ester (**4**)				
Ethyl acetate	4.10 ± 0.13 [a]	2.24 ± 0.78 [b]	3.05 ± 0.10 [c]	4.80 ± 0.02 [a]
Methyl acetate	ND	ND	3.80 ± 0.41 [a]	0.49 ± 0.04 [b]
Ethyl isobutyrate	ND	ND	0.63 ± 0.00	ND
Isoamyl acetate	ND	0.25 ± 0.00 [a]	0.37 ± 0.08 [a]	0.27 ± 0.00 [a]
Aldehyde (**3**)				
Butanal, 3 methyl	0.86 ± 0.06 [a]	0.78 ± 0.05 [ab]	0.21 ± 0.00 [b]	0.74 ± 0.28 [ab]
Butanal, 2 methyl	0.63 ± 0.26 [a]	0.34 ± 0.00 [a]	0.13 ± 0.01 [a]	0.32 ± 0.00 [a]
Benzeneacetaldehyde	0.23 ± 0.02 [a]	0.33 ± 0.10 [a]	ND	ND
HPLC (g/L)				
Glucose	1.08 ± 0.11 [a]	0.75 ± 0.09 [ab]	0.52 ± 0.14 [b]	0.62 ± 0.21 [b]
Fructose	2.52 ± 0.22 [a]	1.78 ± 0.49 [b]	1.26 ± 0.13 [b]	1.47 ± 0.06 [b]
Lactic acid	ND	0.43 ± 0.04 [a]	1.24 ± 0.21 [b]	1.41 ± 0.12 [b]
Ethanol	ND	0.37 ± 0.12 [a]	0.50 ± 0.09 [a]	0.81 ± 0.08 [b]
pH	5.16	4.50	4.17	4.13

HPLC analysis of the liquid fermentation fraction revealed the high presence of lactic acid and ethanol. These compounds were associated with LAB and yeast metabolism, respectively. Lactic acid was the main organic acid produced, reaching a concentration of 1.41 g/L at 24 h (Table 1). This primary metabolite is mainly produced by the central carbon metabolism of homofermentative LAB [35]. The diffusion of acids in the grains can influence the flavor and final quality of the beverage [13,17]. Lactic acid is known to attribute desirable lactic sensory notes and contribute to the acidity and body of the final drink [15]. Ethanol showed a constant increase throughout fermentation (0.8 g/L), indicating a high metabolic activity of non-*Saccharomyces* yeasts [8].

Among the volatiles, GC-MS analyses identified nine compounds, two higher alcohols, four esters and three aldehydes (Table 1). Ethyl acetate was the most abundant compound formed during fermentation. *P. fermentans* YC5.2 was selected based mainly on the production of this metabolite, which plays an important role in the development of aroma for coffee beans [13]. The diffusion and persistence of ethyl acetate in the grain contributes to the development of desirable fruit notes and nuances of grape/cherry in the coffee beverage [3]. Other volatile compounds produced during fermentation, associated with the metabolism of early cultures, include isoamyl acetate and ethyl isobutyrate.

In addition to aromatic function, yeast used in this study as an initial culture has been reported in previous studies as producers of pectinitic enzymes. Thus, it assists in the degradation of pectin present in coffee pulp and mucilage, producing metabolites that spread into coffee beans, favoring the flavor formation of the final beverage [36]. Significantly, the roasted grains of fermentation with the inoculation of YC5.2 and LPBC161 brought these metabolites derived from yeast. Table 2 shows the compounds detected inside the coffee beans during the fermentation process. Twenty-five compounds were identified: three organic acids, seven upper alcohols, eight aldehydes, one terpene, two terpenes, one pyrazine, one ketone and two hydrocarbons. Although the identified compounds fluctuated throughout the fermentation process, acetic acid, isorvalériic acid, butanal, 3-methyl, hexanal, benzeacetaldehyde, benzaldehyde and acetone were observed in high proportions. Interestingly, even though the main compound is identified in the liquid fraction of fermentation, ethyl acetate was only identified in coffee beans after 7 pm. In addition, isoamyl acetate and ethyl isobutyrate were not identified, which may also come from the initial metabolism.

Table 2. Concentration of volatile compounds (area * 10^5) inside coffee beans collected during coffee fermentation process conducted with selected starter cultures (*Pichia fermentans* YC5.2 and *Pediococcus acidilactici* LPBC161). Different letters indicate significant differences over fermentation time.

Compound (Area)	0 h	8 h	19 h	24 h
Organic acid **(3)**				
Acetic acid	5.35 ± 0.04 [a]	6.03 ± 0.86 [a]	9.99 ± 0.33 [b]	9.73 ± 0.84 [b]
Butanoic acid, 3-methyl	0.33 ± 0.03 [a]	0.44 ± 0.24 [ab]	0.58 ± 0.08 [ab]	0.85 ± 0.18 [b]
Isovaleric acid	0.36 ± 0.06 [a]	1.39 ± 0.18 [b]	0.54 ± 0.12 [ac]	0.77 ± 0.04 [c]
Higher alcohols **(7)**				
Propanol, 2-methyl	0.16 ± 0.00 [a]	0.13 ± 0.00 [ab]	0.11 ± 0.00 [b]	0.11 ± 0.02 [b]
1-Octen-3-ol	0.41 ± 0.03 [a]	0.29 ± 0.05 [b]	0.22 ± 0.00 [b]	0.20 ± 0.03 [b]
2-Hexanol, 5 methyl	0.59 ± 0.10 [a]	0.59 ± 0.06 [a]	0.46 ± 0.05 [ab]	0.33 ± 0.01 [b]
1-Butanol, 3-methyl	-	0.60 ± 0.24 [a]	0.28 ± 0.00 [a]	0.59 ± 0.12 [a]
2-Heptanol, 3-methyl	-	-	-	0.13 ± 0.03
1-Butanol, 2-methyl	-	-	-	0.28 ± 0.00
Phenylethyl Alcohol	-	-	-	0.26 ± 0.03
Aldehyde **(8)**				
Butanal, 3 methyl	3.33 ± 0.46 [ab]	4.39 ± 0.33 [b]	2.48 ± 0.18 [c]	1.50 ± 0.79 [c]
Butanal, 2 methyl	0.70 ± 0.07 [ab]	0.76 ± 0.07 [b]	0.26 ± 0.00 [ac]	0.39 ± 0.00 [c]
Hexanal	2.01 ± 0.32 [a]	3.49 ± 0.56 [bc]	3.97 ± 0.52 [c]	2.44 ± 0.27 [ab]
Heptanal	0.20 ± 0.02 [a]	0.07 ± 0.01 [b]	-	-
Benzeacetaldehyde	1.06 ± 0.06 [a]	1.07 ±0.03 [a]	0.94 ± 0.05 [ab]	0.81 ± 0.02 [b]
Benzaldehyde	1.56 ± 0.06 [a]	1.23 ± 0.12 [b]	0.90 ± 0.03 [c]	0.89 ± 0.06 [c]
Pentanal	0.61 ± 0.06 [a]	0.58 ± 0.21 [a]	0.58 ± 0.04 [a]	0.49 ± 0.04 [a]
Methional	0.70 ± 0.00 [a]	0.79 ± 0.22 [a]	0.67 ± 0.12 [a]	0.30 ± 0.04 [b]
Ester **(1)**				
Ethyl acetate	-	-	0.13 ± 0.01 [a]	0.25 ± 0.00 [b]
Terpenes **(2)**				
Linalol	0.65 ± 0.00 [a]	0.71 ± 0.04 [ab]	0.95 ± 0.00 [b]	0.61 ± 0.16 [a]
Limonene	-	-	0.17 ± 0.01	-
Pyrazine **(1)**				
2-Isobuttyl-3-methoxypyrazine	0.23 ± 0.11	-	-	-
Ketones **(1)**				
Acetoin	0.47 ± 0.05 [a]	3.49 ± 0.51 [bc]	4.15 ± 0.08 [c]	3.30 ± 0.09 [b]
Hydrocarbons **(2)**				
Toluene	0.23 ± 0.10 [a]	0.22 ± 0.01 [a]	-	-
Nonane, 3-methyl-5-propyl	0.27 ± 0.00	-	-	-
Furanone **(1)**				
Furan, 2-pentyl	-	0.33 ± 0.07 [a]	0.27 ± 0.06 [a]	0.15 ± 0.03 [a]

The diffusion of volatile compounds in coffee beans is not fully understood. However, Salem et al. [36] demonstrated that the compounds butanal, 2-fenontanol and isoamyl acetate presented different transfer rates from culture medium to coffee beans. The study suggested that the differences observed in the diffusion of compounds in grains may be associated with three processes (i) the parchment layer is a barrier, which "acts" as a filter; (ii) the compounds suffered metabolic reactions, decreasing their amount in coffee beans; (iii) there is an interaction between volatiles and yeast, significantly reducing the transfer of these compounds.

Finally, several other compounds identified in the roasted grains in the fermentation process were detected. These volatiles can be generated mainly during the course of fermentation by biochemical reactions within the bean or even by reactions that occur during the roasting process. It is known that the volatile fraction of coffee beans develops mainly in the form of alcohols, acids, esters and aldehydes [37]. Compounds of these classes are associated with flavor during coffee fermentation.

4. Conclusions

The coffee farm microbiome is composed of a rich microbiome diversity dominated by Saccharomycetales and Enterobacteriaceae. Coffee cherries before de-pulping and depulped beans harbour beneficial microorganisms (yeast and LAB) for the fermentation process. On the other hand, enterobacteria, filamentous fungi and other microbial groups presents in soil, leaves, fruits collected from the ground and over-ripe fruits may transfer unwanted aromas to coffee beans. Therefore, they should be prevented from having access to the fermentation tank. Thus, cleaning procedures should be performed to prevent the growth of these unwanted microbial groups.

The inoculation with high titers of selected yeast and LAB start culture modulates the overall fermentation over wild microbiota, with efficient sugar mucilage consumption and aroma compounds formation. Thus, the results of this study showed that the introduction of starter cultures is essential to control the coffee fermentation process in terms of both kinetics and quality of the resulting product.

Supplementary Materials: The following are available online at https://www.mdpi.com/article/10.3390/fermentation7040278/s1, Table S1: Relative abundance (%) of fungi and bacteria detected in environmental samples collected from Brazil coffee plantations. Table S2: The relative abundance (%) of fungi and bacteria during the inoculated fermentation process.

Author Contributions: Conceptualization, C.R.S.; Data curation, A.d.S.V.; Formal analysis, V.B.P. and D.P.d.C.N.; Funding acquisition, G.V.d.M.P. and C.R.S.; Investigation, G.V.d.M.P. and D.P.d.C.N.; Methodology, A.d.S.V.; Project administration, G.V.d.M.P. and C.R.S.; Writing—original draft, V.B.P. and G.V.d.M.P.; Writing—review & editing, A.d.S.V., D.P.d.C.N. and C.R.S. The authors are responsible for all the aspects of the manuscript. All authors have read and agreed to the published version of the manuscript.

Funding: This work was supported by the Brazilian National Council for Scientific and Technological Development (CNPq) (project number 429560/2018-4).

Institutional Review Board Statement: Not applicable.

Informed Consent Statement: Not applicable.

Acknowledgments: The authors thank the CNPq for fellowships in research (grant numbers 303254/2017-3 and 151885/2019-2).

Conflicts of Interest: The authors declare no conflict of interest.

References

1. ICO. Coffee year production by country. *Data* **2021**, *2021*, 7–9.
2. de Melo Pereira, G.V.; Soccol, V.T.; Brar, S.K.; Neto, E.; Soccol, C.R. Microbial ecology and starter culture technology in coffee processing. *Crit. Rev. Food Sci. Nutr.* **2017**, *57*, 2775–2788. [CrossRef] [PubMed]
3. de Melo Pereira, G.V.; de Carvalho Neto, D.P.; Júnior, A.I.M.; Vásquez, Z.S.; Medeiros, A.B.P.; Vandenberghe, L.P.S.; Soccol, C.R. Exploring the impacts of postharvest processing on the aroma formation of coffee beans–A review. *Food Chem.* **2019**, *272*, 441–452. [CrossRef]

4. de Carvalho Neto, D.P.; de Melo Pereira, G.V.; Tanobe, V.O.A.; Soccol, V.T.; da Silva, B.J.G.; Rodrigues, C.; Soccol, C.R. Yeast diversity and physicochemical characteristics associated with coffee bean fermentation from the Brazilian Cerrado Mineiro region. *Fermentation* **2017**, *3*, 11. [CrossRef]

5. De Bruyn, F.; Zhang, S.J.; Pothakos, V.; Torres, J.; Lambot, C.; Moroni, A.V.; Callanan, M.; Sybesma, W.; Weckx, S.; De Vuysta, L. Exploring the Impacts of Postharvest Processing on the Microbiota and. *Appl. Environ. Microbiol.* **2017**, *83*, 1–16. [CrossRef]

6. de Oliveira Junqueira, A.C.; de Melo Pereira, G.V.; Coral Medina, J.D.; Alvear, M.C.R.; Rosero, R.; de Carvalho Neto, D.P.; Enríquez, H.G.; Soccol, C.R. First description of bacterial and fungal communities in Colombian coffee beans fermentation analysed using Illumina-based amplicon sequencing. *Sci. Rep.* **2019**, *9*, 8794. [CrossRef] [PubMed]

7. Pothakos, V.; De Vuyst, L.; Zhang, S.J.; De Bruyn, F.; Verce, M.; Torres, J.; Callanan, M.; Moccand, C.; Weckx, S. Temporal shotgun metagenomics of an Ecuadorian coffee fermentation process highlights the predominance of lactic acid bacteria. *Curr. Res. Biotechnol.* **2020**, *2*, 1–15. [CrossRef]

8. Elhalis, H.; Cox, J.; Zhao, J. Ecological diversity, evolution and metabolism of microbial communities in the wet fermentation of Australian coffee beans. *Int. J. Food Microbiol.* **2020**, *321*, 108544. [CrossRef]

9. Zhang, S.J.; De Bruyn, F.; Pothakos, V.; Torres, J.; Falconi, C.; Moccand, C.; Weckx, S.; De Vuyst, L. Following coffee production from cherries to cup: Microbiological and metabolomic analysis of wet processing of Coffea arabica. *Appl. Environ. Microbiol.* **2019**, *85*, e02635-18. [CrossRef]

10. da Silva Vale, A.; de Melo Pereira, G.V.; de Carvalho Neto, D.P.; Sorto, R.D.; Goés-Neto, A.; Kato, R.; Soccol, C.R. Facility-specific 'house' microbiome ensures the maintenance of functional microbial communities into coffee beans fermentation: Implications for source tracking. *Environ. Microbiol. Rep.* **2021**, *13*, 470–481. [CrossRef]

11. Carvalho Neto, D.P.; de Melo Pereira, G.V.; Finco, A.M.O.; Letti, L.A.J.; da Silva, B.J.G.; Vandenberghe, L.P.S.; Soccol, C.R. Efficient coffee beans mucilage layer removal using lactic acid fermentation in a stirred-tank bioreactor: Kinetic, metabolic and sensorial studies. *Food Biosci.* **2018**, *26*, 80–87. [CrossRef]

12. de Carvalho Neto, D.P.; Vinícius De Melo Pereira, G.; Finco, A.M.O.; Rodrigues, C.; De Carvalho, J.C.; Soccol, C.R. Microbiological, physicochemical and sensory studies of coffee beans fermentation conducted in a yeast bioreactor model. *Food Biotechnol.* **2020**, *34*, 172–192. [CrossRef]

13. de Melo Pereira, G.V.; Soccol, V.T.; Pandey, A.; Medeiros, A.B.P.; Andrade Lara, J.M.R.; Gollo, A.L.; Soccol, C.R. Isolation, selection and evaluation of yeasts for use in fermentation of coffee beans by the wet process. *Int. J. Food Microbiol.* **2014**, *188*, 60–66. [CrossRef] [PubMed]

14. Elhalis, H.; Cox, J.; Frank, D.; Zhao, J. Microbiological and biochemical performances of six yeast species as potential starter cultures for wet fermentation of coffee beans. *LWT* **2021**, *137*, 110430. [CrossRef]

15. de Melo Pereira, G.V.; de Carvalho Neto, D.P.; Medeiros, A.B.P.; Soccol, V.T.; Neto, E.; Woiciechowski, A.L.; Soccol, C.R. Potential of lactic acid bacteria to improve the fermentation and quality of coffee during on-farm processing. *Int. J. Food Sci. Technol.* **2016**, *51*, 1689–1695. [CrossRef]

16. Ribeiro, L.S.; da Cruz Pedrozo Miguel, M.G.; Martinez, S.J.; Bressani, A.P.P.; Evangelista, S.R.; e Batista, C.F.S.; Schwan, R.F. The use of mesophilic and lactic acid bacteria strains as starter cultures for improvement of coffee beans wet fermentation. *World J. Microbiol. Biotechnol.* **2020**, *36*, 1–15. [CrossRef]

17. Evangelista, S.R.; Silva, C.F.; da Cruz Miguel, M.G.P.; de Sauza Cordeiro, C.; Pinheiro, A.C.M.; Duarte, W.F.; Schwan, R.F. Improvement of coffee beverage quality by using selected yeasts strains during the fermentation in dry process. *Food Res. Int.* **2014**, *61*, 183–195. [CrossRef]

18. Vale, A.; de Melo Pereira, G.V.; de Carvalho Neto, D.P.; Rodrigues, C.; Pagnoncelli, M.G.B.; Soccol, C.R. Effect of Co-Inoculation with pichia fermentans and pediococcus acidilactici on metabolite produced during fermentation and volatile composition of coffee beans. *Fermentation* **2019**, *5*, 67. [CrossRef]

19. Muynarsk, E.S.M.; de Melo Pereira, G.V.; Mesa, D.; Thomaz-soccol, V.; Pagnoncelli, M.G.B.; Soccol, C.R. Draft Genome Sequence of Pediococcus acidilactici Strain. *Microbiol. Resour. Announc.* **2019**, *8*, e00332-19. [CrossRef]

20. Feltrin, V.P.; Sant'Anna, E.S.; Porto, A.C.S.; Torres, R.C.O. Produçao de lactobacillus plantarum em melaço de cana-de-açúcar. *Braz. Arch. Biol. Technol.* **2000**, *43*, 119–124. [CrossRef]

21. de Carvalho Neto, D.P.; de Melo Pereira, G.V.; de Carvalho, J.C.; Soccol, V.T.; Soccol, C.R. High-throughput rRNA gene sequencing reveals high and complex bacterial diversity associated with brazilian coffee bean fermentation. *Food Technol. Biotechnol.* **2018**, *56*, 90–95. [CrossRef]

22. Caporaso, J.G.; Lauber, C.L.; Walters, W.A.; Berg-Lyons, D.; Huntley, J.; Fierer, N.; Owens, S.M.; Betley, J.; Fraser, L.; Bauer, M.; et al. Ultra-high-throughput microbial community analysis on the Illumina HiSeq and MiSeq platforms. *ISME J.* **2012**, *6*, 1621–1624. [CrossRef]

23. Bressani, A.P.P.; Martinez, S.J.; Sarmento, A.B.I.; Borém, F.M.; Schwan, R.F. Organic acids produced during fermentation and sensory perception in specialty coffee using yeast starter culture. *Food Res. Int.* **2020**, *128*, 108773. [CrossRef]

24. Nisiotou, A.A.; Rantsiou, K.; Iliopoulos, V.; Cocolin, L.; Nychas, G.J.E. Bacterial species associated with sound and Botrytis-infected grapes from a Greek vineyard. *Int. J. Food Microbiol.* **2011**, *145*, 432–436. [CrossRef] [PubMed]

25. Pinto, C.; Pinho, D.; Sousa, S.; Pinheiro, M.; Egas, C.; Gomes, A.C. Unravelling the diversity of grapevine microbiome. *PLoS ONE* **2014**, *9*, e85622. [CrossRef]

26. Zhou, S.; Ingram, L.O. Synergistic hydrolysis of carboxymethyl cellulose and acid-swollen cellulose by two endoglucanases (CelZ and CelY) from Erwinia chrysanthemi. *J. Bacteriol.* **2000**, *182*, 5676–5682. [CrossRef] [PubMed]

27. Gueule, D.; Fourny, G.; Ageron, E.; Le Flèche-Matéos, A.; Vandenbogaert, M.; Grimont, P.A.D.; Cilas, C. Pantoea coffeiphila sp. nov., cause of the 'potato taste' of Arabica coffee from the African great lakes region. *Int. J. Syst. Evol. Microbiol.* **2015**, *65*, 23–29. [CrossRef]

28. Zhao, Q.; Xiong, W.; Xing, Y.; Sun, Y.; Lin, X.; Dong, Y. Long-Term Coffee Monoculture Alters Soil Chemical Properties and Microbial Communities. *Sci. Rep.* **2018**, *8*, 6116. [CrossRef]

29. Cabrera-Rodríguez, A.; Trejo-Calzada, R.; García-De la Peña, C.; Arreola-Ávila, J.G.; Nava-Reyna, E.; Vaca-Paniagua, F.; Díaz-Velásquez, C.; Meza-Herrera, C.A. A metagenomic approach in the evaluation of the soil microbiome in coffee plantations under organic and conventional production in tropical agroecosystems. *Emir. J. Food Agric.* **2020**, *32*, 263–270. [CrossRef]

30. Veloso, T.G.R.; da Silva, M.d.C.S.; Cardoso, W.S.; Guarçoni, R.C.; Kasuya, M.C.M.; Pereira, L.L. Effects of environmental factors on microbiota of fruits and soil of Coffea arabica in Brazil. *Sci. Rep.* **2020**, *10*, 14692. [CrossRef]

31. de Melo Pereira, G.V.; De Carvalho Neto, D.P.; Junqueira, A.C.D.O.; Karp, S.G.; Letti, L.A.J.; Magalhães Júnior, A.I.; Soccol, C.R. A Review of Selection Criteria for Starter Culture Development in the Food Fermentation Industry. *Food Rev. Int.* **2020**, *36*, 135–167. [CrossRef]

32. Pereira, G.V.M.; Vale, A.S.; Carvalho Neto, D.P.; Muynarsk, E.S.; Soccol, V.T.; Soccol, C.R. Lactic acid bacteria: What coffee industry should know? *Curr. Opin. Food Sci.* **2020**, *31*, 1–8. [CrossRef]

33. Sun, S.Y.; Gong, H.S.; Zhao, K.; Wang, X.L.; Wang, X.; Zhao, X.H.; Yu, B.; Wang, H.X. Co-inoculation of yeast and lactic acid bacteria to improve cherry wines sensory quality. *Int. J. Food Sci. Technol.* **2013**, *48*, 1783–1790. [CrossRef]

34. Bressani, A.P.P.; Martinez, S.J.; Batista, N.N.; Simão, J.B.P.; Dias, D.R.; Schwan, R.F. Co-inoculation of yeasts starters: A strategy to improve quality of low altitude Arabica coffee. *Food Chem.* **2021**, *361*, 130133. [CrossRef] [PubMed]

35. Dicks, L.M.T.; Endo, A. Taxonomic status of lactic acid bacteria in wine and key characteristics to differentiate species. *S. Afr. J. Enol. Vitic.* **2009**, *30*, 72–90. [CrossRef]

36. Silva, B.L.; Pereira, P.V.; Bertoli, L.D.; Silveira, D.L.; Batista, N.N.; Pinheiro, P.F.; de Souza Carneiro, J.; Schwan, R.F.; de Assis Silva, S.; Coelho, J.M.; et al. Fermentation of Coffea canephora inoculated with yeasts: Microbiological, chemical, and sensory characteristics. *Food Microbiol.* **2021**, *98*. [CrossRef] [PubMed]

37. Gonzalez-Rios, O.; Suarez-Quiroz, M.L.; Boulanger, R.; Barel, M.; Guyot, B.; Guiraud, J.P.; Schorr-Galindo, S. Impact of "ecological" post-harvest processing on coffee aroma: II. Roasted coffee. *J. Food Compos. Anal.* **2007**, *20*, 297–307. [CrossRef]

 *fermentation*

Article

Methods for Oxygenation of Continuous Cultures of Brewer's Yeast, *Saccharomyces cerevisiae*

Timothy Granata [1,*], Cindy Follonier [1,2], Chiara Burkhardt [1,2] and Bernd Rattenbacher [1,2]

[1] Space Science Hub and Space Biology Group, Center for Bio- & Medical Engineering, Lucerne University of Applied Sciences and Arts, Obermattweg 9, 6058 Hergiswil, Switzerland
[2] Biotechnology Space Support Center, Obermattweg 9, 6058 Hergiswil, Switzerland; cindy.follonier@hslu.ch (C.F.); chiara.burkhardt@hslu.ch (C.B.); bernd.rattenbacher@hslu.ch (B.R.)
* Correspondence: timothy.granata@hslu.ch

Abstract: Maintaining steady-state, aerobic cultures of yeast in a bioreactor depends on the configuration of the bioreactor system as well as the growth medium used. In this paper, we compare several conventional aeration methods with newer filter methods using a novel optical sensor array to monitor dissolved oxygen, pH, and biomass. With conventional methods, only a continuously stirred tank reactor configuration gave high aeration rates for cultures in yeast extract peptone dextrose (YPD) medium. For filters technologies, only a polydimethylsiloxan filter provided sufficient aeration of yeast cultures. Further, using the polydimethylsiloxan filter, the YPD medium gave inferior oxygenation rates of yeast compared to superior results with Synthetic Complete medium. It was found that the YPD medium itself, not the yeast cells, interfered with the filter giving the low oxygen transfer rates based on the volumetric transfer coefficient ($K_L a$). The results are discussed for implications of miniaturized bioreactors in low-gravity environments.

Keywords: yeast; continuous cultures; bioreactor; hollow-fiber filters; $K_L a$; oxygen uptake

Citation: Granata, T.; Follonier, C.; Burkhardt, C.; Rattenbacher, B. Methods for Oxygenation of Continuous Cultures of Brewer's Yeast, *Saccharomyces cerevisiae*. *Fermentation* **2021**, *7*, 282. https://doi.org/10.3390/fermentation7040282

Academic Editor: Ronnie G. Willaert

Received: 9 November 2021
Accepted: 22 November 2021
Published: 26 November 2021

Publisher's Note: MDPI stays neutral with regard to jurisdictional claims in published maps and institutional affiliations.

Copyright: © 2021 by the authors. Licensee MDPI, Basel, Switzerland. This article is an open access article distributed under the terms and conditions of the Creative Commons Attribution (CC BY) license (https://creativecommons.org/licenses/by/4.0/).

1. Introduction

Many experiments conducted with Brewer's yeast, *Saccharomyces cerevisiae*, use batch and continuous cultures to rear cells in growth medium. For aerobic cultures, however, oxygenation of media is often problematic and different methods have been proposed for different growth media [1]. Common methods for aerating yeast cultures are bubbling, shaking, or stirring, which promotes oxygen transfer into the culture through the free surface while newer methods use filters for aeration [1]. Some of these methods pose problems, such as the use of commercially available yeast extract peptone dextrose (YPD) medium which foams when bubbled constantly or intermittently and may oxidize compounds in the medium. Anti-foaming agents can be used, but they too may alter the composition of the medium. For situations where the free surface area is reduced or absent, oxygen is not replenished in the head space and stirring and shaking cannot be used. Therefore, other methods are needed to aerate the yeast. Additionally, if cell cultures remove oxygen faster than it can be supplied, which occurs when biomass concentrations and cell respiration rates are very high, then more effective methods are needed to replenish oxygen [2]. One of the main reasons for studying newer oxygen transfer technologies, particularly aeration filters, is to combine these with miniaturized bioreactors and sensors to improve bioreactor design for applications in space. In microgravity, head space, and thus shaking and stirring of cultures, are not feasible, nor is bubbling medium with air or oxygen [3]. Also critical is a small system size since the payload is limited [3].

Recent reviews of bioreactor aeration discuss a variety of methods [4–6]. There are also many variables that can influence aeration including temperature, mixing rate, membranes, biomass concentration, and media [7–10]. In this study, we use a novel sensor array to compare different filter technologies and media to conventional methods to determine if the methods can be used in space to oxygenate yeast cultures.

2. Materials and Methods

2.1. The Bioreactor and Data Analysis

A 70 mL Erlenmeyer flask was used as the bioreactor chamber. The chamber volume was kept constant at 40 mL, leaving a 2 cm head space with a surface area of 7.1 cm^2. The overall volume of the system, including, tubing and filters, was 50 mL. For all experiments, a peristaltic pump (Phamacia LKBP-1, Uppsala, Sweden) with a flow rate of 10 mL h^{-1} was used to transfer the medium to the bioreactor giving a system dilution rate of D = 0.2 h^{-1}. The bioreactor volume was kept constant via an overflow tube connected to a sample bottle. For experiments with filters, a second peristatic pump (Alitea-XV, Stockholm, Sweden) was used to recirculate cells through the filters at a flow rate of 26.7 mL min^{-1}.

The bioreactor was mounted atop an optical array (model SFR vario by PreSens GmbH, Regensburg, Germany) to non-invasively and simultaneously sample time series of optical density (OD) for biomass, dissolved oxygen (DO), and pH in the bioreactor chamber (Figure 1). To sample DO and pH, the array used chemically treated spots that were autoclavable. The DO spots (SP-PSt3-YAU-D7_YOP) reacted with oxygen molecules and emitted a fluorescent signal at 505 nm. The pH spots (SP-PG1-V2-D7-US-SA-2004-1) reacted with hydrogen ions and fluoresced at 470 nm. These two spots were glued to the inside bottom of the bioreactor chamber, aligned with the optical detectors for DO and pH (Figure 1a). Optical density (OD) of yeast in the bioreactor was measured as backscatter from cells with illumination provided at a peak wavelength of 605 nm by an LED and absorption determined over the visible band by a positive intrinsic negative (PIN) photodiode. Additionally, the optical array measured ambient temperature and pressure.

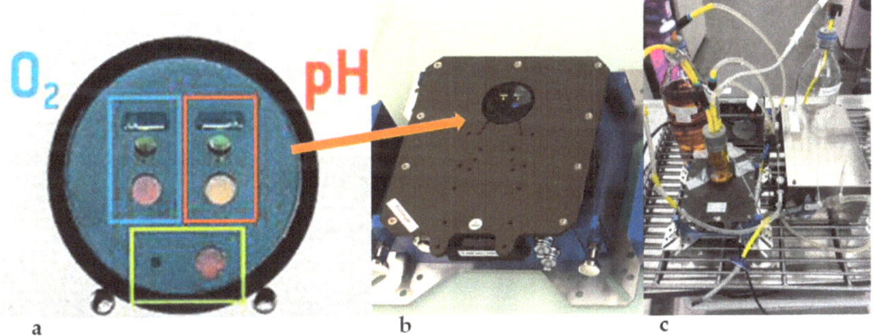

Figure 1. (**a**) the optical window of the sensor array showing the detectors for dissolved oxygen (blue rectangle, top left), pH (red rectangle, top right), and optical density (green rectangle, bottom); (**b**) the optical sensor array with the arrow pointing to the optical window; and (**c**) the complete bioreactor system with the bioreactor flask atop the optical array.

Pre- and post-calibrations of DO and pH were done using standard solutions. Biomass (OD) measured by the optical array was calibrated every 10–12 h from grab samples (2 mL) analyzed on the Eppendorf spectrometer (Eppendorf plus) at 650 nm. Data dropouts were not interpolated and missing points, displayed as zeros, were deleted from the time series. This resulted in gaps in the time series. Aberrant data points were not edited, deleted, or interpolated, resulting in rapid spikes and declines in some time series.

2.2. Culture Conditions

Cultures of wild type *Saccharomyces cerevisiae* (BY4742, MATα his3Δ1 leu2Δ0 lys2Δ0 ura3Δ0) were obtained from Dr. R. Willaert at the University of Brussels, Vrije. For steady-state experiments, cells were grown in 7.5 g L^{-1} glucose in either yeast extract-peptone dextrose (YPD) or synthetic complete (SC) media. Glucose and media were autoclaved separately and combined under sterile conditions.

YPD was made with 1% yeast extract (BD 212750), 2% peptone extract (BD 211677) and either 2% glucose for inocula or 0.75% for steady-state cultures. This is the same composition as in commercial YPD, sold as Gibco® (Thermo Fisher, Waltham, MA, U.S.A.) and NutriSelect®Basic (Sigma-Aldrich, St. Louis, MO, U.S.A.). The SC media was prepared with yeast nitrogen base (YNB) without amino acids, which were added separately.

Since the composition of YPD is seldom analyzed or reported, an example of its composition is provided in Table 1 to compare with the composition of the SC medium, keeping in mind that media can vary from batch to batch depending on the contents of each extract.

Table 1. Formula for the SC medium and an example of the composition of YPD.

Composition [1]	YPD [2] (mg L^{-1})	SC (mg L^{-1})	Composition [1]	YPD [2] (mg L^{-1})	SC (mg L^{-1})
$(NH_4)_2SO_4$	0	5000	Lysine	183	76
YNB	17,000	6700	Methionine	61	76
Amino acids	2611	1216	Phenylalanine	192	76
Alanine	257	76	Proline		
Arginine	292	76	Pyrrolysine [2]		
Aspartic acid			Serine	122	76
Asparagine	102	76	Selenocysteine		
Cysteine		76	Threonine	95.1	76
Glutamine	397	76	Tryptophan		
Glutamic acid			Tyrosine	102	76
Glycine	120	76	Valine	170	76
Histidine	42.5	76	**Nucleic acids**		
Isoleucine			Adenine	ND [3]	18
Leucine	212	176	Uracil	ND [3]	76

[1] Twenty amino acids from nine essential, six conditionally essential, and five non-essential amino acids. [2] Non-essential amino acids normally not in either medium. The YPD composition from [11]. [3] ND—No Data.

Inocula were prepared in YPD batch cultures of 100 mL with 20 g L^{-1} glucose, and cells were grown at 30 °C on a shaker (250 rpm). Within 24 h, the yeast culture was transferred to two, 50 mL centrifuge tubes and centrifuged for 10 min at 4000 rpm to concentrate the inoculum. YPD was decanted and the remaining cell concentrate was washed with either YPD or SC medium (7.5 g L^{-1} glucose), then vortexed for 20 s and centrifuged again. After three washings, the inoculum was prepared by mixing cells in 10 mL of medium, then measuring OD at 650 nm on an Eppendorfplus spectrometer (Eppendorf Co.). The dry weight of the inoculum was estimated based on a predetermined linear relationship (N = 102, r2 = 0.955):

$$\text{Biomass} = 0.1073 + (0.2225 \text{ mg/mL OD}) \times \text{OD}_{\text{measured}} \tag{1}$$

After inoculating the growth chamber, the volume was adjusted to 40 mL, mixed manually to homogenize the cell distribution, and the OD was measured again to determine the initial biomass, t_0, in the system.

2.3. Oxygenation Methods

Since the optical spots were mounted to the bottom of the bioreactor chamber, a stir bar would have interfered with data acquisition and therefore could not be used as a means for oxygen transfer. Consequently, other methods of aeration were tested. Seven experiments were conducted to determine the oxygenation rates of the YPD medium. The ambient conditions during these aeration experiments were relatively constant. Temperatures had a maximum change of 0.5 °C, while the maximum difference in barometric pressure was 19 mbars (Table 2). This resulted in a maximum change in DO_s of 0.2 mg L^{-1}.

Table 2. Operating conditions during the different experiments.

Experiment	Repeated	Medium	Temperature (°C)	Pressure (mbars)	DO_S [1] (mg L^{-1})
1 Bubbling	4	YPD	26.4 ± 1.1	973.8 ± 0.31	7.74
2 Impeller	3	YPD	26.4 ± 1.1	973.8 ± 0.31	7.74
3 H$_2$O$_2$	3	YPD	26.8 ± 0.18	973.8 ± 0.50	7.69
4 Flat Membrane	2 [2]	YPD	26.6 ± 0.10	965.0 ± 2.4	7.65
5 Liqui-Cel HFF	3	YPD	26.6 ± 0.10	965.0 ± 2.4	7.65
6 PDMS HFF	8	YPD	26.3 ± 0.65	957.2 ± 0.11	7.62
7 PDMS HFF	5	SC	26.8 ± 0.23	949.1 ± 0.35	7.54

[1] Dissolved oxygen in the bioreactor (DO_{BR}) and the saturated DO (DO_s) were taken at the end of the experiments. Values were compensated for temperature and pressure. [2] Repeated three times but one file was corrupted.

For experiment 1, the YPD medium was bubbled with sterile air using a 0.2 mm filter in-line with an air pump (Q = 0.2 L min^{-1}). In experiment 2, the bioreactor operated as a continuously stirred tank reactor (CSTR) mixed with an impellor motor (3 V, Motraxx LRE-280RAC-2865) at 13800 rpm to transfer oxygen from the headspace. In experiment 3, DO levels were chemically enhanced by adding 200 mg L^{-1} of hydrogen peroxide (PEROX-AID®) to the bioreactor and mixed by recirculating the bioreactor volume using a peristaltic pump (Q = 26.7 mL min^{-1}). In experiment 4, oxygenation and mixing were performed using a flat membrane filter (Evenflow, PTFE 44) in-line and downstream of the recirculating peristaltic pump. For experiment 5, a Liqui-Cel (G569.75x), hollow-fiber filter (HFF) with a volume of 2.5 mL was substituted for the flat plate membrane.

In experiment 6, yeast cultures were aerated with a polydimethylsiloxan hollow-fiber filter with a volume of 1.5 mL (PDMSXA-100 cm^2 by *Perm*Select Ann Arbor, MI, USA)— hereafter referred to as the PDMS, in-line with the recirculating peristaltic pump (Figure 2). Lastly, in experiment 7, the SC medium was oxygenated using the PDMS downstream of the recirculating pump in order to compare with oxygenation rates of the YPD medium.

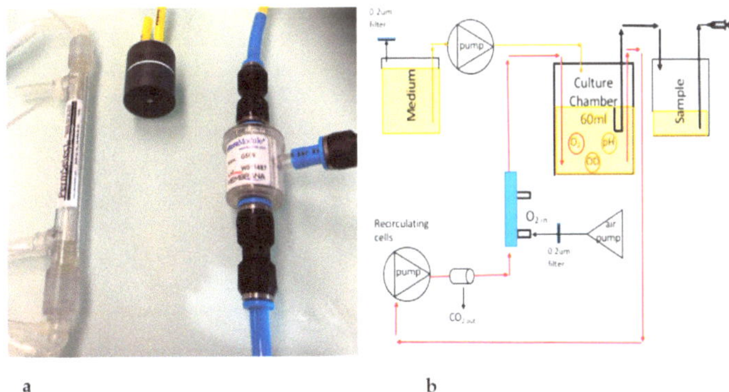

Figure 2. Three filter technologies. (**a**) From left to right: the PDMSXA-100 cm^2 (PDMS) filter, the Evenflow flat membrane filter, and Liqui-Cel filter; (**b**) schematic of the bioreactor system showing the filter location (in blue) downstream of the recirculating pump.

For experiments 4–7 (i.e., filters), the recirculation pump had a constant flow (26.7 mL min^{-1}) that continuously passed cells over the filter to provide aeration. The retention time in the growth chamber was 1.9 min, which was sufficient to keep yeast suspended without causing cell lysis. Initially, the air pump was connected to the filters and a CO$_2$ trap was placed between the filters and recirculation pump (Figure 2b), but these were omitted since they had no impact on gas transfer in the cultures.

To determine how fast the filters could replenish oxygen-free YPD and SC media, 10 mg mL^{-1} of Na_2SO_3 was prepared in stock media and 20 mL added to the sample chamber, after removing 20 mL of oxygenated medium [4]. As a control, filtered de-ionized water (FDIW) was aerated with the filters and DO was monitored without and with the addition of Na_2SO_3.

Generally, oxygen transfer rate to the medium, from gas to dissolved state, was quantified as the coefficient:

$$K_L a = \frac{dDO}{dt} \frac{1}{(DO_s - DO_{Br})} \tag{2}$$

where DO_s is the saturated oxygen concentration and DO_{Br} is the oxygen concentration in the bioreactor. For the impeller (treated as a CSTR), K_{La} was defined as:

$$K_L a = \frac{-ln\left(\frac{DO_s - DO_{BrN}}{DO_s - DO_{Br0}}\right)}{t_N - t_0} \tag{3}$$

where the subscripts 0 and N are for the beginning and end time of the measurement, respectively. For the membrane and hollow-fiber filters, the K_{La} was calculated as:

$$K_L a = \frac{QDO_{Br}}{A \, ln\left(1 - \frac{DO_{Br}}{DO_S}\right)} \tag{4}$$

where Q is the fluid flow rate through the filters in mL min^{-1}. The PDMS filter volume was 2.1 mL and its lumen surface area was 100 cm^2. For all K_{La} measurements, the mass transfer rate has units of min^{-1}.

Oxygen utilization rate (OUR) by the yeast was determined as:

$$\frac{dDO}{dt} = K_L a(DO_s - DO_{Br}) - q_{DO} * X_{biomass} \tag{5}$$

$$\frac{dDO}{dt} = K_L a(DO_s - DO_{Br}) - OUR \tag{6}$$

Such that $OUR = q_{DO}(X_{biomass})$ where q_{DO} is the specific respiration rate per unit biomass and $X_{biomass}$ is the biomass concentration.

Solving for OUR gives,

$$OUR = K_L a(DO_S - DO_{Br}) - \frac{dDO_{Br}}{dt} \tag{7}$$

For steady-state conditions $d/dt = 0$ and Equation (7) reduces to

$$OUR = K_L a(DO_S - DO_{Br}) \tag{8}$$

For all experiments, DO_{Br} and DO_S were corrected for changes in temperature and atmospheric pressure (Table 2).

Analysis of variance (ANOVA) was conducted on data from the same time series and separate time series using SPSS v27 and a Tukey test was used to discriminate differences between means. Statistical tests were run for unequal variance with alpha values set to 0.01. Mean values, plus and minus one standard deviation, are used throughout the paper and in all tables where the sample size is designated by N.

3. Results

3.1. The Bubbling Method

Filtered air was bubbled in the YPD medium of the culture chamber using an air stone (i.e., sparge) connected to an air pump (Figure 3a). However, bubbles adhered to the DO and pH spots interfering with the optical measurements (Figure 3a).

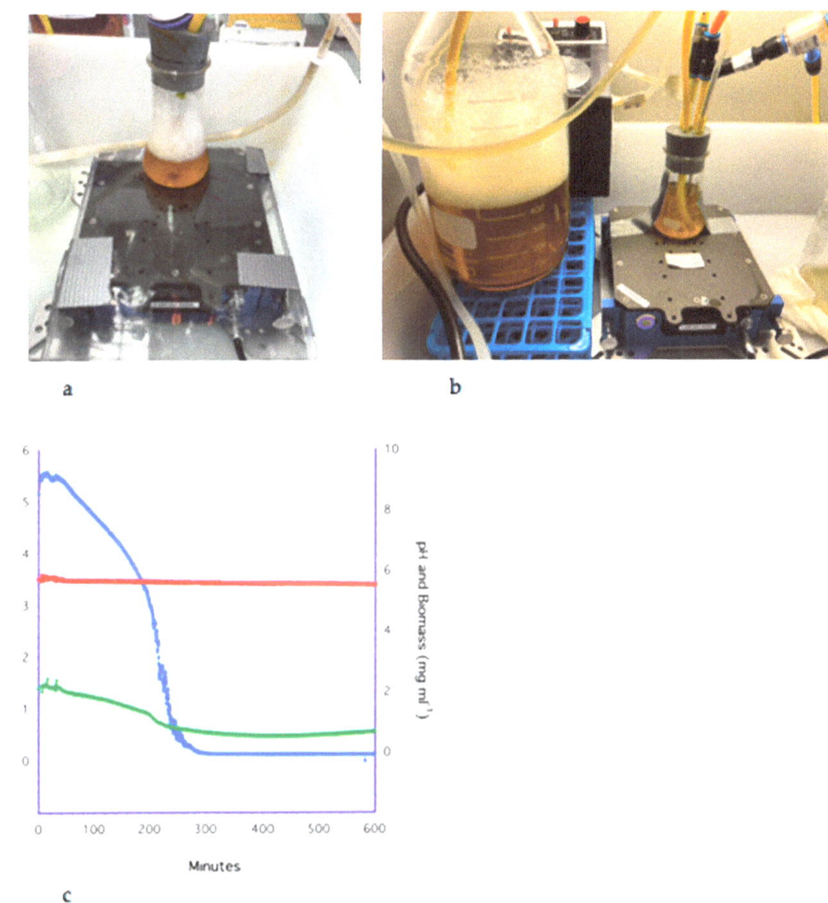

Figure 3. Bubbled YPD medium in: (**a**) the culture chamber; (**b**) the medium reservoir to the left of the culture chamber; and (**c**) time series of aerated medium delivered at D = 0.2 h^{-1}. Colors are blue for DO, red for pH, and green for biomass.

As an alternative, filtered air was bubbled into the medium reservoir (Figure 3b) and pumped into the growth chamber at the dilution rate of 0.2 hr^{-1} (Q = 10 mL hr^{-1}). The average rate of decrease in dissolved oxygen was -0.23 ± 0.03 mg L^{-1} min^{-1}. The decrease in DO indicated that oxygen was consumed by the yeast (i.e., respired) at a greater rate than the medium could replenish it. This is illustrated in Figure 3c.

3.2. The Impeller Method

The impeller method (Figure 4a) was very effective for transferring oxygen over the air-medium interface. The average rate of aeration was 0.15 ± 0.16 mg L^{-1} min^{-1} and DO remained high for all replicated experiments. A typical time series is shown in Figure 4b, where DO in the YPD culture was initially 6.8 mg L^{-1} and biomass was 2.8 mg mL^{-1} (the pH sensor failed). After diluting the culture with low DO medium (0.5 mg mL^{-1}

oxygen), DO was reduced from 6.8 to 3.8 mg L^{-1}, however, mixing rapidly increased DO to 5.7 mg L^{-1} within 10 min, and equilibrated at approximately 6 mg L^{-1}.

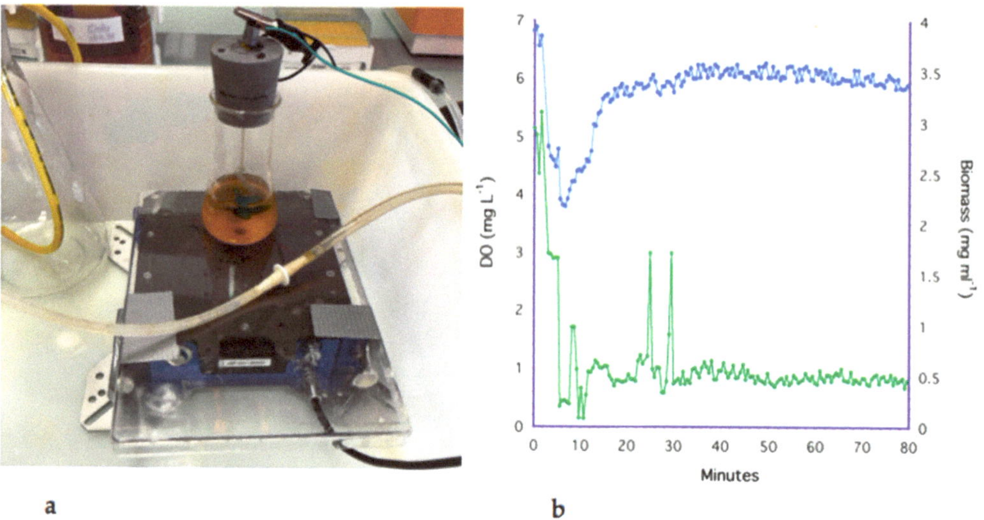

a

b

Figure 4. (**a**) Set-up of stirred bioreactor; and (**b**) a time series of the bioreactor mixed with an impeller and YPD medium. Colors are blue for DO and green for biomass.

3.3. The H$_2$O$_2$ Method

YPD medium was chemically oxygenated by adding hydrogen peroxide, which gave a mean aeration rate of 0.40 ± 0.15 mg L^{-1} min^{-1}. However, for all replicated experiments, dissolved oxygen decreased over time to anoxic levels. This is illustrated by the time series in Figure 5 where the anoxic medium increased from DO < 0.5 mg L^{-1} to 12 mg L^{-1} within 5 min but was reduced to 1.1 mg L^{-1} in less than 30 min. Coincident with the addition of H$_2$O$_2$, biomass deceased from 4 mg mL^{-1} to a steady-state value of 1.2 mg L^{-1}, which may have been a response of cells to the strong oxidizing potential of peroxide (O$_2^{2-}$), which is toxic for the yeast.

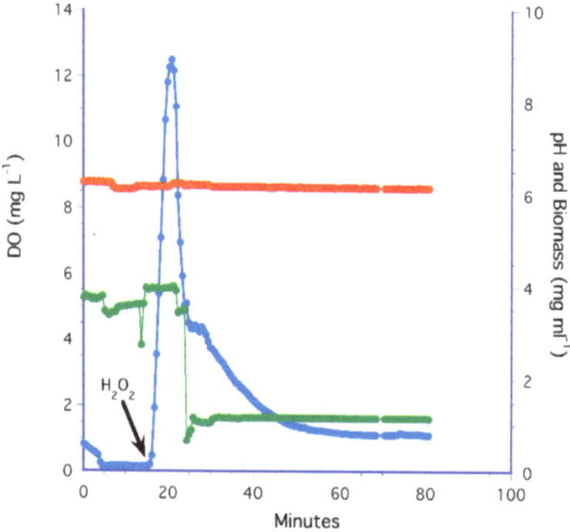

Figure 5. Hydrogen peroxide addition to YPD medium in the bioreactor (arrow). Colors are blue for DO, red for pH, and green for biomass.

3.4. Flat Membrane Filter Method

For the flat plate filter, low DO medium was inoculated with yeast and circulated over the membrane using the recirculating pump. The average aeration rate was -0.21 ± 0.35 mg L^{-1} min^{-1}, indicating that the membrane could aerate the cultures. Figure 6 shows that DO was unable to increase in concentration and remained low even when the biomass was diluted from 8.9 to 2.3 mg mL^{-1} with low DO medium. After shaking the flask (at t = 77 min) to mechanically aerate the bioreactor, the DO increased slightly to 2 mg L^{-1} but rapidly decreased again.

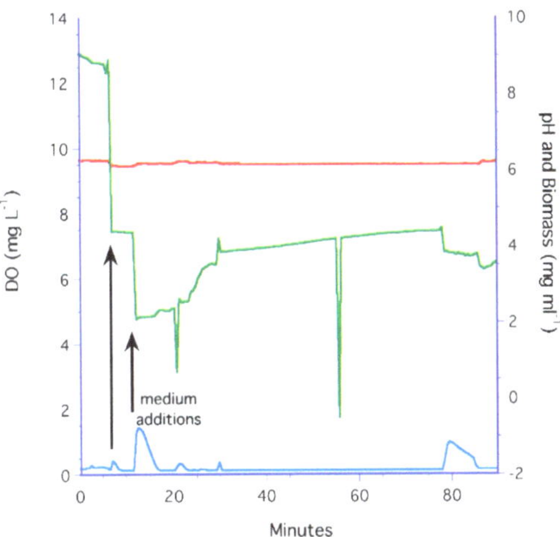

Figure 6. Oxygenation of the bioreactor with the flat membrane and YPD medium. Colors are blue for DO, red for pH, and green for biomass.

3.5. The LiquiCel Hollow-Fiber Filter (HFF) Method

The Liqui-Cel HFF had a negative aeration rate with an average of -0.26 ± 0.13 mg L^{-1} min^{-1}. A typical run is shown in Figure 7.

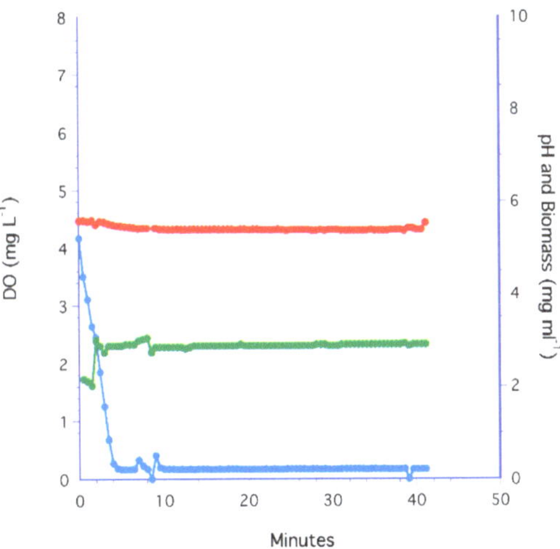

Figure 7. Liqui-Cel HFF and YPD medium. Colors are blue for DO, red for pH, and green for biomass.

The DO decreased from 4.1 mg L^{-1} to a nearly constant level of 0.18 mg L^{-1} as a result of yeast respiration being higher than the aeration rate. The pH was fairly constant between and 5.5 and 5.3. The biomass initially increased from 2.1 to 2.8 mg mL^{-1} as a result of initial cell mixing, but thereafter remained constant.

3.6. The PDMS Hollow-Fiber Filter (HFF) Method

Aeration of yeast cultures with the PDMS filter and YPD was not effective over the long term. Overall, DO declined at an average rate of -0.14 ± 0.29 mg L^{-1} min^{-1}. For example, in Figure 8a DO was initially high at 6.8 mg L^{-1} but within 240 min, DO decreased to a low of 0.18 mg L^{-1} as a result of yeast respiration.

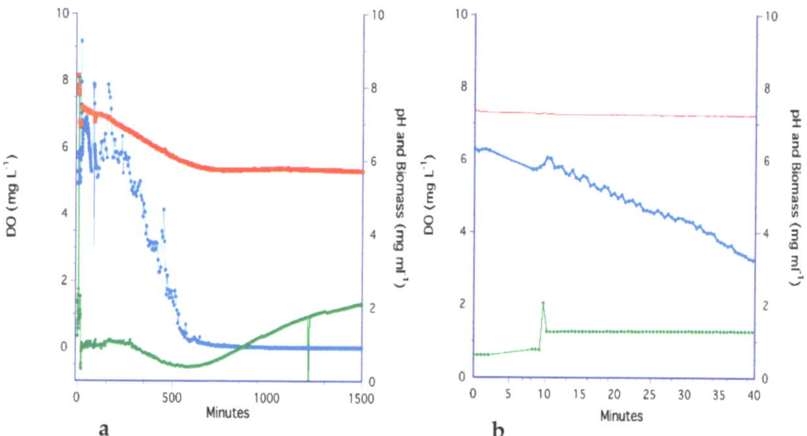

Figure 8. PDMS hollow-fiber filter with YPD medium for: (**a**) longer term experiment and (**b**) after cleaning the PDMS filter. Colors are blue for DO, red for pH, and green for biomass.

Biomass was initially steady at 1 mg mL^{-1} when DO was high, but as oxygen was depleted, biomass fell to a minimum of 0.35 mg mL^{-1} over a 10 h period. Biomass increased again during anoxic conditions to a high of 2.1 mg mL^{-1}. The pH decreased linearly from 7.4 to 5.8 as respired CO$_2$ increased. Thereafter, pH was nearly constant at 5.8.

The experiment was repeated after cleaning the PDMS by flushing the 100 mL of 70% ethanol, rinsing with 100 mL of filtered de-ionized water (FDIW), and re-inoculating the YPD. However, even after cleaning the filter, the DO in the culture progressively decreased Figure 8b).

Using the same PDMS filter (after cleaning), the experiment was repeated using the SC medium. In all runs with SC medium, the PDMS filter kept the yeast cultures aerated at an average rate of 0.26 ± 0.13 mg L^{-1} min^{-1}. The final time series in Figure 9a illustrates how DO recovered each time Na$_2$SO$_3$ was added. Initially, DO was high and steady at 7.6 mg L^{-1}. After the first addition of Na$_2$SO$_3$, at $t = 16$ min, the DO in the bioreactor was reduced to 0.33 mg L^{-1} within 14 min, at a rate of -0.24 mg L^{-1} min^{-1}. After a 2-fold dilution of Na$_2$SO$_3$ solution, the DO rose to 1.2 mg L^{-1} before the next addition at $t = 116$ min as Na$_2$SO$_3$. Despite the data dropouts in the time series between minutes 229 and 334, there was an increase in the DO concentration to 4.2 mg L^{-1} over this time period. This increase in DO was not an artifact as demonstrated by: (1) the rapid increase in DO from 3.4 to 6.6 mg L^{-1} after the addition of a small volume of Na$_2$SO$_3$ at 366 min; and (2) a longer time series (2.1 days) where both DO and pH increased slightly and biomass remained constant (Figure 9b). For this longer time series, DO was near 4.1 mg L^{-1}, which was sufficient to maintain aerobic cell respiration.

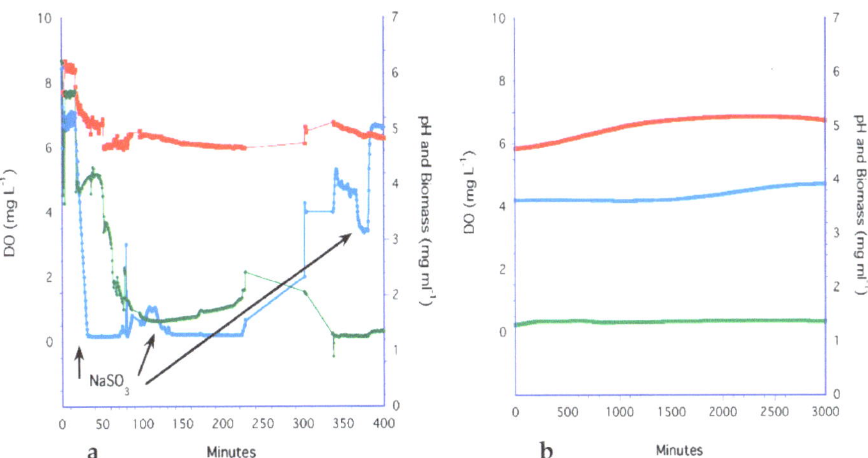

Figure 9. (**a**) PDMS hollow-fiber filter using SC medium with additions of Na_2SO_3; (**b**) longer term, time series of the steady-state culture using the PDMS bioreactor with SC but no Na_2SO_3 additions. Colors are blue for DO, red for pH, and green for biomass.

3.7. Analysis of Methods

Of the seven aeration methods, four could not effectively aerate cultures. Negative trends in aeration (i.e., DO over time = dDO/dt) occurred in the YPD medium for bubbling, flat membrane, Liqui-Cel, and PDMS methods, indicating that these methods could not replenish DO at rates equal to or greater than respiration rates of the yeast cultures.

The three methods that did provide sufficient aeration for cultures were the PDMS filter using SC medium and the impeller, and H_2O_2, both using YPD medium. While the H_2O_2 trend was only transient, both the impeller/YPD and the PDMS/SC methods did reach a steady state, however, DO trends were not statistically different (N = 3, $|t| = 2.01$, $p = 0.098$).

To determine the effectiveness of the PDMS filter to oxygenate the two media, several time series were collected using only the YPD and SC media without yeast. In an anoxic YPD medium, DO never recovered (for example in Figure 10a where DO remained low and constant below 0.22 mg L^{-1}). Even after cleaning the filter and repeating the experiment with oxygenated YPD the maximum DO of 5 mg L^{-1} progressively decreased despite continued aeration through the PDMS filter (Figure 10b). This negative aeration rate represents a negligible oxygen transfer rate, K_La.

To determine the K_La for the SC medium, Na_2SO_3 was added periodically and the increase in DO was monitored. The result was an average aeration rate of 0.34 ± 0.13 mg L^{-1} min^{-1}. As a control, FDIW was substituted for SC resulting in an average aeration rate of 0.37 ± 0.12 mg L^{-1} min^{-1}. Both these processes are illustrated in Figure 11. DO of the SC medium was initially high and constant at 7.6 mg L^{-1} which decreased to 7.1 mg L^{-1} after the first addition of Na_2SO_3 (Figure 11a). A second addition reduced DO to 1.2 mg L^{-1} at a rate of -0.05 mg L^{-1} min^{-1}, but the medium rapidly reached a DO of 7.6 mg L^{-1} at a rate of 0.69 mg L^{-1} min^{-1}. The same trend occurred using FDIW, where the DO was initially high and constant at approximately 8 mg L^{-1} (Figure 11b). After Na_2SO_3 was added to the FDIW, DO decreased to <1 mg L^{-1} but was soon re-aerated by the filter.

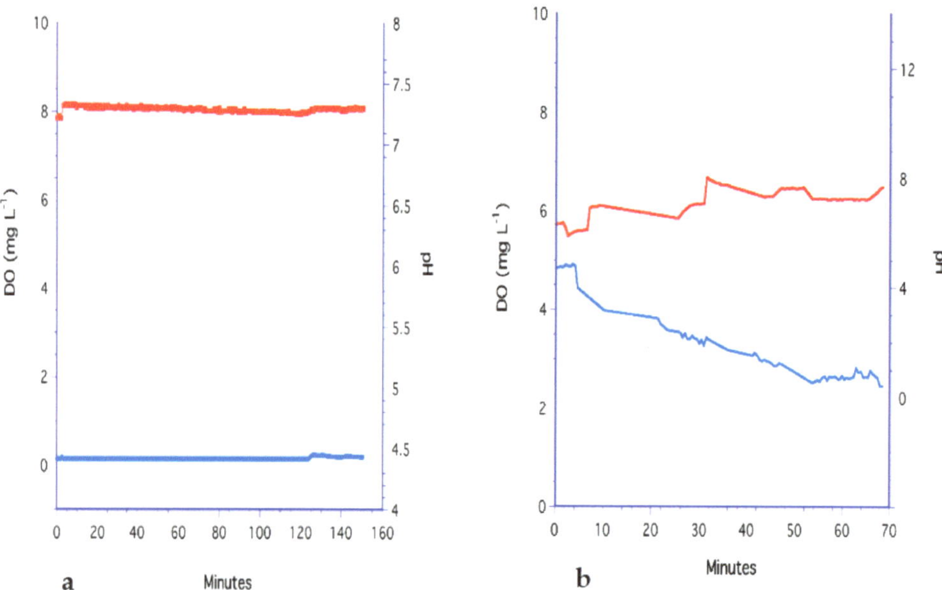

Figure 10. (**a**) YPD aeration with the PDMS hollow-fiber filter; (**b**) PDMS washed with 100 mL of FDIW to remove YPD.

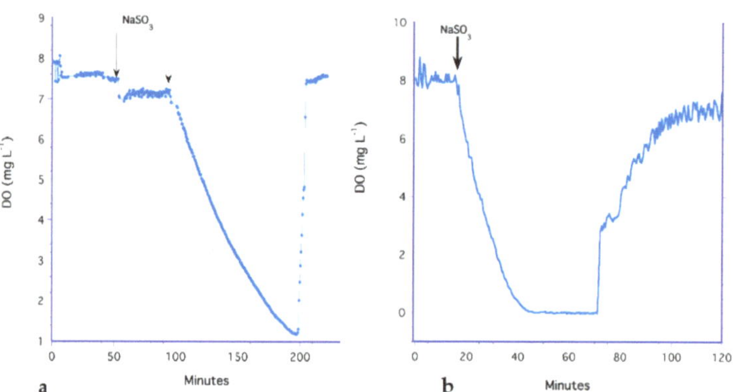

Figure 11. Oxygenation using the PDMS hollow-fiber filter with (**a**) SC medium without yeast and (**b**) filtered de-ionized water.

To determine if the mean $K_L a$ values for the PDMS filter with SC and FDIW (i.e., no yeast) are representative of other bioreactors, they were compared to literature values of other aeration methods (Table 3). An ANOVA of the $K_L a$ data indicate the methods were not statistically different (df = 33, F = 0.525, p = 0.597). Additionally, when YPD was used with the PDMS filter, the aeration rate was near zero, and YPD without glucose gave the same result (Table 3).

Table 3. $K_L a$ from experiments and the literature.

Aeration Process	Medium	$K_L a$ (min^{-1})	N [1]	References
Bubbled	Broth	0.30−2.4	4	[12,13]
Bubbled	DIW	1.5−3.4	2	[14]
Impeller (CSTR)	Broth	0.09−1.7	4	[2,15]
Impeller-Sparged	Broth	0.30−2.4	15	[16–30]
Impeller-Sparged	DIW	0.24−5.5	8	[5,12,23,28]
Membranes	WW [2], Broth	0.13−8.1	3	[9,31,32]
Various HFF	DIW	0.03−2.7	8	[9,10,33]
PDMS 2500	DIW	0.33−7.8	2	[10]
PDMS 100 [1]	FDIW	1.3−1.7	5	This study
PDMS 100 [1]	SC	1.1−1.5	4	This study
PDMS 100 [1]	YPD	≤ -0.09	4	This study
PDMS 100 [1]	YPD (no glucose)	≤ -0.02	3	This study

[1] N—sample size. [2] WW—wastewater.

Few papers report the oxygen uptake rate (OUR) of yeast in continuous cultures and those that do seldom provide $K_L a$ values for their aeration method. However, some values for *S. cerevisiae* are shown in Table 4 along with our experimental results. Statistically, there is a difference in the mean OUR values (df = 24, F = 4.93, p = 0.006), where PDMS/SC was greater than literature values, most of which were for sparged-impellers.

Table 4. Oxygen uptake rates of yeast [1].

Method	Biomass (mg mL^{-1})	$K_L a$ (min^{-1})	OUR [2] (mg O$_2$ L^{-1} min^{-1})	q_{O2} (mg O$_2$ gDwt^{-1} min^{-1})	N [3]
H$_2$O$_2$/YPD	2.3−3.5	2.3 ± 1.2	2.3 ± 0.29	3.7 ± 5.7	3
Impeller/YPD	0.3−2.67	0.87 ± 0.02	6.0 ± 3.1	2.3 ± 1.2	3
PDMS/YPD	0.6−1.2	negligible	−0.18 ± 0.32	0.06 ± 0.09	4
PDMS/ SC	1.4−38	1.3 ± 0.17	5.6 ± 4.3	3.0 ± 4.1	4
Sparged-Impeller [20,23–27,29,30,34]	1.4−50	1.3 ± 0.17	0.88 ± 1.0	4.8 ± 5.7	13

[1] All yeast are *S. cevevisiae* except for one, *Schizosaccharomyces pomb* [25]. [2] For negligible $K_L a$, OUR was calculated based on dDO/dt. OUR is reported as positive uptake. [3] N is the sample size.

To resolve if this difference was a result of different biomass concentrations, q_{O2} means were calculated by standardizing OUR by biomass for those data sets where both OUR and biomass were available. Interestingly, none of the q_{O2} means were statistically different (df = 24, F = 0.714, p = 0.558), indicating that biomass could not account for the high oxygen uptake rates of the PDMS/SC method.

4. Discussion

The $K_L a$ values for the steady-state impeller/YPD and the PDMS SC methods were in the same range as most bioreactor aeration processes using yeast media and de-ionizer water. The most surprising result was the negligible $K_L a$ coefficients for PDMS/YPD method. Some studies have suggested that the viscosity of the broth plays the dominant role in reducing gas transfer to cultures as a result of substances in the solution, for example, sugars, especially in high concentrations [4,19]. Yet glucose concentrations were the same in the YPD and SC media (7.5 g L^{-1}) for all the experiments. However, additional experiments using the PDMS filter and YPD with and without glucose confirmed that the YPD medium, and not the glucose, inhibited oxygen transfer in the hollow-fiber filter. The main differences of the YPD recipe from SC were the concentrations of the yeast nitrogen base and various amino acids (Table 1). Thus, one possibility is that amino acids in the medium could have clogged filter surfaces. For example, arginine, isoleucine, and phenylalanine were determined to be "sticky" in comparison to lysine and glutamic acid [35]. While it is conceivable that amino acid concentrations may have accounted for the difference in K_{La} values between the two media, it is also possible that other constituents in the exacts contributed to this effect.

It is known that K_La can be enhanced with higher mixing rates [5,12] as demonstrated by the impeller experiment. However, increased mixing increases shear stress on cells and can potentially damage them [36]. The highest K_La for a bioreactor, which was nearly 10 times higher than the average values in Table 3, was achieved with a PDMS filter at a high flow rate. However, increased flow through these filters also equates to increased shear stress on cells. The low flow rate (i.e., recirculation rate) in our PDMS/SC experiments not only achieved adequate aeration rates but kept cells growing at a constant specific growth rate of $0.2 \ \text{h}^{-1}$.

The measurement of DO also affects the calculation of K_La. Aroniada et al. [5] caution that the use of membrane DO probes have slow response times at high DO levels which increase when DO tends to zero. The use of the optical array to measure DO eliminated this variability, as evidenced by the immediate response of the DO spot to the addition of Na_2SO_3 to the media and FDIW. However, air bubbling in the bioreactor chamber is not compatible with this non-invasive sensor array since bubbles often adhered to the optical spots.

The oxygen utilization rate (OUR) using the PDMS filter differed depending on the medium in the bioreactor, which was also found for other filters [7]. For the continuous cultures, the lowest oxygen uptake rate (OUR) was in the YPD medium coincident with the lowest biomass. OUR progressively increased as biomass increased (not shown). The fact q_{O2} levels were not different indicates that biomass cannot account for the higher OUR values of the PDMS/SC method, signifying that the PDMS/SC method is superior to sparged-impeller aeration in maintaining highly oxygenated cultures.

5. Conclusions

Overall, the PDMS hollow-fiber filter in combination with the SC medium maintained near saturating DO levels for most cultures. This was essential since at low dilution rates the volume of the aerated medium being fed to the bioreactor was inadequate to supply cells with sufficient oxygen. Overall, the PDMS/SC method is easy to use, provides sufficient aeration of the yeast culture at lower mixing and dilution rates for moderate biomass concentrations. Combined with the novel PreSens sensor array, the system provides non-invasive, continuous sampling of aerobic yeast respiration in small sample volumes and thus may be suitable for space applications.

Author Contributions: Conceptualization, T.G., C.F. and B.R.; methodology, T.G. and C.F., formal analysis, T.G.; B.R.; data curation, C.F. and B.R.; writing—original draft preparation, T.G.; writing—review and editing, T.G., C.F., C.B. and B.R.; supervision, T.G. and B.R.; project administration, T.G.; funding acquisition, B.R. All authors have read and agreed to the published version of the manuscript.

Funding: This research was funded by the European Space Agency's PRODEX program under the guidance of Martin Haag.

Institutional Review Board Statement: Not applicable.

Informed Consent Statement: Not applicable.

Data Availability Statement: Data are available from the PRODEX program via the corresponding author.

Acknowledgments: We thank Marcel Egli, for securing the original project for the yeast bioreactor. Finally, our thanks to the three reviews whose critical comments and suggestions helped to better this paper.

Conflicts of Interest: The authors declare no conflict of interest. It should be noted that the authors have collaborated on yeast production with Ronnie Willaert.

References

1. Montes, F.J.; Catalán, J.; Galán, M.A. Prediction of kLa in yeast broths. *Process. Biochem.* **1999**, *34*, 549–555. [CrossRef]
2. Emery, A.N.; Jan, D.C.-H.; Al-Rubeai, M. Oxygenation of intensive cell-culture. *Appl. Microbiol. Biotechnol.* **1995**, *43*, 1028–1033. [CrossRef] [PubMed]
3. Walther, I. Space bioreactors and their applications. *Adv. Space Biol. Med.* **2002**, *8*, 197–213. [CrossRef]
4. Garcia-Ochoa, F.; Gomez, E. Bioreactor scale-up and oxygen transfer rate in microbial processes: An overview. *Biotechnol. Adv.* **2009**, *27*, 153–176. [CrossRef]
5. Aroniada, M.; Maina, S.; Koutinas, A.; Kookos, I.K. Estimation of volumetric mass transfer coefficient (kLa)—Review of classical approaches and contribution of a novel methodology. *Biochem. Eng. J.* **2020**, *155*, 107458. [CrossRef]
6. I Betts, J.; Baganz, F. Miniature bioreactors: Current practices and future opportunities. *Microb. Cell Factories* **2006**, *5*, 21. [CrossRef]
7. Cruz, A.J.G.; Silva, A.S.; Araujo, M.L.G.C.; Giordano, R.C.; Hokka, C.O. Estimation of the volumetric oxygen transfer coefficient (KLa) from the gas balance and using a neural network technique. *Braz. J. Chem. Eng.* **1999**, *16*, 179–183. [CrossRef]
8. Badino, A.; Facciotti, M.; Schmidell, W. Volumetric oxygen transfer coefficients (kLa) in batch cultivations involving non-Newtonian broths. *Biochem. Eng. J.* **2001**, *8*, 111–119. [CrossRef]
9. Marrot, B.; Barrios-Martinez, A.; Moulin, P.; Roche, N. Experimental Study of Mass Transfer Phenomena in a Cross Flow Membrane Bioreactor: Aeration and Membrane Separation. *Eng. Life Sci.* **2005**, *5*, 409–414. [CrossRef]
10. Orgill, J.J.; Atiyeh, H.; Devarapalli, M.; Phillips, J.R.; Lewis, R.S.; Huhnke, R. A comparison of mass transfer coefficients between trickle-bed, hollow fiber membrane and stirred tank reactors. *Bioresour. Technol.* **2013**, *133*, 340–346. [CrossRef]
11. Martínez-Force, E.; Benítez, T. Effects of varying media, temperature, and growth rates on the intracellular concentrations of yeast amino acids. *Biotechnol. Prog.* **2005**, *11*, 386–392. [CrossRef]
12. García-Ochoa, F.; Santos, V.E.; Casas, J.A.; Gomez, E. Xanthan gum: Production, recovery, and properties. *Biotechnol. Adv.* **2000**, *18*, 549–579. [CrossRef]
13. Russell, A.B.; Thomas, C.R.; Lilly, M.D. Oxygen transfer measurements during yeast fermentations in a pilot scale airlift fermenter. *Bioprocess Eng.* **1995**, *12*, 71–79. [CrossRef]
14. Zedníková, M.; Orvalho, S.; Fialová, M.; Ruzicka, M.C. Measurement of Volumetric Mass Transfer Coefficient in Bubble Columns. *ChemEngineering* **2018**, *2*, 19. [CrossRef]
15. Kim, H.J.; Jong, H.K.; Hyuck, J.O.; Chul, S.S. Morphological control of Monascus cells and scale-up of pigment fermentation. *Process Biochem.* **2002**, *38*, 649–655. [CrossRef]
16. Schwippel, J.; Votruba, J. Oxygen and/or glucose limitation in a chemostat culture of *Candida utilis* mathematical model identification. *Bioprocess Eng.* **1995**, *13*, 133–140. [CrossRef]
17. Rieger, M.; Käppeli, O.; Fiechter, A. The Role of Limited Respiration In The Incomplete Oxidation Of Glucose By *Saccharomyces Cerevisiae*. *J. Gen. Microbiol.* **1983**, *129*, 653–661. [CrossRef]
18. Banti, D.C.; Tsali, A.; Mitrakas, M.; Samaras, P. The Dissolved Oxygen Effect on the Controlled Growth of Filamentous Microorganisms in Membrane Bioreactors. *Environ. Sci. Proc.* **2020**, *2*, 39. [CrossRef]
19. Al-Mayah, A.M.R.; Muallah, S.K.; Al-Jabbar, A.A. Prediction of Oxygen Mass Transfer Coefficients in Stirred Bioreactor with Rushton Turbine Impeller for Simulated (Non-Microbial) Medias. *Al-Khwarizmi Eng. J.* **2014**, *10*, 1–14.
20. Raghavendran, V.; Webb, J.P.; Cartron, M.L.; Springthorpe, V.; Larson, T.R.; Hines, M.; Mohammed, H.; Zimmerman, W.; Poole, R.K.; Green, J. A microbubble-sparged yeast propagation–fermentation process for bioethanol production. *Biotechnol. Biofuels* **2020**, *13*, 104. [CrossRef]
21. Terasaka, K.; Shibata, H. Oxygen transfer in viscous non-Newtonian liquids having yield stress in bubble columns. *Chem. Eng. Sci.* **2003**, *58*, 5331–5337. [CrossRef]
22. Caşcaval, D.; Galaction, A.-I.; Turnea, M. Comparative analysis of oxygen transfer rate distribution in stirred bioreactor for simulated and real fermentation broths. *J. Ind. Microbiol. Biotechnol.* **2011**, *38*, 1449–1466. [CrossRef] [PubMed]
23. Chisti, Y.; Jauregui-Haza, U.J. Oxygen transfer and mixing in mechanically agitated airlift bioreactors. *Biochem. Eng. J.* **2002**, *10*, 143–153. [CrossRef]
24. Galaction, A.-I.; Cascaval, D.; Oniscu, C.; Turnea, M. Prediction of oxygen mass transfer coefficients in stirred bioreactors for bacteria, yeasts and fungus broths. *Biochem. Eng. J.* **2004**, *20*, 85–94. [CrossRef]
25. Klein, T.; Schneider, K.; Heinzle, E. A system of miniaturized stirred bioreactors for parallel continuous cultivation of yeast with online measurement of dissolved oxygen and off-gas. *Biotechnol. Bioeng.* **2013**, *110*, 535–542. [CrossRef]
26. Jouhten, P.; Rintala, E.; Huuskonen, A.; Tamminen, A.; Toivari, M.; Wiebe, M.; Ruohonen, L.; Penttilä, M.; Maaheimo, H. Oxygen dependence of metabolic fluxes and energy generation of Saccharomyces cerevisiae CEN.PK113-1A. *BMC Syst. Biol.* **2008**, *2*, 60. [CrossRef]
27. Lee, J.H.; Lim, Y.B.; Park, K.M.; Lee, S.W.; Baig, S.Y.; Shin, H.T. Factors Affecting Oxygen Uptake by Yeast Issatchenkia orientalis as Microbial Feed Additive for Ruminants. *Asian Australas. J. Anim. Sci.* **2003**, *16*, 1011–1014. [CrossRef]
28. Özbek, B.; Gayik, S. The studies on the oxygen mass transfer coefficient in a bioreactor. *Process. Biochem.* **2001**, *36*, 729–741. [CrossRef]

29. Peddie, F.L.; Simpson, W.J.; Kara, B.V.; Robertson, S.C.; Hammond, J.R.M. Measurement of Endogenous Oxygen Uptake Rates of Brewers' Yeasts. *J. Inst. Brew.* **1991**, *97*, 21–25. [CrossRef]
30. Ju, L.-K.; Sundararajan, A. The effects of cells on oxygen transfer in bioreactors. *Bioprocess Eng.* **1995**, *13*, 271–278. [CrossRef]
31. Yildiz, E.; Keskinler, B.; Pekdemir, T.; Akay, G.; Nuhoglu, A. High strength wastewater treatment in a jet loop membrane bioreactor: Kinetics and performance evaluation. *Chem. Eng. Sci.* **2005**, *60*, 1103–1116. [CrossRef]
32. Millward, H.; Bellhouse, B.; Sobey, I. The vortex wave membrane bioreactor: Hydrodynamics and mass transfer. *Chem. Eng. J.* **1996**, *62*, 175–181. [CrossRef]
33. Ahmed, T.; Semmens, M.J.; Voss, M.A. Oxygen transfer characteristics of hollow-fiber, composite membranes. *Adv. Environ. Res.* **2004**, *8*, 637–646. [CrossRef]
34. Kuriyama, H.; Kobayashi, H. Effects of oxygen supply on yeast growth and metabolism in continuous fermentation. *J. Ferment. Bioeng.* **1993**, *75*, 364–367. [CrossRef]
35. Levy, E.D.; De, S.; Teichmann, S. Cellular crowding imposes global constraints on the chemistry and evolution of proteomes. *Proc. Natl. Acad. Sci. USA* **2012**, *109*, 20461–20466. [CrossRef]
36. Stafford, R.A. Yeast Physical (Shear) Stress: The Engineering Perspective. In *Brewing Yeast Fermentation Performance*, 2nd ed.; Smart, K., Ed.; Wiley: New York, NY, USA, 2003.

 *fermentation*

MDPI

Article

The Influence of Yeast Strain on Whisky New Make Spirit Aroma

Christopher Waymark and Annie E. Hill *

The International Centre for Brewing and Distilling, Heriot-Watt University, Edinburgh EH14 4AS, UK; chriswaymark@yahoo.co.uk
* Correspondence: A.Hill@hw.ac.uk; Tel.: +44-(0)131-451-3458

Abstract: Flavour in Scotch malt whisky is a key differentiating factor for consumers and producers alike. Yeast (commonly *Saccharomyces cerevisiae*) metabolites produce a significant amount of this flavour as part of distillery fermentations, as well as ethanol and carbon dioxide. Whilst yeast strains contribute flavour, there is limited information on the relationship between yeast strain and observed flavour profile. In this work, the impact of yeast strain on the aroma profile of new make spirit (freshly distilled, unmatured spirit) was investigated using 24 commercially available active dried yeast strains. The contribution of alcoholic, fruity, sulfury and sweet notes to new make spirit by yeast was confirmed. Generally, distilling strains could be distinguished from brewing and wine strains based on aroma and ester concentrations. However, no statistically significant differences between individual yeast strains could be perceived in the intensity of seven aroma categories typically associated with whisky. Overall, from the yeast strains assessed, it was found that new make spirit produced using yeast strains marketed as 'brewing' strains was preferred in terms of acceptability rating.

Keywords: yeast; distillery fermentation; spirit flavour; whisky; new make spirit; aroma

Citation: Waymark, C.; Hill, A.E. The Influence of Yeast Strain on Whisky New Make Spirit Aroma. *Fermentation* **2021**, *7*, 311. https://doi.org/10.3390/fermentation7040311

Academic Editor: Ronnie G. Willaert

Received: 14 October 2021
Accepted: 16 November 2021
Published: 14 December 2021

Publisher's Note: MDPI stays neutral with regard to jurisdictional claims in published maps and institutional affiliations.

Copyright: © 2021 by the authors. Licensee MDPI, Basel, Switzerland. This article is an open access article distributed under the terms and conditions of the Creative Commons Attribution (CC BY) license (https://creativecommons.org/licenses/by/4.0/).

1. Introduction

The production of Scotch malt whisky is both beautifully simple and extraordinarily complex. Simplicity lies in the use of only three ingredients and the same basic processing features. However, no two production sites or operational parameters are alike and the myriad choices, from barley variety and malting specification, fermentation regime, still design and operation, through to cask selection and maturation time, lead to immense product diversity. The final flavour and aroma of whisky is due to the interaction of more than 1000 flavour-active compounds termed congeners. The major congener classes that were identified in Scotch whisky are higher alcohols, esters, acids, phenols, lignin-degradation products, and lactones. Studies of processed raw materials, fermentation products, spirit distillate and matured whisky indicate the likely production stage (s) of flavour congeners (Figure 1) and highlight the role of yeast in the production of floral, fruity, and sweet attributes [1].

It is noted that some congener compounds could be generated in multiple stages of the production process. Furfurals and pyrazines from Maillard reactions that provide biscuity, chocolatey or marzipan flavours could arise from conditions in mashing, distillation and cask preparation as well as kilning [2]. Esters are primarily yeast metabolites but are also produced, albeit more slowly, as a result of condensation reactions in the cask [3], or, to a lesser extent, in the still [4]. Phenols are characteristic of peated malt, but also emerge in fermentation [5] and as oak decomposition products in the cask [3]. Lactones are principally formed as oak extractives [3] but were also identified in fermentation [6].

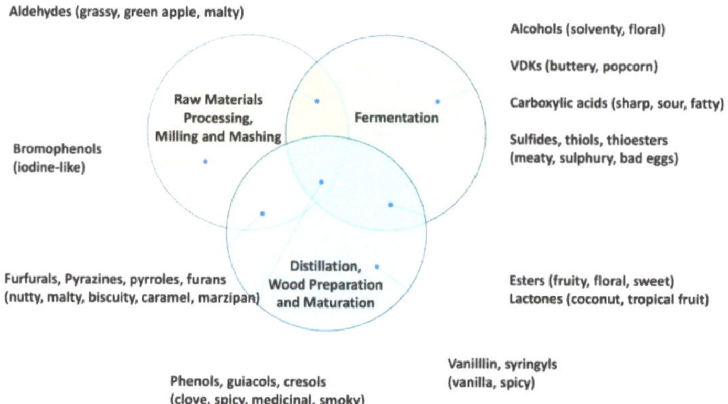

Figure 1. Congeners categorized by Scotch whisky production stage (VDKs: Vicinal Diketones) [1].

Under The Scotch Whisky Regulations (2009) [7], for a whisky to be denoted as 'Scotch whisky', it must be distilled and matured in oak casks in Scotland where malted barley and/or other cereal ingredients are processed, converted, and fermented at a Scottish distillery. The stipulation surrounding fermentation is simply that it is 'by yeast'. Yeast strains used in industrial distillery fermentations are predominantly those belonging to *Saccharomyces cerevisiae*. Throughout much of the 20th century, spent brewing yeast was employed in both baking and whisky production, prior to the initial development of specialised leavening yeast strains for baking and subsequent specialised distilling strains with improved characteristics more suited to distillery fermentations. The typical commercial strains that are currently used derive from a hybrid between ale yeast and a wild *S. cerevisiae* strain (previously known as *S. diastaticus*) with amylolytic properties [8]. Such strains are high yielding in terms of ethanol and produce consistent concentrations of desired congeners. As a result, both grain and most malt whisky distilleries predominantly use a sole strain of distilling yeast, with the practice of adding spent brewing yeast having fallen out of favour.

The focus on spirit yield and consistency in congener production is a sensible commercial strategy. However, a number of distillers are seeking a greater diversity in flavour profile in order to either enhance specific attributes or to expand the palette for blending. One route to achieve this is through yeast choice.

In this work, the link between yeast strain selection and new make spirit aroma profile was investigated using 24 commercially available active dried yeast strains. The influence of yeast application type in terms of designation as 'distilling', 'brewing' or 'wine' in predicting new make spirit flavour is discussed.

2. Materials and Methods

2.1. Wort Preparation

A concentrated batch of unboiled fresh wort (SG of 1.0726) was produced from Distillers Malt (Muntons, Stowmarket, UK) using the International Centre for Brewing and Distilling (ICBD) pilot brewing kit at Heriot-Watt University, Edinburgh. A weight of 55 kg of malt was milled using a two-roller mill (Fraser Agricultural, Inverurie, GBR) then mashed with 150 L of water at 64 °C for 1 h prior to separation using a lauter tun. The grain bed was then sparged continuously with 81 °C water to obtain the desired volume and gravity. Fresh wort was frozen in aliquots and stored at −20 °C. When required, the wort was defrosted and diluted to the desired gravity (1.0612 ± 0.0018). An amount of 850 mL of defrosted unboiled wort was diluted with 150 mL of autoclaved water. Gravities were measured using a benchtop Anton Paar DMA 4500 (Anton Paar, St. Albans, UK).

2.2. Yeast Viability

Yeast viability assessments were carried out using the ABER Countstar^{TM} automatic yeast counter (ABER Instruments, Aberystwyth, UK) using methylene blue stain. In a 1.5 mL Eppendorf tube, 10 µL of yeast slurry was pipetted into 990 µL of sodium acetate (0.1 M) (Fisher Scientific, Loughborough, UK) EDTA (10 mM) (Sigma Aldrich, Milwaukee, WI, USA) solution (pH 4.5) and agitated. Into another 1.5 mL Eppendorf tube, 500 µL of the resulting 1000 µL of solution was then transferred into 500 µL of methylene blue staining solution (Fisher Scientific, Loughborough, UK) and agitated. An amount of 20 µL of the stained solution (dilution factor of 200) was then pipetted into a chamber on an ABER Countstar^{TM} slide.

2.3. Yeast Rehydration, Pitching and Fermentation

A weight per volume pitching rate of 1 g active dried yeast (ADY) per 1 L of wort was used. For rehydration, ADY was slurried with autoclaved water (30 °C) then placed in an unagitated water bath (Grant SUB36; Grant Instruments, Shepreth, UK) at 30 °C for 60 min prior to pitching.

Fermentations using individual yeast strains (Table 1) were performed. Fermentation time was set at 72 h, and temperature at 35 °C, with fermentations carried out in triplicate for each strain. Fermentations (300 mL) were carried out in 500 mL conical flasks (Fisher Scientific, Loughborough, UK). The triplicate fermentations per yeast strain were consolidated after 72 h to give 900 mL of wash. Wash samples were centrifuged for five minutes (1006.2 g) to remove yeast from suspension (Denley, Guangzhou, China). Wash alcohol by volume (ABV%) calculations were then conducted by first measuring the specific gravity of a centrifuged wash sample using a benchtop Anton Paar DMA 4500 (Anton Paar, St. Albans, UK). The ABV% of the sample was then estimated using the following formula: Alcohol by volume (ABV%) = (Original Gravity − Final Gravity) × 135.25.

2.4. Distillation

Wash distillations were carried out using 2 L copper pot stills (operating capacity of 1 L) and associated worm-tub condensers (Copper-Alembic, Gandra, Portugal). Stills were heated using electric hotplates (Duronic, Romford, UK). 900 mL of wash was charged per yeast strain. 500 µL of silicon-based antifoam was added to each 900 mL wash distillation. The collection of low wines was carried out until the distillate coming off the still was 1.5% alcohol by volume (ABV).

Low wines distillations were carried out using 2 L copper pot stills (operating capacity of 1 L) and associated worm-tub condensers (Copper-Alembic, Gandra, Portugal). Approximately 200–300 mL of low wines was charged per yeast strain, and 2.5 mL of foreshots were collected per spirit distillation. After the collection of foreshots, a 3.5 mL sample of distillate (spirit or 'hearts cut') was collected and the ABV (%) recorded. Spirit was collected until the distillate coming off the still was 70% ABV. Feints (distillate produced post spirit cut) were not collected in this experiment.

Alcohol by volume (ABV%) and density values were determined using a handheld Anton Paar DMA35 density meter (Anton Paar, St. Albans, UK).

Table 1. Yeast strains used in this study.

Number	Name	Yeast Strain	Application
1	Pinnacle MG+	*Saccharomyces cerevisiae*	Whisk(e)y
2	SafSpirits M1	*S. cerevisiae*	Whisk(e)y
3	SafSpirits USW6	*S. cerevisiae*	Bourbon/Whiskey
4	Pinnacle M	*S. cerevisiae*	Whisk(e)y
5	Distilamax MW	*S. cerevisiae*	Whisk(e)y
6	Distilamax GW	*S. cerevisiae*	Whisk(e)y
7	Pinnacle S	*S. cerevisiae*	Whisk(e)y
8	Pinnacle G	*S. cerevisiae*	Whisk(e)y
9	Kerry M	*S. cerevisiae*	Whisk(e)y
10	Distilamax XP	*S. cerevisiae var. diastaticus*	Whisk(e)y
11	Safale T58	*S. cerevisiae*	English/Belgian ale
12	Safale WB06	*S. cerevisiae*	Wheat beer
13	Safale BE-256	*S. cerevisiae*	Belgian ale
14	Safale S-04	*S. cerevisiae*	US/English ale
15	Saflager 189	*S. pastorianus*	Swiss lager
16	Safale K-97	*S. cerevisiae*	German/Belgian ale
17	Safale S-33	*S. cerevisiae*	Belgian/English ale
18	Safale BE134	*S. cerevisiae var diastaticus*	Belgian-Saison
19	Hothead	*S. cerevisiae*	Norwegian ale
20	Kveik	*S. cerevisiae*	Norwegian ale
21	LalvinV116	*S. cerevisiae*	Ice wine
22	Lalvin EC1118	*S. cerevisiae var. bayanus*	Champagne
23	Lalvin ICV OKAY	*S. cerevisiae*	Wine
24	Exotics SPH	*Hybrid S. cerevisiae/S. paradoxus*	Red wine

2.5. Gas Chromatography Mass Spectrometry (GC/MS)

18 new make spirit (NMS) samples were analysed to determine specific ester concentrations using a solid-phase microextraction (SPME) GC/MS method developed within Heriot-Watt University's International Centre for Brewing and Distilling (ICBD). Specific ester concentrations were determined using a Shimadzu QP2010 Ultra (Shimadzu, Kyoto, Japan) GC/MS integrated system with split/splitless injector and Shimadzu AOC 5000 autosampler. SPME was conducted using a 65 mm polydimethylsiloxane/divinylbenzene coated fibre (Supleco, Bellefonte, PN, USA). Samples were prepared by adding 1 mL of NMS to a 15 mL vial and 5 mL of distilled water then added to the NMS and mixed. Data were handled using Shimadzu GC/MS Solutions data handling systems. For instrument details see Table 2.

Table 2. GC/MS Instrument conditions.

SPME	Conditioning: 5 min at 70 °C Extraction: 5 min Desorption: 1 min
Column	DB Wax UI (30 m × 0.25 mm × 0.25 µm) (Agilent, Santa Clara, CA, USA)
Carrier gas	Helium (BOC; 5.0)
Internal standard	Methyl heptanoate
Oven	40 °C for 3 min; ramp to 100 °C at 10 °C/min; ramp to 160 °C at 4 °C/min; ramp to 220 °C at 10 °C/min. Hold for 100 min. Ramp to 250 °C at 70 °C/min. Hold for 3 min.

2.6. Sensory Analysis

Quantitative descriptive analysis: Aroma Intensity Rating. A form of quantitative descriptive analysis (QDA) was carried out on the NMS samples. Sensory analysis was carried out in two phases: (1) Language familiarisation, (2) NMS nosing, undiluted (still strength).

The language familiarisation sessions utilised descriptors from the Scotch whisky research institute (SWRI) whisky wheel [9] and aroma standards using the Whisky aroma kit (The Aroma Academy, Aberdeen, UK). A total of 200 min of language familiarisation was carried out with the panel ($n = 5$).

To avoid the introduction of bias derived from numerical scales, the NMS nosing ballot constructed for this experiment was an adaptation of a graphic line scale [10]. In place of a line scale it consists of three tick boxes, each with a corresponding word anchor relating to the intensity of the descriptor. The word anchors were, from left to right: 'Not present', 'Mild/Subtle' and 'Strongly present'. Unbeknownst to the assessors, the tick boxes represented a scale of 0–1.0. The assessor responses were converted to their numerical value by the panel leader after the conclusion of the sensory session.

2.7. Hedonic Assessment: Acceptability Rating

An acceptability test using the nine-point hedonic scale was also performed. The nine-point hedonic scale was historically used for the hedonic assessment of food, beverages and non-food products [11].

Assessors ($n = 5$) were asked to nose samples individually and then check only one of the phrases from the nine presented on a ballot that best described their opinion of the sample. The phrases represented a scale of 1–9 where 9 represented 'like extremely' [11]. This assessment was performed in order to narrow down the initial selection of yeast strains [12].

2.8. Statistical Analysis

Kolmogorov–Smirnov, Analysis of Variance (ANOVA), Friedman and Games Howell multiple comparison tests were performed using the IBM SPSS statistics software package (IBM, Armonk, NY, USA).

The Kolmogorov–Smirnov test compares an observed sample distribution and a specified theoretical distribution and is sometimes referred to as a normality test [13]. Analysis of variance (ANOVA) is a statistical technique that tests to determine if significant differences exist between three or more groups and this includes one group with multiple variations [14] such as the yeast used in this work. The Friedman test in a non-parametric one-way ANOVA [15].

The Games Howell multiple comparison test is a pairwise comparison procedure, designed for use with unequal sample sizes, that can determine which specific groups are statistically significantly different [16,17].

3. Results

3.1. Yeast Viability and Wash Alcohol Volume

The yeast viability, measured using methylene blue staining following 60 min of rehydration in water at 30 °C, was found to be less than 90% for all strains tested. However, no significant difference was found in the average final-wash alcohol by volume (ABV). Ten of the yeast strains assessed achieved average final-wash ethanol concentrations of 8% ABV or over. This included all of the yeast strains marketed as 'distilling' strains (1–10), with the exception of yeast strain 7 (Pinnacle S).

3.2. Aroma Intensity Ratings

The aroma intensity ratings experiment intended to investigate whether the yeast strain used to produce the new make spirit had an effect on the perceived intensity of seven common aromas associated with whisky new make spirit. Seven individual Friedman tests were carried out.

From the individual Friedman tests, it was found that there was no statistically significant difference in the perceived intensity of the fruity (fresh) aroma depending on the yeast strain used to produce the new make spirit (χ^2 (22) = 25.26, $p = 0.285$). This was also the case for all of the other aroma categories; fruity (2), sulphury, floral, sweet, cereal and feinty (χ^2 (22) = 27.03, $p = 0.210$, $\chi^2 = 27.49$, $p = 0.193$, $\chi^2 = 13.75$, $p = 0.910$, $\chi^2 = 31.63$, $p = 0.084$, $\chi^2 = 25.14$, $p = 0.290$ and $\chi^2 = 30.31$, $p = 0.11$, respectively).

Distilling yeast strain 2 achieved the highest average intensity rating for the fruity (fresh) aroma category (0.7; Figure 2a), with both distilling and wine yeast receiving an average score of 0.45.

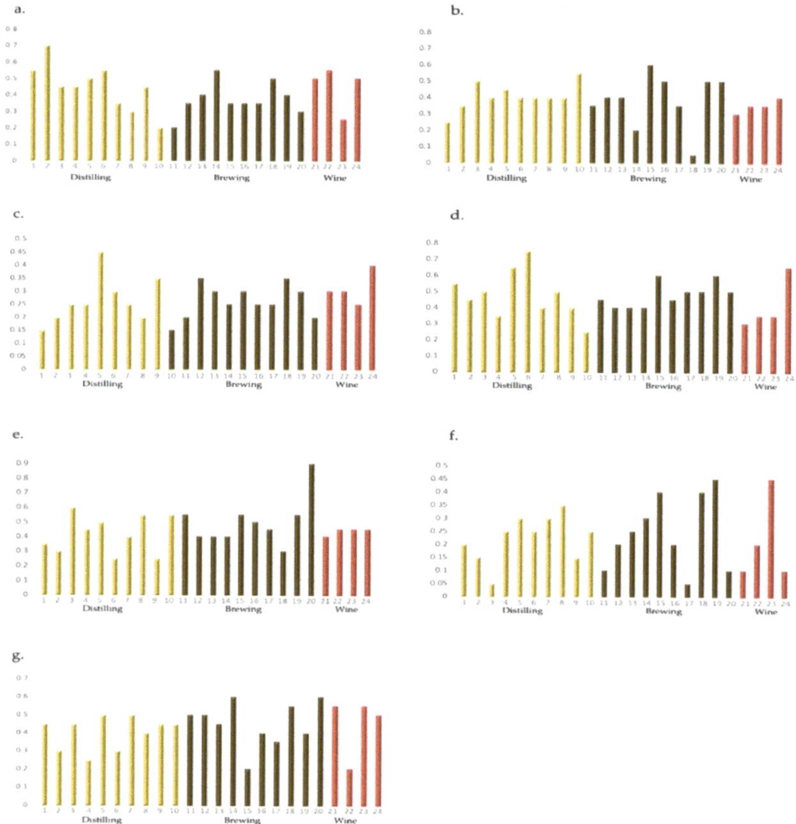

Figure 2. Average aroma intensity rating from 0 (absent) to 1.0 (most intense) for seven common aromas associated with whisky new make spirit: (**a**) Fruity (Fresh), (**b**) Fruity (Dried), (**c**) Floral, (**d**) Sweet, (**e**) Cereal, (**f**) Sulfury, (**g**) Feinty. Ratings are grouped by yeast type: Distilling, Brewing, or Wine; $n = 5$.

This suggests that the NMS spirit produced using these strains possesses the fruity characteristics of the solvent-like and fresh fruit variety. As described in the Scotch Whisky Research Institute (SWRI) flavour wheel [9].

The 'Fruity (dried)' category relates to the dried fruit aroma category described on the SWRI whisky flavour wheel [9]. Brewing yeast strain 18 achieved the lowest average rating for this attribute (0.05) (Figure 2b). Regardless of classification, the average intensity rating was below 0.5 in relation to the Fruity (dried) attribute. This suggests an overall lower perception of this attribute by the panel in the given samples. In addition, only NMS samples produced using distilling yeast 10 and brewing yeast 15 could be considered as exhibiting mild-to-strong fruity characteristics of the dried fruit variety (Figure 2b).

In relation to the floral attribute, which is associated with fresh flower-like aromas, all strains achieved average intensity ratings below 0.5 (Figure 2c). More specifically, with the exception of yeast strain 5 (0.45), all of the new make spirit produced achieved average aroma intensity ratings or 0.4 or lower, which suggests that the NMS produced using the yeast strains included in this study possessed weak floral aroma characteristics.

The sweet aroma category was associated with honey, vanilla and caramel-like aroma [9]. Six yeast strains achieved average aroma intensity ratings above 0.5, with yeast strain 6 achieving the highest average rating of 0.75 (Figure 2d). With the exceptions of yeast strains 4, 10, 21, 22 and 23, all of the assessed yeast strains achieved average intensity ratings of 0.4 or above (Figure 2d) suggesting that the NMS produced using these yeast strains exhibits mild-to-strong sweet aroma characteristics. Interestingly, three of the four wine yeast strains achieved scores below 0.4, with the average score of 0.41 caused by a single higher-scoring strain. Overall, distilling and brewing strains were ranked higher with an average score of 0.48.

Brewing yeast strain 20 scored significantly high in relation to cereal notes (0.9; Figure 2e), which include toasted, malt and husky aromas. On average the brewing yeast scored higher (0.5) than both the distilling and wine yeast (0.42 and 0.43, respectively).

Yeast strains 3, 11, 17, 20, 21 and 24, achieved average intensity ratings of 0.1 or lower in relation to the sulfury aroma category. This suggests a particularly low perception of this attribute by the assessors in these NMS samples (Figure 2f). Yeast strains 19 and 23 on the other hand both achieved the highest average rating for this attribute at 0.45 (Figure 2f). Overall, the mean average intensity ratings did not exceed 0.3 across all of the yeast strains assessed suggesting that sulfury characteristics in the NMS were minimal.

The average aroma intensity rating for feinty notes, which include leathery, tobacco and sweaty aromas, was consistent across all yeast strains with brewing and wine yeast averaging 0.45 and distilling yeast 0.4. Of the brewing yeast, strains 14 and 20 were the highest ranking.

3.3. Ester Profiles

The concentration of seven esters was assessed in new make spirit samples from eight distilling yeast strains, eight brewing yeast strains and two wine yeast strains used in the aroma intensity test. Ethyl hexadecanoate was not detected in any of the samples. The results for ethyl hexanoate, ethyl lactate, ethyl decanoate, phenylethyl acetate, ethyl dodecanoate and ethyl octanoate are shown in Table 3.

Table 3. Average ester concentration in new make spirit per yeast type: Distilling ($n = 8$), Brewing ($n = 8$), Wine ($n = 2$).

			Average Concentration per Yeast Type (mg/L)		
Ester	**Aroma**	**Aroma Threshold (mg/L) [18]**	**Distilling**	**Brewing**	**Wine**
Ethyl hexanoate	Aniseed, apple-like	0.076	1.71	0.52	1.60
Ethyl lactate	Buttery, butterscotch, artificial strawberry	>14	1.30	23.90	7.97
Ethyl decanoate	Floral, soapy	1.1	1.09	0.79	1.96
Phenylethyl acetate	Roses, honey, waxy	0.7	1.63	0.92	2.47
Ethyl dodecanoate	Soapy, estery	0.64	0.25	0.56	0.92
Ethyl octanoate	Sour apple	0.24	0.75	0.71	1.41
		Total	6.73	27.4	16.33

All samples exceeded the threshold value for ethyl hexanoate with distilling yeast generating new make spirit with the highest concentration. Samples from brewing yeast contained an average of 23.90 mg/L ethyl lactate, exceeding the aroma threshold of 14 mg/L.

Samples from wine yeast contained the highest concentration of ethyl decanoate, phenylethyl acetate, ethyl dodecanoate, and ethyl octanoate, with each case exceeding threshold levels. Threshold levels were not reached for either distilling or brewing strains for ethyl decanoate and ethyl dodecanoate.

The total ester concentration, by addition of each of the six individual averages, places brewing strains highest followed by wine and then distilling yeast (Table 3).

3.4. Hedonic Assessment: Acceptability Ratings

New make spirit samples produced using the 24 different yeast strains used in this study were assessed individually and assigned an acceptability rating using a ballot constructed based on the 9-point hedonic scale.

It is generally not recommended to perform statistical analyses of data derived from hedonic assessments such as the 9-point hedonic scale, as the type of data collected is not mathematically suitable. However, mean ratings can be collected and ratings of seven or higher are considered indicative of a product of a highly acceptable sensory quality [19].

As shown in Table 4, seven of the twenty-four yeast strains achieved an average acceptability rating of 7 or above.

Table 4. Average mean acceptability ratings of new make spirit samples that achieved a mean rating of 7 or above (out of 9).

Yeast Strain	Mean Acceptability Rating
14	7.75
2	7.5
13	7.5
20	7.5
11	7.25
16	7
17	7

Yeast strain 14 (brewing yeast) achieved the highest average acceptability rating. Yeast strains 2, 13 and 20 (distilling, brewing, and brewing yeast, respectively) jointly achieved the second highest average acceptability rating (Table 4). Thirdly, these were followed by yeast strain 11 (brewing yeast) and then joint fourth by yeast strains 16 and 17 (both brewing yeast).

From these data and the selection of yeast strains used in this study, it appears that, with the exception of spirit produced using yeast strain 2, NMS produced using 'brewing' strains were found to be the more acceptable products by the panel.

4. Discussion

Multiple studies identified that different yeast strains produce different congener profiles and different aromas in beer [20–22]. Furthermore, the mechanisms for production of higher alcohols and esters were well studied [23,24]. The difference in congener profiles between different strains, (even within species), depends on differences in the metabolism of sugars, proteins and amino acids and autolysis of yeast [25].

Overall, although statistical analysis produced no credible evidence from this study that the yeast strain had a statistically significant impact on the perceived intensity of the aroma categories assessed, it is suggested from the results of the panel responses that there were differences in the overall perceived aroma profiles of the assessed new make spirit (NMS) profiles and that a large selection of the yeast strains used in this study produced NMS with distinctly fresh-fruit-like, sweet and non-sulphury characters.

Commercially, yeasts are commonly split and marketed into groups related to their intended industrial application. To look more broadly at the yeast strains involved in this study, the strains were separated into three groups: distilling strains, brewing strains, or wine strains, based on either the industrial application or fermentation medium specified by the yeast manufacturer. The categorisation of strains based on application (distilling, brewing, or wine) highlighted distinct patterns. Generally, distilling yeast was regarded as fruitier than brewing or wine yeast. Brewing yeast was regarded as having cereal, sulfury, and feinty notes, and wine yeast was rated alongside distilling yeast for fresh fruit aroma and highest for floral aroma (although intensity ratings for floral aroma were low across all samples). These results compare with analytical studies of ester concentration, with wine yeast containing, on average, higher concentrations of ethyl decanoate, phenylethyl acetate, ethyl dodecanoate, and ethyl octanoate, compared to distilling and brewing yeast.

The aroma characteristics of these esters are floral, rose, estery, and sour apple, respectively. These results also concur with data from the literature and unpublished industry studies; higher alcohols, acetate esters and ethyl esters are more commonly cited in fermentation studies with distilling yeast in comparison with brewing or wine yeasts [1,12].

That distilling yeast was perceived as fruitier than brewing or wine yeast contrasts with the total ester concentration (Table 3). However, the analysis of aroma compounds and the perceived character of matured Scotch whisky demonstrates that the relative levels of individual esters within the headspace may be low but the interactions with other components within the spirit matrix leads to an increase in the overall 'estery' character [26].

The results from the hedonic assessment clearly indicate that brewing strains produce new make spirit that is 'most acceptable'. Six of the top seven scoring strains were brewing strains, with each strain also ranking highly in at least one attribute within the aroma intensity ratings. On average brewing strains also produced a spirit with a significantly higher concentration of ethyl lactate, a compound associated with buttery, butterscotch and artificial strawberry characteristics. Traditionally, brewing strains were employed within distillery fermentations either as the main production yeast, or in co-fermentation at additions of up to 50%. The results from this study highlight the positive influence that brewing strains of *S. cerevisiae* have on new make spirit aroma.

Borneman et al. [27] warns that high levels of genetic similarity exist between commercial *S. cerevisiae* strains which potentially reduce the scope for genetic, metabolic, and thus new congeneric differences. The lack of statistically significant differences between individual strains in both aroma intensity ratings and in ester production in this study reflects this theory. Lambrechts and Pretorius [28] reviewed the production of higher alcohols and esters in *Saccharomyces* strains and found them to be strain-dependent. For alcohols this is potentially due to differences in amino acid utilisation in the Ehrlich pathway. These metabolic differences were found to be greater in non-*Saccharomyces* yeasts [29]. For ester synthesis, the precursor alcohol and acyl-CoA concentrations, as well as the expression of the Eeb1 and Eht1 genes [30], were shown to be key [31]. Saerens et al. [32] later reported that yeasts could be chosen for the production of desired alcohols and esters based on the expression levels of genes involved in their biosynthesis. The production of congener types by non-conventional yeasts, such as *Hanseniaspora guillermondii*, is cited commonly in the literature, with multiple studies demonstrating higher concentrations of higher alcohols and esters in comparison to brewing and wine yeast strains [33,34]. As such, there is a great potential in their application within distillery fermentations.

5. Conclusions

Developments in yeast isolation, propagation techniques, and format have greatly expanded the diversity of the yeast strains available to Scotch whisky distillers. It is also no secret in the distilling industry that using alternative yeast strains, compared to the 'industry standards', is a perfectly viable production strategy. There are already a number of products available on the Scotch whisky market that are produced with yeast strains that are either proprietary or strains that were not initially developed and marketed for whisky fermentations.

This work highlights the aroma characteristics imparted by yeast, the variation in overall aroma perception, and consequently the importance of yeast strain selection. There is clear scope to increase the palette of flavour attributes in new make spirit giving distillers an opportunity to innovate in this very traditional process.

Author Contributions: Conceptualization, A.E.H.; methodology, A.E.H.; investigation, C.W.; writing—original draft preparation, A.E.H.; writing—review and editing, C.W.; supervision, A.E.H.; project administration, A.E.H.; funding acquisition, A.E.H. All authors have read and agreed to the published version of the manuscript.

Funding: This research was part-funded by Innovate UK and The Port of Leith Distillery.

Institutional Review Board Statement: The study was conducted according to the guidelines of Heriot-Watt University and approved by the Ethics Committee of the School of Engineering and Physical Sciences (approval number 19/EA/AH/1, 9 July 2016).

Informed Consent Statement: Informed consent was obtained from all subjects involved in the study.

Data Availability Statement: The data presented in this study are available on request from the corresponding author. The data are not publicly available due to commercial sensitivity.

Acknowledgments: Many thanks to Victoria Muir-Taylor for practical work and Maarten Gorseling for analytical support.

Conflicts of Interest: The authors declare no conflict of interest.

References

1. Waymark, C. How Well Can the Choice of Yeast Strain Predict Scotch Malt Whisky New Make Spirit flavour profile? Ph.D. Thesis, Heriot-Watt University, Edinburgh, UK, 13 August 2021.
2. Piggott, J.R. Whisky, Whiskey and Bourbon: Composition and Analysis of Whisky. In *Encyclopedia of Food and Health*; Caballero, B., Finglas, P.M., Toldrá, F., Eds.; Academic Press: Oxford, UK, 2016; pp. 514–518. [CrossRef]
3. Mosedale, J.R.; Puech, J.-L. Wood maturation of distilled beverages. *Trends Food Sci. Technol.* **1998**, *9*, 95–101. [CrossRef]
4. Nicol, D. Batch Distillation. In *Whisky Technology, Production and Marketing*, 2nd ed.; Russell, I., Stewart, G., Eds.; Elsevier Ltd.: Amsterdam, The Netherlands, 2014; pp. 155–178.
5. Dzialo, M.C.; Park, R.; Steensels, J.; Lievens, B.; Verstrepen, K.J. Physiology, ecology and industrial applications of aroma formation in yeast. *FEMS Microbiol. Rev.* **2017**, *41* (Suppl. S1), S95–S128. [CrossRef] [PubMed]
6. Wanikawa, A.; Yamamoto, N.; Hosoi, K. The influence of brewers' yeast on the quality of malt whisky. In *Distilled Spirits: Tradition and Innovation*; Nottingham University Press: Nottingham, UK, 2004; pp. 95–101.
7. The Scotch Whisky Regulations 2009. Available online: https://www.legislation.gov.uk/uksi/2009/2890/contents/made (accessed on 7 October 2021).
8. Campbell, I. Yeast and Fermentation. In *Whisky: Technology, Production and Marketing*; Russell, I., Bamforth, C., Stewart, G.G., Eds.; Elsevier Ltd.: Amsterdam, The Netherlands, 2003; pp. 117–145.
9. Jack, F. Sensory Analysis. In *Whisky Technology, Production and Marketing*, 2nd ed.; Russell, I., Stewart, G., Eds.; Elsevier Ltd.: Amsterdam, The Netherlands, 2014; pp. 229–242.
10. Stone, H.; Bleibaum, R.N.; Thomas, H.A. Descriptive Analysis. In *Sensory Evaluation Practices*, 4th ed.; Elsevier Academic Press: Amsterdam, The Netherlands, 2012; pp. 233–289.
11. Lawless, H.T. Acceptance Testing. In *Sensory Evaluation of Food*, 2nd ed.; Heldman, D., Ed.; Springer: New York, NY, USA, 2010; pp. 325–342.
12. Muir-Taylor, V.G. Yeast Selection for Scotch Malt Whisky Fermentations: An Assessment of Commercially Available Active Dried Yeast Strains. Ph.D. Thesis, Heriot-Watt University, Edinburgh, UK, 4 March 2021.
13. Hoffman, J.I.E. Categorical and Cross-classified data: McNemar's Test, Kolmogorov-Smirnov Tests, Concordance. In *Biostatistics for Medical and Biomedical Practitioners*; Hoffman, J.I.E., Ed.; Elsevier: Amsterdam, The Netherlands, 2015; pp. 221–237.
14. MacFarland, T.W. Oneway Analysis of Variance (ANOVA). In *Introduction to Data Analysis and Graphical Presentation in Biostatistics with R Statistics in the Large*; Macfarland, T.W., Ed.; Springer: Berlin/Heidelberg, Germany, 2014; pp. 73–97.
15. Scheff, S.W. Nonparametric statistics. In *Fundamental Statistical Principles for the Neurobiologist a Survival Guide*; Scheff, S.W., Ed.; Academic Press: Cambridge, MA, USA, 2016; pp. 157–182.
16. Saunder, D.C.; DeMars, C.E. An Updated Recommendation for Multiple Comparisons. *Adv. Methods Pract. Psychol. Sci.* **2019**, *2*, 26–44. [CrossRef]
17. Midway, S.; Robertson, M.; Shane, F.; Kaller, M. Comparing multiple comparisons: Practical guidance for choosing the best multiple comparisons test. *PeerJ* **2020**, *8*, 1–26. [CrossRef] [PubMed]
18. Salo, P.; Nykänen, L.; Suomalainen, H. Odor Thresholds and Relative Intensities of Volatile Aroma Components in an Artificial Beverage Imitating Whisky. *J. Food Sci.* **1972**, *37*, 394–398. [CrossRef]
19. Everitt, M. Consumer-targeted sensory quality. In *Global Issues in Food Science and Technology*; Barbosa-Canovas, G., Buckle, K., Colonna, P., Lineback, D., Mortimer, A., Speiss, W., Eds.; Academic Press: Cambridge, MA, USA, 2009; pp. 117–128.
20. ÄYräpää, T. Formation of Higher Alcohols by Various Yeasts. *J. Inst. Brew.* **1968**, *74*, 169–178. [CrossRef]
21. Nykänen, L.; Nykänen, I. Production of Esters by Different Yeast Strains in Sugar Fermentations. *J. Inst. Brew.* **1977**, *83*, 30–31. [CrossRef]
22. Peddie, H.A.B. Ester formation in brewery fermentations. *J. Inst. Brew.* **1990**, *96*, 327–331. [CrossRef]
23. Pires, E.; Teixeira, J.; Brányik, T.; Vicente, A. Yeast: The soul of beer's aroma—A review of flavour-active esters and higher alcohols produced by the brewing yeast. *Appl. Microbiol. Biotechnol.* **2014**, *98*, 1937–1949. [CrossRef] [PubMed]

24. Stewart, G.G. Esters–the most important group of flavour-active compounds in alcoholic beverages. In *Distilled Spirits: Production, Technology and Innovation*; Bryce, J.H., Piggot, J.R., Stewart, G.G., Eds.; Nottingham University Press: Nottingham, UK, 2004; Volume 2, pp. 43–250.
25. Fleet, G.H. Wine yeasts for the future. *FEMS Yeast Res.* **2008**, *8*, 979–995. [CrossRef] [PubMed]
26. Steele, G.M.; Fotheringham, R.N.; Jack, F.R. Understanding flavour development in Scotch Whisky. In *Distilled Spirits: Tradition and Innovation*; Bryce, J.H., Stewart, G.G., Eds.; Nottingham University Press: Nottingham, UK, 2008; pp. 161–167.
27. Borneman, A.R.; Forgan, A.H.; Kolouchova, R.; Fraser, J.A.; Schmidt, S.A. Whole genome comparison reveals high levels of inbreeding and strain redundancy across the spectrum of commercial wine strains of *Saccharomyces cerevisiae*. *G3 Genes Genomes Genet.* **2016**, *6*, 957–971. [CrossRef] [PubMed]
28. Lambrechts, M.G.; Pretorius, I.S. Yeast and its Importance to Wine Aroma-A Review. *S. Afr. J. Enol. Vitic.* **2000**, *21*, 97–129. [CrossRef]
29. Herraiz, T.; Reglero, G.; Herraiz, M.; Martin-Alvarez, P.J.; Cabezudo, M.D. The influence of the yeast and type of culture on the volatile composition of wines fermented without sulfur dioxide. *Am. J. Enol. Vitic.* **1990**, *41*, 313–318.
30. Saerens, S.M.G. The *Saccharomyces cerevisiae* EHT1 and EEB1 Genes Encode Novel Enzymes with Medium-chain Fatty Acid Ethyl Ester Synthesis and Hydrolysis Capacity. *J. Biol. Chem.* **2006**, *281*, 4446–4456. [CrossRef] [PubMed]
31. Calderbank, J.; Hammond, J.R.M. Influence of Higher Alcohol Availability on Ester Formation by Yeast. *J. Am. Soc. Brew. Chem.* **1994**, *52*, 84–90. [CrossRef]
32. Saerens, S.M.G.; Verbelen, P.J.; Vanbeneden, N.; Thevelein, J.M.; Delvaux, F.R. Monitoring the influence of high-gravity brewing and fermentation temperature on flavour formation by analysis of gene expression levels in brewing yeast. *Appl. Microbiol. Biotechnol.* **2008**, *80*, 1039–1051. [CrossRef] [PubMed]
33. Bourbon-Melo, N.; Palma, M.; Rocha, M.P.; Ferreira, A.; Bronze, M.R.; Elias, H.; Sá-Correia, I. Use of *Hanseniaspora guilliermondii* and *Hanseniaspora opuntiae* to enhance the aromatic profile of beer in mixed-culture fermentation with *Saccharomyces cerevisiae*. *Food Microbiol.* **2021**, *95*, 103678. [CrossRef] [PubMed]
34. Gamero, A.; Belloch, C.; Querol, A. Genomic and transcriptomic analysis of aroma synthesis in two hybrids between *Saccharomyces cerevisiae* and *S. kudriavzevii* in winemaking conditions. *Microb. Cell Factories* **2015**, *14*, 128. [CrossRef] [PubMed]

fermentation

MDPI

Article

Saccharomyces cerevisiae Dehydrated Culture Modulates Fecal Microbiota and Improves Innate Immunity of Adult Dogs

Karine de Melo Santos [1], Larissa Wünsche Risolia [1], Mariana Fragoso Rentas [1], Andressa Rodrigues Amaral [2], Roberta Bueno Ayres Rodrigues [1], Maria Isabel Gonzalez Urrego [1], Thiago Henrique Annibale Vendramini [1], Ricardo Vieira Ventura [1], Júlio César de Carvalho Balieiro [1], Cristina de Oliveira Massoco [1], João Paulo Fernandes Santos [1], Cristiana Fonseca Ferreira Pontieri [3] and Marcio Antonio Brunetto [1,2,*]

[1] Pet Nutrology Research Center, Nutrition and Production Department, School of Veterinary Medicine and Animal Science, University of São Paulo, Pirassununga 13635-900, SP, Brazil; kdemelosantos@yahoo.com.br (K.d.M.S.); larissa.risolia@usp.br (L.W.R.); mariana.rentas@usp.br (M.F.R.); roberta_barodrigues@hotmail.com (R.B.A.R.); migu@usp.br (M.I.G.U.); thiago.vendramini@usp.br (T.H.A.V.); rvventura@usp.br (R.V.V.); balieiro@usp.br (J.C.d.C.B.); cmassoco@gmail.com (C.d.O.M.); joao_paulo@zootecnista.com.br (J.P.F.S.)

[2] Veterinary Nutrology Service, Veterinary Teaching Hospital, School of Veterinary Medicine and Animal Science, University of São Paulo, Cidade Universitária, Sao Paulo 05508-270, SP, Brazil; andressa.rodrigues.amaral@usp.br

[3] Nutrition Development Center, Grandfood Industria e Comercio LTDA (Premier Pet), Dourado 13590-000, SP, Brazil; cristiana@premierpet.com.br

* Correspondence: mabrunetto@usp.br; Tel.: +55-19-3565-4226

Citation: Santos, K.d.M.; Risolia, L.W.; Rentas, M.F.; Amaral, A.R.; Rodrigues, R.B.A.; Urrego, M.I.G.; Vendramini, T.H.A.; Ventura, R.V.; Balieiro, J.C.d.C.; Massoco, C.d.O.; et al. *Saccharomyces cerevisiae* Dehydrated Culture Modulates Fecal Microbiota and Improves Innate Immunity of Adult Dogs. *Fermentation* **2022**, *8*, 2. https://doi.org/10.3390/fermentation8010002

Academic Editor: Ronnie G. Willaert

Received: 22 November 2021
Accepted: 13 December 2021
Published: 23 December 2021

Publisher's Note: MDPI stays neutral with regard to jurisdictional claims in published maps and institutional affiliations.

Copyright: © 2021 by the authors. Licensee MDPI, Basel, Switzerland. This article is an open access article distributed under the terms and conditions of the Creative Commons Attribution (CC BY) license (https:// creativecommons.org/licenses/by/ 4.0/).

Abstract: *Saccharomyces cerevisiae* yeast culture can be dehydrated, and it has a potential prebiotic effect. This study evaluated the effects of supplementing increasing levels of dehydrated yeast culture (DYC) of *Saccharomyces cerevisiae* (Original XPC™, Diamond V, Cedar Rapids, IA, USA) on fecal microbiota, nutrient digestibility, and fermentative and immunological parameters of healthy adult dogs. Eighteen adult male and female dogs with a mean body weight of 15.8 ± 7.37 kg were randomly assigned to three experimental treatments: CD (control diet), DYC 0.3 (control diet with 0.3% DYC) and DYC 0.6 (control diet with 0.6% DYC). After 21 days of acclimation, fecal samples were collected for analysis of nutrient digestibility, microbiota and fecal fermentation products. On the last day, the blood samples were collected for the analysis of immunological parameters. The microbiome profile was assessed by the Illumina sequencing method, which allowed identifying the population of each bacterial phylum and genus. The statistical analyses were performed using the SAS software and the Tukey test for multiple comparison ($p < 0.05$). Our results suggest that the addition of DYC increased the percentage of the phyla Actinobacteria and Firmicutes ($p = 0.0048$ and $p < 0.0001$, respectively) and reduced that of the phylum Fusobacteria ($p = 0.0008$). Regardless of the inclusion level, the yeast addition promoted reduction of the genera *Allobaculum* and *Fusobacterium* ($p = 0.0265$ and $p = 0.0006$, respectively) and increased ($p = 0.0059$) that of the genus *Clostridium*. At the highest prebiotic inclusion level (DYC 0.6), an increase ($p = 0.0052$) in the genus *Collinsella* and decrease ($p = 0.0003$) in *Prevotella* were observed. Besides that, the inclusion of the additive improved the apparent digestibility of the crude fiber and decreased the digestibility of crude protein, nitrogen-free extract and metabolizable energy ($p < 0.05$). There were no significant changes in the production of volatile organic compounds. However, an increase in propionate production was observed ($p = 0.05$). In addition, the inclusion of yeast resulted in an increased phagocytosis index in both treatments ($p = 0.01$). The addition of 0.3 and 0.6% DYC to the diet of dogs wase able to modulate the proportions of some phyla and genera in healthy dogs, in addition to yielding changes in nutrient digestibility, fermentative products and immunity in healthy adult dogs, indicating that this additive can modulate fecal microbiota and be included in dog nutrition.

Keywords: bacteria; beta-glucan; fermentation; illumina; mannan oligosaccharides; prebiotic

1. Introduction

Prebiotics are substrates that are selectively used by host microorganisms and promote health benefits [1]. By modulating the intestinal microbiota, prebiotics can alter the host physiology and improve the fight against innumerable metabolic and immune infections and diseases, such as obesity, diabetes, and inflammatory bowel disease [2]. Yeasts are an example of prebiotics used in nutrition, such as *Saccharomyces cerevisiae*.

Structurally, the yeast cell wall of *S. cerevisiae* is composed of two fractions: one formed by beta-1,3/1,6-glucans and chitin, and another comprising mannoproteins partially formed by mannan oligosaccharides (MOS) [3]. *S. cerevisiae* can be dehydrated and is used as the commercial product Original XPC™ (Diamond V, Cedar Rapids, IA, USA). Its final composition includes MOS, beta-glucans, nucleotides, organic acids, polyphenols, amino acids, vitamins and minerals.

The MOS present in the *Saccharomyces cerevisae* cell wall is able to increase *lactobacilli* populations [4] and fecal *bifidobacterial* populations [5], which are considered beneficial bacteria to the host, and it appears to preserve the integrity of the gut absorption surface [6]. Another component, the beta-glucan, can selectively stimulate the growth of *lactobacilli* populations in a rat model [7], which also suggests a prebiotic activity [8].

Some studies have evaluated the supplementation of prebiotics with similar compositions of *S. cerevisiae* for modulation of fermentative products in dogs and obtained lower concentrations of phenols and indoles [9], reduced fecal ammonia excretion [10], and an increase in short chain fatty acids (SCFA) [11].

Regarding immunological parameters, supplementation with MOS present in the yeast cell wall can induce an increase in white cell blood concentrations, stimulating the immune response against pathogens [12], in addition to reducing inflammatory activity and improving innate immunity [11,13].

Considering these potential benefits, the aim of this experiment was to evaluate the effect of a food enriched with increasing levels of a commercial product called Original XPC™ (Diamond V, Cedar Rapids, IA, USA) (OXPC) composed of dehydrated yeast culture on fecal microbiota, nutrient digestibility, and fermentative and immunological parameters in healthy adult dogs.

2. Materials and Methods

2.1. Ethics

All experimental procedures were approved by the Ethics Research Committee for Animal Welfare of the School of Veterinary Medicine and Animal Science, University of São Paulo (protocol number 9148270415).

2.2. Animals

The experiment was carried out at the Nutritional Development Center and Bromatology Laboratory of the Premier Pet company, located in Dourado, São Paulo, Brazil. Eighteen adult male and female dogs, with a mean weight of 15.8 ± 7.37 kg and mean age of 3.3 ± 1.58 years, were included in this investigation (Table 1). The health status was confirmed prior to the beginning of the experiment. During the experiment, the dogs were housed individually in kennels with dimensions of 2.0 × 5.60 m and solarium of 2.0 × 4.90 m.

2.3. Experimental Design

The animals were randomly assigned to one of three experimental treatments, resulting in 6 replicates per treatment. The whole experiment last 25 days, which included an acclimation period of 14 days for diet adaptation; after this period, total fecal samples were collected for analysis of nutrient digestibility every day for a week (7 days). On the next 3 days, feces were collected immediately after defecation with sterile gloves for fecal pH, fecal fermentation products, and microbiota analysis. On the last day, the blood samples were collected for analysis of immunological parameters.

Table 1. Descriptive information of the animals included in the study.

Animal	Diet	Sex	Breed	Weight (Kg)	Age (Years)
Animal 1	DC	Male	Whippet	11.6	1.5
Animal 2	DC	Female	English Setter	24.4	1.7
Animal 3	DC	Female	Cocker Spaniel	10.5	6.0
Animal 4	DC	Male	Beagle	12.0	5.0
Animal 5	DC	Female	Beagle	12.8	5.0
Animal 6	DC	Male	Golden Retriever	29.0	3.3
Animal 7	OXPC 0.3	Male	Beagle	12.2	4.0
Animal 8	OXPC 0.3	Female	Beagle	10.4	4.0
Animal 9	OXPC 0.3	Female	Whippet	9.1	1.2
Animal 10	OXPC 0.3	Female	English Bulldog	12.6	2.6
Animal 11	OXPC 0.3	Male	West Highland White Terrier	8.4	1.1
Animal 12	OXPC 0.3	Male	Labrador Retriever	30.4	4.2
Animal 13	OXPC 0.6	Male	Beagle	14.5	5.7
Animal 14	OXPC 0.6	Female	Beagle	7.8	1.5
Animal 15	OXPC 0.6	Male	French Bulldog	15.2	2.4
Animal 16	OXPC 0.6	Male	English Setter	26.9	1.6
Animal 17	OXPC 0.6	Male	Beagle	11.4	4.0
Animal 18	OXPC 0.6	Female	Golden Retriever	25.0	4.5

2.4. Diets

A control diet was formulated to meet the requirements of AAFCO [14] for adult dogs under maintenance, and the prebiotic was included in different concentrations. The additive used in this study was Original XPC™ (Diamond V, Cedar Rapids, Iowa, USA) (OXPC), composed of dehydrated yeast culture of *S. cerevisiae*, with approximately 11.50% moisture, 14.90% crude protein, 1.30% fat, 25.20% crude fiber, and 8.50% ash. The OXPC is produced through the fermentation of selected liquids and cereal grains and raw ingredients with *S. cerevisiae*. After this process, the entire culture medium is dried without destroying the yeast factors, B-vitamins, and other nutritional fermentation products to form the final product.

Experimental treatments included increasing levels of the additive, as follows: DC (control diet without OXPC), OXPC 0.3 (control diet with 0.3% OXPC), and OXPC 0.6 (control diet with 0.6% OXPC). All diets were isonutritive and formulated with the same ingredients, differing only by the addition and concentration of OXPC, which was proportionally compensated by starch between diets (Table 2). Diets were extruded at the Pet Unit of Premier Pet, Dourado—SP (Brazil), and all ingredients were obtained from a single batch in order to avoid variability among treatments.

All animals were fed sufficient amounts of calories according to the National Research Council's energy requirement [15] for the maintenance of adult dogs, calculated as $95 \text{ kcal} \times (BW)^{0.75}$ per day, with water offered ad libitum. The daily total amount of the food was divided into two equal portions, offered at 07:00 a.m. and 03:30 p.m. The feeders were removed 30 min after offering the diets. Consumption and food leftovers were measured and recorded throughout the experiment.

Table 2. Ingredient and chemical composition of the experimental diets with and without the additive Original XPC™ (OXPC).

Item	Diets [1]		
	DC	OXPC 0.3	OXPC 0.6
Ingredients (%)			
Starch	1.00	0.70	0.40
Dehydrated yeast culture	–	0.30	0.60
Corn grain	20.91	20.91	20.91
Poultry viscera meal	36.00	36.00	36.00
Broken rice	30.00	30.00	30.00
Poultry fat	8.20	8.20	8.20
Liquid palatability enhancers	2.00	2.00	2.00
Powdered palatability enhancers	0.50	0.50	0.50
Potassium chloride	0.43	0.43	0.43
Common salt	0.30	0.30	0.30
Premix mineral/vitamin [2]	0.52	0.52	0.52
Antifungal	0.10	0.10	0.10
Antioxidant	0.04	0.04	0.04
Chemical composition (% of dry matter)			
Ash	6.65	5.92	6.09
Crude protein	35.66	31.25	33.37
Fat	15.63	16.38	14.36
Nitrogen-free extract [3]	35.35	39.09	39.58
Crude fiber	6.71	7.36	6.59
Gross energy (kcal/g)	5.23	5.22	5.14

[1] DC (control diet), OXPC 0.3 (control diet with 0.3% OXPC) and OXPC 0.6 (control diet with 0.6% OXPC); [2] Addition per kilogram of product: Iron 100 mg, Copper 10 mg, Manganese 10 mg, Zinc 150 mg, Iodine 2 mg, Selenium 0.3 mg, Vitamin A 18000IU, Vitamin D 1200IU, Vitamin E 200IU, Thiamine 6 mg, Riboflavin 10 mg, Pantothenic Acid 40 mg, Niacin 60 mg, Pyridoxine 6 mg, Folic Acid 0.30 mg, Vitamin B12 0.1 mg, and Choline 2000 mg; [3] Nitrogen-free extract was calculated by the formula NFE = 100 − (Moisture + crude protein + fat + crude fiber + ash).

2.5. Determination of the Coefficients of Apparent Digestibility of Nutrients

The apparent digestibility coefficients (ADC) of the nutrients were determined by the total fecal collection method, according to AAFCO [14] recommendations. Individual food consumption was recorded daily, as well as the quantities offered and rejected. Feces were collected within a 24 h period for 7 days, subsequently weighed and conditioned in individual plastic bags, previously identified, closed, and stored in a freezer (-15 °C) for further analysis. At the end of the collection period, they were thawed and homogenized, forming a single sample per animal (feces pool). The content of dry matter (DM), crude protein (CP), ethereal extract in acid hydrolysis, a.k.a. fat (EEAH), ashes, and crude fiber (CF) from food and feces were analyzed according to the methodology described by AOAC [16]. All analyses were conducted in duplicate and repeated when the coefficient of variation was greater than 5%.

Nitrogen-free extract (NFE) was calculated by the difference between DM and the sum of CP, EEAH, CF, and ashes. The gross energy (GE) of food and feces was determined on a calorimetric pump (1281, Parr Instrument Company, Moline, IL, USA). Based on the results obtained in the laboratory, ADC of DM, organic matter (OM), CP, EEAH, CF, and NFE of the diets were calculated. These calculations were performed with the following formula: CDC of the nutrient (%) = [ingested nutrient (g) − excreted nutrient (g)]/(ingested nutrient (g))] × 100.

2.6. Determination of Fecal Score, Fecal pH, and Ammoniacal Nitrogen

The fecal score was evaluated according to grading scores from 0 to 5, of which 0 = liquid stools; 1 = pasty and shapeless stools; 2 = soft, malformed stools that assume the shape of the collection container; 3 = soft, formed, and moist stools that mark the floor;

4 = well-formed and consistent stools that do not mark the floor; 5 = well-formed, hard, and dry stools. Values between 3 and 4 were considered as ideal fecal score [17].

For determination of fecal pH, a 2 g sample of feces was separated and diluted in 18 mL of distilled water within 30 min after defecation. Determination was carried out with a digital benchtop pH meter (Digimed, DM-20, São Paulo, SP, Brazil), according to the methodology adapted from Walter et al. [18]. For quantification of fecal ammoniacal nitrogen, stool samples were collected within 30 min after defecation. The sampling process, as well as the distillation, were performed according to Sá et al. [19].

2.7. Evaluation of Volatile Organic Compounds in Feces

Fresh feces were collected from the animals within 30 min after defecation and were quickly homogenized; 0.5 g of sample was placed in a sealed 20 mL glass vial with a leak-proof metal cap and double-sided silicone/Teflon. Samples were stored and maintained at $-20\,°C$. The samples were evaluated by gas chromatography coupled to mass spectrometry (GC-MSD) (Agilent Technologies, Santa Clara, CA, USA) using an Agilent 7890 A gas chromatograph (CG) and an Agilent 5975C mass sorting detector (MSD), according to an adapted methodology [20,21]. The NIST mass spectra library of 2008 was used to identify the compounds detected.

2.8. Determination of Short-Chain (SCFA) and Short Branched Chain Fatty Acids (SBCFA) and Lactic Acid in Feces

Stool samples were collected within 30 min after defecation. Subsequently, they were homogenized and weighed for the quantification of SCFA and SBCFA. Three grams of feces were acidified with 9 mL of 16% formic acid in a 15 mL falcon tube. The determination of the short and short branched chain fatty acids was performed by gas chromatography (Shimadzu Corporation, Kyoto, Japan) according to Erwin et al. [22].

Lactic acid was measured according to the methodology described by Pryce et al. [23], by the spectrophotometry method at 565 nm (500 to 570 nm). After collection, the feces were homogenized and mixed with 9 mL of distilled water (1: 3 v/v).

2.9. Determination of Biogenic Amines

Five grams of fresh stool were collected in duplicate within 30 min after defecation and stored in 7 mL of 5% trichloroacetic acid and refrigerated. Subsequently, the samples were centrifuged at $10,000\times g$ for 20 min at 4 °C, and the supernatant was filtered on qualitative filter paper. The residue was extracted two more times using 7 mL and 6 mL of 5% trichloroacetic acid. The supernatants were combined for further determination of the biogenic amines. The determination and separation of the biogenic amines were performed by high-performance liquid chromatography by reverse phase column ion pairing and subsequently quantified by fluorimetry after post-column derivation with second ophthalaldehyde [24]. The amines were identified by comparing the retention time of the peaks found in the samples with those of the amines of the standard solution, according to the methodology described by Gomes et al. [25].

2.10. Determination of Fecal Bacteria Concentration by Illumina Sequencing Technology

After a 7-day digestibility period, fecal samples were immediately and aseptically collected for 3 days for microbiota determination. The samples did not have contact with any other surface besides sterile gloves. The DNA extraction was performed using the Mobio Power Soil kit (MO BIO Laboratories, Carlsbad, CA, USA) according to the methodology described by McInnes et al. [26]. After extraction, the DNA concentration was quantified using a Qubit® 2.0 Fluorometer (Life Technologies, Grand Island, NY, USA).

Amplifications of the 16S rRNA gene were generated using a Fluidigm Access matrix (Fluidigm Corporation, South San Francisco, CA, USA) in combination with a Roche High Fidelity Fast Start Kit (Roche, Indianapolis, IA, USA). For this step, primers 515F (5′-GTGCCAGCMGCCGCGGTAA-3′) and 806R (5′-GGACTACHVGGGTWTCTAAT-3′)

291 bp-fragment of the V4 region were used [27]. After this, Fluidigm primer specific forward (CS1 tag) and inverse (CS2 tag) were added according to the Fluidigm protocol. To confirm the quality of the regions and the sizes of the amplicons, the fragment analyzer (Advanced Analytics, Ames, IA, USA) was used.

A pool of DNA was generated by combining equimolar amounts of the amplified fragments from each sample. The pooled samples were selected by gel size from 2% E-gel agarose (Life Technologies, Grand Island, New York, NY, USA) and extracted using Qiagen gel purification kit (Qiagen, Valencia, CA, USA). To confirm the appropriate profile and average size, the sorted and cleaned clustered products were run on an Agilent Bioanalyzer.

The characterization of the microbial community through the Illumina sequencing was performed on a MiSeq using V3 reagents (Illumina Inc., San Diego, CA, USA) at the W.M. Keck Center for Biotechnology at the University of Illinois. Fluidigm tags were removed using FASTX-Toolkit (version 0.0.13); to process the resulting sequence data, QIIME 1.8.1 was used according to Caporaso et al. [28]. The sequence data were imported from demultiplexed fastq files, and we filtered out low-quality sequencing reads considering a quality score threshold of 25. After that, the sequences were grouped into operational taxonomic units (OTUs) using OTU open reference and were compared against the reference database OTU Greengenes 13_8 using a 97% similarity threshold.

An even sampling depth of sequences per sample was used for assessing alpha and beta diversity measures. A total of 917,433 reads were obtained, with an average of 50,436 reads (range = 17,666–93,602) per sample. Rarefaction curves based on observed species, Chao1, and phylogenetic distance (PD) whole tree measures plateaued, suggesting enough sequencing depth. The dataset was rarified to 16,300 reads for analysis of diversity and species richness. Principal coordinate analysis (PCoA) was performed, using both weighted and unweighted unique fraction metric (UniFrac) distances, to measure the phylogenetic distance between sets of taxa in a phylogenetic tree as the fraction of the branch length of the tree, on the 97% OTU composition and abundance matrix [29]. The unweighted distance checks for the presence or absence of different taxa of microbial communities between/among samples, whereas the weighted distance investigates proportional changes in the microbial community.

2.11. Phagocytosis and Oxidative Burst Test

In order to perform this assay, blood leukocytes (lymphocytes, neutrophils, and monocytes) were incubated with a fluorescent reagent that indicated the production of reactive oxygen species in the basal state, and after carrying out phagocytosis of *Staphylococcus aureus* bacteria, which indicated the percentage and intensity of phagocytosis. Cells were incubated with the DCFH-DA reagent in phosphate buffered saline (PBS) and DCFH and fluorochrome-labeled (propidium iodide) labeled *S. aureus* and maintained at 37 °C for 20 min. After this period, the red cells were disrupted with a lysis solution and washed with PBS until a clear-looking sample was obtained. This sample was then read on a FACS Calibur (Becton & Dickinson, Franklin Lakes, NJ, USA) flow cytometer.

2.12. Lymphoproliferation Test

The assay was performed in 96-well microtiter plates with U-shaped bottom. Blood lymphocytes were obtained by separation into iron particles and, after purification and washes in RPMI-1640 medium, were added to the wells in a concentration equivalent to 1×10^5 cells/well in 200 µL/well. The mitogens used were Concanavalin A and Phytohemagglutinin. The plates were incubated for 72 h at 37 °C in 5% CO_2 atmosphere. After 72 h of incubation, cells were collected and evaluation of proliferation was performed on a FACS Calibur (Becton & Dickinson, Franklin Lakes, NJ, USA) flow cytometer. For the analysis of fluorescence data, the values of the percentage of cell divisions and the index of cell proliferation were considered. In order to obtain and analyze the results, we used the software Cell Queste Flow Jo (TreeStar).

2.13. Immunophenotyping of Lymphocytes

The number of CD4/CD8+ CD4+ CD8+ lymphocytes (CD3/CD4+/CD45R−) and CD4/CD8+ lymphocytes (CD3/CD8+/CD45R−) was assessed. Mononuclear cells (2×10^5 cells/mL) were incubated in microtubes (1.5 mL) with CD3 (1:100), CD4 (1:10), CD8 (1:20), CD21 (1:100), and CD45R (1:100) (Serotec Antibody, Biolegend and eBioscience), diluted in 100 μL of cytometry buffer (PBS containing 0.5% bovine serum albumin and 0.02% sodium azide). The isotypic antibodies for background definition were included in the assay. Cells were incubated for 20 min at 4 °C, protected from light. At the end of the incubation period, the samples were washed twice with buffer for cytometry in a volume of 1000 μL/microtube. Finally, the cells were resuspended in 500 μL of phosphate buffer. The population of cells with low size (FSC) and low complexity (SSC) according to the delimited gate was selected as the lymphocyte population. From this selection, the different populations of lymphocytes were determined. The acquisition and analysis of 10,000 cells were performed using the flow cytometry technique.

2.14. Statistical Analysis

The results were analyzed using the Statistical Analysis System (SAS Institute Inc., Cary, NC, USA, 2004). The normality of the residues was verified by the Shapiro–Wilk test (PROC UNIVARIATE) and the variances compared by the F test. The statistical assumptions underwent logarithmic or square root transformation, and then analysis of variance was performed by PROC GLM of the SAS with the means compared by the Tukey test at 5% of significance, as well as by simple polynomial regression, considering 2 degrees of freedom (linear and deviation).

The abundances observed for phyla and genera of each animal were evaluated by means of a generalized linear model, considering binomial distribution and a logit link function. The model included the fixed effect of the treatments (OXPC levels: 0.0, 0.3% or 0.6%) in addition to the random effect of the residue. The Tukey multiple comparison test was performed to identify which specific means differed at 5% significance level. All analyses were performed using the PROC GENMOD of the SAS procedure from the Statistical Analysis System, version 9.4 (SAS Institute Inc., Cary, NC, USA).

3. Results

There was no difference between the treatments in relation to the average daily consumption of DM, NM, OM, CP, EEAH, CF, ash, and GE ($p > 0.05$; Table 3). The inclusion levels assessed for OXPC did not influence the ADC of DM, MO, ash, EEAH, and GE ($p > 0.05$; Table 3). The presence of additive decreased the digestibility of NFE ($p = 0.04$), CP ($p = 0.01$), and the metabolizable energy of the diets ($p < 0.01$), but increased the digestibility of CF ($p < 0.001$).

There were no differences for fecal production in organic matter, dry matter, and fecal score ($p > 0.05$; Table 4).

The presence of OXPC increased the propionic acid amount compared to that in the control group ($p = 0.05$). The other variables of intestinal fermentation did not differ between the treatments in this study ($p > 0.05$; Table 4). There was no difference among treatments regarding VOCs ($p > 0.05$; Table 5).

Regarding fecal microbiota, the alpha diversity was measured to determine the number of OTUs and then to give a basic measure of the bacterial diversity within each sample. All samples showed similar rarefaction curves regardless of treatment, indicating that these samples had similar diversity and no treatment effect (Figure 1). The principal coordinate analysis (PCoA) measures the overall bacterial genera relatedness, where the samples with similar bacterial communities are localized in similar positions in the diagram. Figures 2 and 3 suggest that OXPC supplementation at both levels (0.3% and 0.6%) did not have a beta-diversity effect on fecal microbiota.

Table 3. Intake of nutrients, apparent digestibility coefficients of nutrients, and metabolizable energy from experimental diets with different doses of the additive Original XPC™ (OXPC) given to the adult dogs.

Item	Diets [1]			SEM [2]	*p* Value
	CO	OXPC 0.3	OXPC 0.6		
	Consumption (g/day)				
Natural matter	197.33	170.67	198.00	15.07	0.75
Dry matter	187.95	156.28	180.54	14.46	0.69
Organic matter	184.21	160.56	185.94	14.54	0.76
Crude protein	70.37	53.34	66.08	5.37	0.45
Ethereal extract in acid hydrolysis	30.83	27.95	28.44	2.38	0.89
Crude fiber	13.25	12.55	13.05	1.05	0.97
Ash	13.12	10.11	12.06	0.99	0.50
Nitrogen-free extract	69.76	66.72	78.37	5.91	0.74
Gross energy (Kcal/day)	1031.29	891.74	1018.82	80.56	0.77
	Apparent digestibility coefficient (%)				
Dry matter	85.84	85.09	85.43	0.72	0.76
Organic matter	89.10	88.44	88.35	0.58	0.62
Ash	40.08	31.82	40.41	3.13	0.12
Crude protein	90.56 [a]	86.90 [c]	88.62 [b]	0.74	0.01
Crude fiber	60.95 [b]	74.41 [a]	72.13 [a]	2.03	<0.001
Ethereal extract in acid hydrolysis	94.44	95.78	95.72	0.79	0.42
Nitrogen-free extract	91.02 [a]	89.24 [b]	88.15 [b]	0.72	0.04
Gross energy	89.69	88.78	88.60	0.56	0.36
Metabolizable energy (Kcal/kg of food consumed)	4100 [a]	4003 [b]	3902 [c]	33.38	<0.01

[1] DC (control diet), OXPC 0.3 (control diet with 0.3% OXPC) and OXPC 0.6 (control diet with 0.6% OXPC); [2] SEM, standard error of the mean. [a,b,c] mean in the lines followed by the same letters do not differ by Tukey test ($p > 0.05$).

The predominant fecal phyla present in all dogs included Firmicutes, Fusobacteria, and Bacteriodetes (Table 6). Together, Firmicutes and Fusobacteria constituted about 85–88% of the bacterial sequences, and Bacteroidetes contributed 8–12% of the sequences. An increase ($p = 0.0048$) in the abundance of fecal Actinobacteria was observed as the dose of the OXPC diet increased (Table 6). Besides that, the concentration of Firmicutes increased ($p < 0.0001$) while Fusobacteria decreased with the additive inclusion ($p = 0.0008$).

Table 4. Fecal quality and production, concentration of lactic acid, short and branched chain fatty acids, and biogenic amines of feces from dogs fed with different doses of the additive Original XPC™ (OXPC).

Item	Diets [1]			SEM [2]	*p* Value
	CD	OXPC 0.3	OXPC 0.6		
Fecal production, fecal quality and lactic acid					
Fecal production g MN/dog/day	85.34	73.54	88.54	14.46	0.75
Fecal production g MS/dog/day	26.16	23.11	26.37	3.86	0.80
Fecal score	3.97	3.88	3.97	0.05	0.42
Fecal pH	6.66	6.56	6.44	0.10	0.34
N ammoniacal	130.53	150.72	165.67	24.33	0.60
Lactic acid	13.22	16.30	11.69	3.56	0.62
Short chain fatty acids. mmol/Kg of dry matter					
Acetic acid	55.02	86.47	84.17	10.00	0.07
Propionic acid	25.77 [b]	42.98 [a]	40.80 [a]	4.94	0.05
Butyric acid	9.41	12.04	13.00	2.16	0.49
SCFA [3] total	90.21	141.5	137.98	10.05	0.07
Branched chain fatty acids. mmol/Kg of dry matter					
Valeric acid	–	–	–	–	–
Iso-valeric acid	2.20	2.58	3.40	0.59	0.36
Iso-butyric acid	1.95	2.00	2.55	0.39	0.50
SBCFA [4] total	4.16	4.58	5.96	0.55	0.42
Total fatty acids	94.36	146.08	143.94	16.46	0.07
Biogenic amines. mg/Kg of feces in the of natural matter					
Tyramine	80.38	12.47	65.91	43.88	0.53
Putrescin	130.54	92.14	106.35	26.33	0.59
Cadaverine	54.42	18.31	37.29	24.57	0.56
Spermidine	41.22	34.80	41.25	5.67	0.65
Phenylethylamine	–	4.25	3.04	–	–
Tryptamine	1.81	3.18	2.44	–	-
Total amines	271.32	155.35	223.95	41.25	0.56

[1] DC (control diet), OXPC 0.3 (control diet with 0.3% OXPC) and OXPC 0.6 (control diet with 0.6% OXPC); [2] SEM, standard error of the mean; [3] SCFA. short chain fatty acids; [4] SBCFA. short branched chain fatty acids. [a,b] mean in the lines followed by the same letters do not differ by Tukey test ($p > 0.05$).

Table 5. Mean percentage of the peak area of the most abundant volatile organic compounds present in feces from dogs fed with different doses of the additive Original XPC™ (OXPC).

Item	Diets [1]			SEM [2]	*p* Value
	DC	OXPC 0.3	OXPC 0.6		
Acetic acid	12.88	13.78	12.56	3.19	0.95
Butanoic acid	4.52	6.26	8.33	1.47	0.28
Ethanol	1.49	3.66	7.08	1.48	0.08
Indol	6.80	14.10	7.82	2.45	0.09
Phenol	1.59	4.18	4.26	1.10	0.24
Propanoic	11.73	15.01	14.76	2.29	0.54
2-piperidinone	2.35	2.87	1.86	0.53	0.43

[1] DC (control diet), OXPC 0.3 (control diet with 0.3% OXPC) and OXPC 0.6 (control diet with 0.6% OXPC); [2] SEM, standard error of the mean.

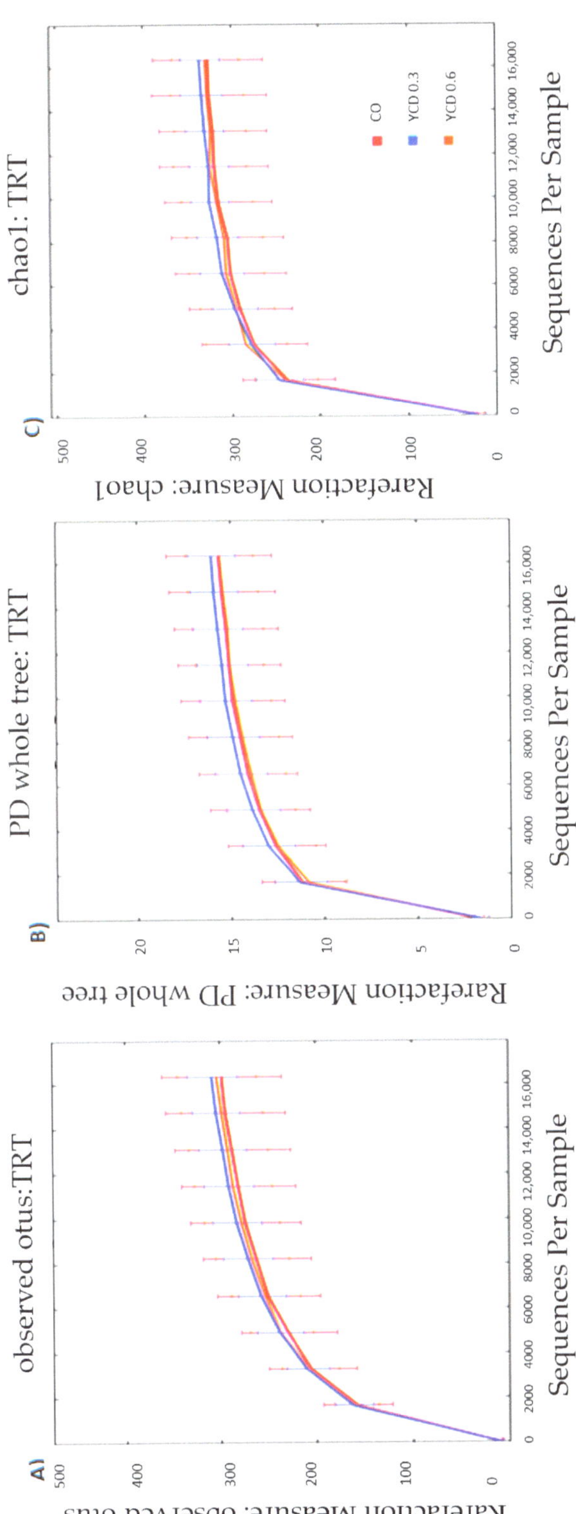

Figure 1. Alpha diversity: Rarefaction curves based on observed operational units (observed OTU) (**A**), phylogenetic distances (PD whole tree) (**B**) and metrics (Chao1) (**C**) according to the diet consumed by the animals. X axis represents the sequence depth (16,300 readings/sample), lines represent the mean of each group (red = control diet, blue = control diet with 0.3% OXPC, and orange = control diet with 0.6% OXPC).

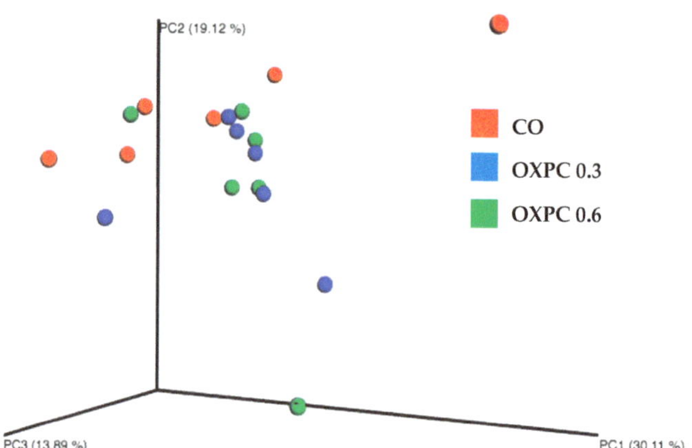

Figure 2. Beta diversity: Principal coordinate analysis (PCoA) of the unweighted portion of the unique metric fraction (Unifrac), according to the diet consumed by the animals. The plot showing clustering of microbial communities from feces of dogs fed with 0% (red), 0.3% (blue), and 0.6% (green) of OXPC. The closer the items, the more similar the microbial communities in the samples.

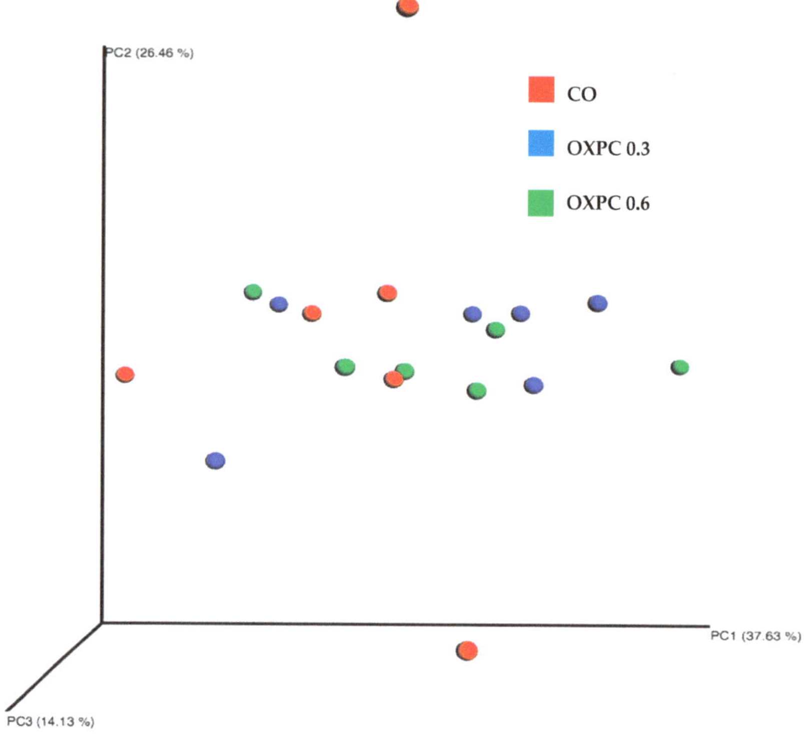

Figure 3. Beta diversity: Principal coordinate analysis (PCoA) of the weighted portion of the unique metric fraction (Unifrac), according to the diet consumed by the animals. The plot showing clustering of microbial communities from feces of dogs fed with 0% (red), 0.3%(blue), and 0.6% (green) OXPC. The closer the items, the more similar the microbial communities in the samples.

Table 6. Prominent bacterial phyla (expressed as percentage of total sequences) in feces of dogs fed with different doses of the additive Original XPC™ (OXPC).

Item	Diets [1]			SEM [2]	*p* Value
	CO	OXPC 0.3	OXPC 0.6		
Unassigned bacteria	0.011 ± 0.0004	0.010 ± 0.0004	0.019 ± 0.0006	0.0078	0.9886
Actinobacteria	0.160 [b] ± 0.0016	0.703 [a,b] ± 0.0034	1.869 [a] ± 0.0055	0.3311	0.0048
Bacteroidetes	11.990 ± 0.0133	9.370 ± 0.0119	8.150 ± 0.0112	30.550	0.0768
Deferribacteres	0.006 ± 0.0003	0.019 ± 0.0006	0.0101 ± 0.0004	0.0088	0.9754
Firmicutes	60.050 [b] ± 0.0200	70.380 [a] ± 0.0186	70.760 [a] ± 0.0186	5.937	<0.0001
Fusobacteria	25.500 [a] ± 0.0178	17.960 [b] ± 0.0157	17.620 [b] ± 0.0156	41.330	0.0008
Proteobacteria	2.280 ± 0.0061	1.560 ± 0.0051	1.570 ± 0.0051	0.5619	0.5709

[1] DC (control diet), OXPC 0.3 (control diet with 0.3% OXPC), and OXPC 0.6 (control diet with 0.6% OXPC); [2] SEM, standard error of the mean. [a,b] mean in the lines followed by the same letters do not differ by Tukey test (*p* > 0.05).

The predominant fecal bacterial genera were *Clostridium* (18–25%), *Fusobacterium* (16–25%), and *Blautia* (7–11%) (Table 7). The fecal concentrations of *Prevotella*, *Allobaculum*, and *Fusobacterium* were lower after including OXPC (*p* = 0.0003, *p* = 0.0265, and *p* = 0.0006 respectively; Table 7). The *Clostridium* proportion increased with OXPC inclusion (*p* = 0.0059; Table 7) and the *Collinsella* proportion was also greater when the highest prebiotic level was supplemented (*p* = 0.0052; Table 7).

Table 7. Prominent bacterial genera (expressed as percentage of total sequences) in feces of dogs fed different doses of the additive Original XPC™ (OXPC).

Item	Diets [1]			SEM [2]	*p* Value
	DC	OXPC 0.3	OXPC 0.6		
Unassigned bacteria	0.012 ± 0.0004	0.010 ± 0.0004	0.020 ± 0.0006	0.0078	0.9886
Actinobateria					
Bifidobacterium	0.056 ± 0.0010	0.090 ± 0.0012	0.016 ± 0.0005	0.062	0.8422
Collinsella	0.050 [b] ± 0.009	0.427 [b] ± 0.0027	1.441 [a] ± 0.0049	0.256	0.0052
Slackia	0.054 ± 0.0009	0.186 ± 0.0018	0.412 ± 0.0026	0.093	0.3835
Bacteroidetes					
Bacteroides	6.946 ± 0.0104	7.392 ± 0.0107	6.465 ± 0.0100	2.356	0.8188
Parabacteroides	0.126 ± 0.0014	0.070 ± 0.0011	0.079 ± 0.0011	0.044	0.9446
Prevotella	3.044 [a] ± 0.0070	0.584 [b] ± 0.0031	0.561 [b] ± 0.0031	0.355	0.0003
S24-7	0.102 ± 0.0013	0.009 ± 0.0004	0.007 ± 0.0003	0.036	0.6529
Other [Paraprevotellaceae]	0.441 ± 0.0027	0.419 ± 0.0026	0.234 ± 0.0020	0.162	0.7954
[Paraprevotellaceae]	0.045 ± 0.0009	0.035 ± 0.0008	0.017 ± 0.0005	0.019	0.9581
[Paraprevotellaceae] [Prevotella]	1.287 ± 0.0070	0.863 ± 0.0031	0.789 ± 0.0031	0.355	0.6505
Deferribacteres					
Mucispirillum	0.006 ± 0.0003	0.019 ± 0.0006	0.011 ± 0.0004	0.009	0.9754
Firmicutes					
Lactobacillus	0.003 ± 0.0002	0.402 ± 0.0026	0.003 ± 0.0002	0.230	0.0841
Streptococcus	0.015 ± 0.0005	0.061 ± 0.0010	0.007 ± 0.0003	0.033	0.8391
Turicibacter	0.108 ± 0.0013	0.017 ± 0.0005	0.010 ± 0.0004	0.047	0.6865
Other Clostridiales	0.290 ± 0.0022	0.417 ± 0.0026	0.183 ± 0.0017	0.137	0.7529
Clostridiales	0.433 ± 0.0027	0.368 ± 0.0025	0.457 ± 0.0028	0.092	0.9696

Table 7. *Cont.*

Item	Diets [1]			SEM [2]	*p* Value
	DC	OXPC 0.3	OXPC 0.6		
Other Clostridiaceae	0.175 ± 0.0017	0.209 ± 0.0019	0.305 ± 0.0023	0.043	0.8914
Clostridiaceae	2.641 ± 0.0065	2.730 ± 0.0067	2.530 ± 0.0064	0.348	0.9768
Clostridium	18.185 [b] ± 0.0158	24.513 [a] ± 0.018	25.043 [a] ± 0.0177	3.931	0.0059
Pseudoramibacter Eubacterium	0.041 ± 0.0008	0.020 ± 0.0006	0.004 ± 0.0003	0.024	0.9024
Other Lachnospiraceae	0.547 ± 0.0030	0.840 ± 0.0037	0.801 ± 0.0036	0.191	0.8033
Lachnospiraceae	3.769 ± 0.0078	3.402 ± 0.0074	4.975 ± 0.0089	0.754	0.3603
Blautia	7.122 ± 0.0105	10.870 ± 0.013	9.698 ± 0.0121	1.573	0.0667
Coprococcus	0.073 ± 0.0011	0.134 ± 0.0015	0.198 ± 0.0018	0.039	0.8358
Dorea	2.404 ± 0.0063	3.790.0078	3.899 ± 0.0079	0.620	0.2579
Roseburia	0.031 ± 0.0007	0.007 ± 0.0004	0.017 ± 0.0005	0.009	0.9543
[Ruminococcus]	2.930 ± 0.0069	4.378 ± 0.0084	4.314 ± 0.0083	0.127	0.3235
Peptococcus	0.956 ± 0.0040	1.299 ± 0.0046	0.700 ± 0.0034	0.622	0.5734
Peptostreptococcaceae	0.538 ± 0.0030	0.237 ± 0.0020	0.365 ± 0.0025	0.192	0.6946
Ruminococcaceae	1.639 ± 0.0052	1.488 ± 0.0049	1.526 ± 0.0050	0.277	0.9762
Fecalibacterium	5.828 ± 0.0096	4.650 ± 0.0086	5.355 ± 0.0092	1.162	0.6535
Ruminococcus	0.300 ± 0.0022	0.190 ± 0.0018	0.098 ± 0.0013	0.127	0.7235
Megamonas	0.665 ± 0.0033	0.980 ± 0.0040	1.970 ± 0.0057	0.685	0.1052
Phascolarctobacterium	0.276 ± 0.0021	0.220 ± 0.0019	0.151 ± 0.0016	0.103	0.8935
[Mogibacteriaceae]	0.038 ± 0.0008	0.001 ± 0.0001	0.0003 ± 0.0001	0.011	0.8000
Erysipelotrichaceae	2.825 ± 0.0068	2.383 ± 0.0062	2.199 ± 0.0060	0.525	0.7750
Allobaculum	6.905 [a] ± 0.0104	3.850 [b] ± 0.079	3.996 [b] ± 0.0080	3.318	0.0265
Catenibacterium	0.455 ± 0.0027	1.014 ± 0.0041	1.016 ± 0.0041	0.349	0.4300
Clostridium	0.024 ± 0.0006	0.018 ± 0.0006	0.020 ± 0.0006	0.011	0.9975
Coprobacillus	0.056 ± 0.0010	0.183 ± 0.0017	0.016 ± 0.0005	0.041	0.5815
[Eubacterium]	0.781 ± 0.0036	1.697 ± 0.0053	0.906 ± 0.0039	0.390	0.2840
Fusobacteria					
Other Fusobacteriaceae	0.611 ± 0.0032	0.643 ± 0.0033	0.692 ± 0.0034	0.070	0.9849
Fusobacterium	24.888 [a] ± 0.0177	17.319 [b] ± 0.015	16.928 [b] ± 0.0153	4.106	0.0006
Proteobacteria					
Sutterella	0.894 ± 0.0038	0.487 ± 0.0028	0.300 ± 0.0022	0.277	0.3770
Campylobacter	0.000 ± 0.0000	0.051 ± 0.0009	0.016 ± 0.0005	0.028	0.8022
Succinivibrionaceae	0.031 ± 0.0007	0.004 ± 0.0003	0.278 ± 0.0022	0.103	0.2653
Anaerobiospirillum	1.328 ± 0.0047	0.865 ± 0.0038	0.956 ± 0.0040	0.371	0.7132
Enterobacteriaceae	0.024 ± 0.0006	0.148 ± 0.0016	0.015 ± 0.0005	0.084	0.6069

[1] DC (control diet), OXPC 0.3 (control diet with 0.3% OXPC), and OXPC 0.6 (control diet with 0.6% OXPC); [2] SEM, standard error of the mean. [a,b] mean in the lines followed by the same letters do not differ by Tukey test ($p > 0.05$).

There was no difference for CD4+ and CD8+ lymphocytes, CD4+/CD8+ ratio, oxidative burst (baseline and SAPI and PMA induced), and lymphocyte proliferative response ($p > 0.05$; Table 8). However, the phagocytosis index was higher with inclusion of OXPC compared to that of control ($p = 0.01$).

Table 8. Results of lymphocyte immunophenotyping, phagocytosis test, proliferation, and oxidative burst of dogs fed different doses of the additive Original XPC™ (OXPC).

Item	Diets [1]			SEM [2]	*p* Value
	DC	OXPC 0.3	OXPC 0.6		
CD4+ % (T helper cells)	28.66	28.33	28.83	1.28	0.96
CD8+ % (cytotoxic T cells)	15.16	14.33	16.83	0.87	0.15
CD4+/CD8+	6.26	2.16	1.56	1.77	0.16
Basal oxidative burst	216.17	206.50	196.83	8.00	0.26
Oxidative burst SAPI [3]	497.83	575.83	507.00	29.45	0.15
Oxidative burst PMA [4]	764.33	778.00	774.67	60.51	0.98
Phagocytosis index	264.67 [b]	295.17 [a]	297.17 [a]	7.38	0.01
Proliferation index	310.33	355.50	356.17	16.14	0.10

[1] DC (control diet), OXPC 0.3 (control diet with 0.3% OXPC), and OXPC 0.6 (control diet with 0.6% OXPC); [2] SEM, standard error of the mean; [3] SAPI, *Staphylococcus aureus* conjugated with propidium iodide; [4] PMA, phorbol 12-myristate 13-acetate. [a,b] mean in the lines followed by the same letters do not differ by Tukey test ($p > 0.05$).

4. Discussion

Few studies have evaluated nutrient digestibility in dogs fed with diets supplemented with additives similar to OXPC, such as MOS mixed with basal diet [10], spray-dried yeast cell wall [30], *S. cerevisiae* live yeast [31], or *S. cerevisiae* fermentation product [6], among others. Different results were found with regard to nutrient digestibility that mentioned increases, decreases, and even non-alteration. Therefore, there is no consensus on the influence of the addition of this additive on digestibility parameters. Nonetheless, in this study, there was a decrease in ADC of CP and NFE and an increase in the ADC of the CF. The increase in intestinal microbial biomass caused by the inclusion of prebiotics in the diet may reduce ADC of CP [30]. Fecal bacterial mass enhances fecal protein content, which implicates in less ADC compared to the control group. Ideally, the true digestibility coefficient of the protein should be evaluated, in order to avoid considering fecal microbial protein content [32]. In regard to NFE digestibility, the OXPC consisted of dehydrated yeast culture, which has soluble fiber in its composition. That may have increased the viscosity of the bolus and impaired the interaction of pancreatic enzymes with the substrate, thereby decreasing the rate of carbohydrate digestion by pancreatic amylase [33].

Although the ADC of CF increased, this result must be evaluated with caution. The methodology used to determine CF was not entirely satisfactory. The laboratory technique is deficient because it yields low estimates of the fiber fraction present in the samples, destroying all of the soluble fraction and part of the insoluble fraction [34]. The main limitation is related to the fact that it does not separate cellulose from hemicellulose and causes loss of lignin (which is not considered carbohydrate) and hemicellulose. This method provides values that may change due to very drastic digestion, which leads to the loss of some components, and therefore, the values and differences obtained in our study may not be accurate [35]. The differences found in the metabolizable energy content of the foods may be actually a reflection of the small variations in the crude energy of the diets and levels and types of fibers.

Regarding fecal pH, no differences were observed among treatments, as well as in the study developed by Swanson et al. [4], who supplemented dogs with 2 g of FOS plus 1 g MOS. It is known that lactate produced by lactobacillus can lower fecal pH [36], and in this study, the authors attributed the non-detection of some bacterial species that consume lactate to this absence of differences in the results of fecal pH. Besides that, the SCFAs are absorbed quickly in the intestine, and may not be possible to identify in large amounts in feces that could have masked minor effects on pH [4].

In our study, the addition of OXPC at the concentrations of 0.3 and 0.6% were not capable of altering this genus population, which may have been implicated in the lack of lactic acid alteration. In addition, in a study by Vickers et al. [37], the authors evaluated the fermentation characteristics of different substrates found in canine diets and could observe an increase in lactate with the use of FOS; however, when MOS was used, this product had its concentration decreased. *Saccharomyces cerevisae* processing or concentration also may explain the differences in these results.

Volatile organic compounds (VOCs) reflect differences in diet, intestinal microbiota, and exposure to chemical contaminants, as they are usually generated by the metabolism of intestinal microorganisms [38]. In this study, the main fecal VOCs did not change among treatments, which can be considered a positive effect, once it is related to lower stool odor [39]. In dogs, the presence of fecal VOCs was identified with the inclusion of prebiotics [40].

Total SCFAs did not change with OXPC addition, which can be explained by their rapid absorption by colonocytes, reducing fecal detection by presenting smaller quantities in feces [4]. These findings corroborate other studies [9,12,31]. SCFAs are associated with cell proliferation due to their role in the energy metabolism of colonocytes and are among the products generated in prebiotic fermentation [41]. Among them, propionic acid showed a higher fecal concentration in dogs treated with OXPC. The MOS present in OXPC reduces colonization by pathogenic bacteria in the intestine by competitive exclusion [30], which may result in an increase in propionic acid.

Biogenic amines are putrefactive compounds that can cause damage to intestinal health [42]. The extra source of energy promoted by fermentable carbohydrates, undigested protein, and their metabolites are used by bacteria for protein synthesis, decreasing the fecal concentration of protein-derived fermentation compounds [43]. No difference between treatments was observed for biogenic amines, which also corroborates the results of Swanson et al. [4].

According to Slavin [44], the level of inclusion of the prebiotic, as well as its source and time of use, can influence its effect. In a recent study, Perini et al. [45] compared the efficacy of prebiotics over 30 and 60 days of supplementation, and observed some changes in fermentation products over time. Therefore, the period of 21 days may not have been long enough to observe the effects of these prebiotics on the fermentative products evaluated in this study.

The microbial balance in the gastrointestinal tract is mostly determined by diet, and prebiotics can influence the gastrointestinal microbiota [46]. Likewise, evaluating the effect of adding a fermented *S. cerevisae* dry product in vitro, Possemiers et al. [47] did not find strong changes in the microbial community composition of the mucosal associated microbiota.

The predominant microbial phyla in the canine and feline gut, reported by previous investigations are *Firmicutes, Bacteroidetes, Proteobacteria, Fusobateria,* and *Actinobacteria. Fusobacteria* is one of the predominant phyla in the intestinal microbiome (intestine or feces). It frequently represents 10% or more of the genera sequences that inhabit the intestine [32]. This characteristic was also observed in this study, where the predominant fecal phyla present in all dogs included *Firmicutes, Fusobacteria, Bacteriodetes, Proteobacteria, Actinobacteria,* and *Deferribacteres.*

Middelbos et al. [48] phylogenetically characterized the fecal microbiota of healthy dogs with 454 pyrosequencing for the first time. The dominant phyla were Fusobacteria (23–40% of the readings), followed by *Firmicutes* (14–28% of the readings), *Bacteroidetes* (31–34% of the readings), *Actinobacteria* (0.8–1.4% of the readings), and *Proteobacteria* (5–7% of readings). Although *Fusobacteria* was not the most dominant, it was among the phyla with the highest proportion. It was also observed in the study of Beloshapka et al. [46], in which *Fusobacteria* and *Firmicutes* constituted about 75–80% of the bacterial sequences, with *Bacteroidetes, Proteobacteria,* and *Actinobacteria* contributing only about 10–15%, 5%, and 2–3% of the sequences, respectively.

An increase in *lactobacilli, bifidobacterium,* and aerobic bacteria was reported when supplementing the diet of healthy dogs with a combination of 2 g FOS and 1 g MOS [4], as FOS is preferentially fermented by lactic acid-producing bacteria [49]. This effect may have been evidenced by the type of additive used and its processing, and for this reason, this result was not observed in our study.

The abundance of fecal *Actinobacteria* increased with the highest dose of OXPC in the diet and has also been observed in cats fed a diet containing a prebiotic [50]. This increase was correlated to microbiota adaptation of the prebiotic, which led to the increase in a genus belonging to the phylum *Actinobacteria*. The same may have occurred in this study, where the presence of OXPC stimulated the increase in the genus *Collinsella*, which belongs to the phylum *Actinobacteria* that was responsible for this increase. In the same way, an increase in *Clostridium* was observed, which may be correlated with the increase in the *Firmicutes* phylum, to which it belongs.

The members of the phylum *Firmicutes*, especially those of the genus *Clostridium*, can provide some benefits to the animals and are positively correlated with the oxidation of carbohydrates [51]. Although clostridial species are not all considered negative [52], an increase in the population of a potentially pathogenic bacterial genus could be considered a disadvantage of adding OXPC to dog's diet. Despite this, the animals did not shown clinical signs of infection, so it is reasonable to associate the increase in *Clostridium* with non-pathogenic strains.

Allobaculum also belongs to the *Firmicutes* phylum and is associated with weight regulation and regulation of hormones known to influence energy homeostasis (e.g., leptin) [53]. *Fusobacterium* reduction was observed with the inclusion of OXPC in the diet of the animals. This effect is considered beneficial since this genus is associated with gastrointestinal diseases [54,55].

Previous studies have shown an association of the genus *Prevotella* with diets containing high concentrations of carbohydrates [56,57]. However, the addition of OXPC reduced the concentration of this genus to the detriment of the others, which highlights the importance of conducting investigations on the interactions between bacterial populations and dietary substrates. We hypothesized that a greater inclusion of OXPC would allow the capture of evident differences among the bacterial groups.

Among the immunological tests performed in the study, the phagocytosis index presented greater activity in the animals that were supplemented with OXPC (0.3 and 0.6) compared to control. Few studies have assessed the effect of including this prebiotic on the immunity of pets; the pioneers were Middelbos et al. [3], who did not find differences in the immunity of dogs with the use of blends containing beet pulp, cellulose or blends of cellulose, fructooligosaccharides, and yeast cell wall at 2.5% in the diet. However, another study that evaluated MOS supplementation observed an increase in the total percentage of white blood cells [4]. A more recent study demonstrated that the inclusion of 1.0% of a commercial blend containing MOS, FOS, GOS, and beta-glucan in healthy dogs increased the polymorphonuclear cell count, phagocytosis index, and oxidative burst in supplemented animals compared to the control group [58]. Finally, Lin et al. [59] observed that supplementation of 0.2% yeast cell wall fractions to dogs tended to increase fecal IgA concentrations. All of the aforementioned suppliers concluded that this finding is related to positive modulation of the immune system.

Studies in other species show that the beta-glucans and MOS contained in *S. cerevisiae* have been identified as agents capable of triggering strong antigenic stimuli and immune responses. Beta-glucan is designated as an immunological response modifier, for when recognized by the organism, it has the ability to trigger a series of events in the immune response. Kubala et al. [60] reported that modulation of cellular activity by beta-glucan begins with the activation of macrophages, endothelial and dendritic cells, and B and T cells. In addition, they involve the specific immune response by inducing the expression of various cytokines such as TNF, IL-6, IL-8, and IL-12 [61]. Therefore, the composition of the OXPC treatment (0.3 and 0.6) explains the improvement in the index of phagocytosis in supplemented dogs.

5. Conclusions

According to the results, the addition of 0.3 and 0.6% OXPC in the diet of dogs was able to alter some phyla and genera abundances to increase propionic acid production and the phagocytosis index in healthy adult dogs with minor alteration in digestibility. Other studies should evaluate higher doses of OXPC supplementation and its effects on the intestine of healthy dogs and those suffering from gastrointestinal disorders.

Author Contributions: Investigation, K.d.M.S., L.W.R., M.F.R., A.R.A., R.B.A.R., M.I.G.U., T.H.A.V., C.d.O.M. and J.P.F.S.; Methodology, M.A.B., K.d.M.S., J.P.F.S., L.W.R., M.F.R., A.R.A., R.B.A.R., M.I.G.U., T.H.A.V., R.V.V., J.C.d.C.B. and R.V.V.; Resources, C.F.F.P. and M.A.B.; Writing—original draft, K.d.M.S., L.W.R., M.F.R., A.R.A., R.B.A.R., T.H.A.V. and M.A.B.; Writing—review and editing, T.H.A.V., C.F.F.P. and M.A.B. All authors have read and agreed to the published version of the manuscript.

Funding: This research received no external funding.

Institutional Review Board Statement: All experimental procedures were approved by the Ethics Research Committee for Animal Welfare of the School of Veterinary Medicine and Animal Science, University of São Paulo (protocol number 9148270415).

Informed Consent Statement: Not applicable.

Data Availability Statement: The data presented in this study are available on request from the corresponding author.

Acknowledgments: We would like to thank Premier Pet (Grandfood Ind. Com. LTDA) for the construction and maintenance of Pet Nutrology Research Center; and the first author acknowledges the scholarship granted by FAPESP—Fundação de Amparo à Pesquisa do Estado de São Paulo (process 2015/05493-0).

Conflicts of Interest: The authors declare no conflict of interest.

References

1. Gibson, G.R.; Hutkins, R.; Sanders, M.E.; Prescott, S.L.; Reimer, R.A.; Salminen, S.J.; Scott, K.; Stanton, C.; Swanson, K.S.; Cani, P.D.; et al. Expert consensus document: The International Scientific Association for Probiotics and Prebiotics (ISAPP) consensus statement on the definition and scope of prebiotics. *Nat. Rev. Gastroenterol. Hepatol.* **2017**, *14*, 491–502. [CrossRef]
2. Bindels, L.B.; Delzenne, N.M.; Cani, P.D.; Walter, J. Opinion: Towards a more comprehensive concept for prebiotics. *Nat. Rev. Gastroenterol. Hepatol.* **2015**, *12*, 303–310. [CrossRef]
3. Middelbos, I.S.; Fastinger, N.D.; Fahey, G.C. Evaluation of fermentable oligosaccharides in diets fed to dogs in comparison to fiber standards. *J. Anim. Sci.* **2007**, *85*, 3033–3044. [CrossRef] [PubMed]
4. Swanson, K.S.; Grieshop, C.M.; Flickinger, E.A.; Bauer, L.L.; Healy, H.-P.; Dawson, K.A.; Merchen, N.R.; Fahey, G.C., Jr. Supplemental Fructooligosaccharides and Mannanoligosaccharides Influence Immune Function, Ileal and Total Tract Nutrient Digestibilities, Microbial Populations and Concentrations of Protein Catabolites in the Large Bowel of Dogs. *J. Nutr.* **2002**, *132*, 980–989. [CrossRef]
5. Grieshop, C.M.; Flickinger, E.A.; Bruce, K.J.; Patil, A.R.; Czarnecki-Maulden, G.L.; Fahey, G.C. Gastrointestinal and immunological responses of senior dogs to chicory and mannan-oligosaccharides. *Arch. Anim. Nutr.* **2004**, *58*, 483–494. [CrossRef] [PubMed]
6. Roberfroid, M. Functional food concept and its application to prebiotics. *Dig Liver Dis.* **2002**, *34* (Suppl. 2), S105–S110. [CrossRef]
7. Snart, J.; Bibiloni, R.; Grayson, T.; Lay, C.; Zhang, H.; Allison, G.E.; Laverdiere, J.K.; Temelli, F.; Vasanthan, T.; Bell, R.; et al. Supplementation of the diet with high-viscosity beta-glucan results in enrichment for *lactobacilli* in the rat cecum. *Appl. Environ. Microbiol.* **2006**, *72*, 1925–1931. [CrossRef]
8. Hughes, S.A.; Shewry, P.R.; Gibson, G.R.; McCleary, B.V.; Rastall, R.A. In vitro fermentation of oat and barley derived β-glucans by human faecal microbiota. *FEMS Microbiol. Ecol.* **2008**, *64*, 482–493. [CrossRef]
9. Lin, C.-Y.; Celesye, A.; Steelman, A.J.; Warzecha, C.M.; Godoy, M.R.C.; Swanson, K.S. Effects of a *Saccharomyces cerevisiae* fermentation product on fecal characteristics, nutrient digestibility, fecal fermentative end-products, fecal microbial populations, immune function, and diet palatability in adult dogs. *J. Anim. Sci.* **2019**, *97*, 1586–1599. [CrossRef] [PubMed]
10. Zentek, J.; Marquart, B.; Pietrzak, T. Intestinal Effects of Mannanoligosaccharides, Transgalactooligosaccharides, Lactose and Lactulose in Dogs. *J. Nutr.* **2002**, *132*, 1682S–1684S. [CrossRef]
11. Propst, E.L.; Flickinger, E.A.; Bauer, L.L.; Merchen, N.R.; Fahey, G.C. A dose-response experiment evaluating the effects of oligofructose and inulin on nutrient digestibility, stool quality, and fecal protein catabolites in healthy adult dogs. *J. Anim. Sci.* **2003**, *81*, 3057–3066. [CrossRef]

12. Pawar, M.M.; Pattanaik, A.K.; Sinha, D.K.; Goswami, T.K.; Sharma, K. Effect of dietary mannanoligosaccharide supplementation on nutrient digestibility, hindgut fermentation, immune response and antioxidant indices in dogs. *J. Anim. Sci. Technol.* **2017**, *59*, 11. [CrossRef] [PubMed]

13. de Souza, T.S.; Putarov, T.C.; Tiemi, C.; Volpe, L.M.; de Oliveira, C.A.F.; de Abreu, G.M.B.; Carciofi, A.C. Effects of the solubility of yeast cell wall preparations on their potential prebiotic properties in dogs. *PLoS ONE* **2019**, *14*, e0215332.

14. Association of American Feed Control Official. AAAFCO methods for substantiating nutritional adequacy of dog and cat food. In *AAFCO Dog Cat Food Nutrient Profiles*; Association of American Feed Control Officials: Champaign, IL, USA, 2016.

15. National Research Council. *Nutrient Requirements of Dogs and Cats*; National Academies Press: Washington, DC, USA, 2006.

16. Association of Official Agricultural Chemists. *Official Methods of Analysis of AOAC International*, 17th ed.; Association of Official Analysis Chemists International: Arlington, VA, USA, 2006.

17. Carciofi, A.C.; Takakura, F.S.; De Oliveira, L.D.; Teshima, E.; Jeremias, J.T.; Brunetto, M.A.; Prada, F. Effects of six carbohydrate sources on dog diet digestibility and post-prandial glucose and insulin response. *J. Anim. Physiol. Anim. Nutr.* **2008**, *92*, 326–336. [CrossRef]

18. Walter, M.; Silva, L.P.; Perdomo, D.M.X. Biological response of rats to resistant starch Resposta biológica de ratos ao amido resistente. *Rev. Inst. Adolfo Lutz.* **2005**, *64*, 252–257.

19. Sá, F.C.; Vasconcellos, R.S.; Brunetto, M.A.; Filho, F.O.R.; Gomes, M.O.S.; Carciofi, A.C. Enzyme use in kibble diets formulated with wheat bran for dogs: Effects on processing and digestibility. *J. Anim. Physiol. Anim. Nutr.* **2013**, *97*, 51–59. [CrossRef]

20. Valente, A.L.P.; Augusto, F. As espessuras dos recobrimentos L. *Quim. Nova* **2000**, *23*, 523–530. [CrossRef]

21. Silva, B.J.G.; Lanças, F.M.; Queiroz, M.E.C. In-tube solid-phase microextraction coupled to liquid chromatography (in-tube SPME/LC) analysis of nontricyclic antidepressants in human plasma. *J. Chromatogr. B Anal. Technol. Biomed. Life Sci.* **2008**, *862*, 181–188. [CrossRef] [PubMed]

22. Erwin, E.S.; Marco, G.J.; Emery, E.M. Volatile Fatty Acid Analyses of Blood and Rumen Fluid by Gas Chromatography. *J. Dairy Sci.* **1961**, *44*, 1768–1771. [CrossRef]

23. Pryce, J.D. A modification of the Barker-Summerson method for the determination of lactic acid. *Analyst* **1969**, *94*, 1151–1152. [CrossRef]

24. Vale, S.R.; Glória, B.A. Determination of biogenic amines in cheese. *J. AOAC* **1997**, *80*, 1006–1012. [CrossRef]

25. de Gomes, M.O.S.; Beraldo, M.C.; Putarov, T.C.; Brunetto, M.A.; Zaine, L.; Glória, M.B.A.; Carciofi, A.C. Old beagle dogs have lower faecal concentrations of some fermentation products and lower peripheral lymphocyte counts than young adult beagles. *Br. J. Nutr.* **2011**, *106*, 187–190. [CrossRef]

26. McInnes, P.; Cutting, M. Manual of Procedures for Human Microbiome Project: Core Microbiome Sampling Protocol A HMP Protocol # 07-001, Version 12.0. 2010. Available online: http://hmpdacc.org/resources/tools_protocols.php (accessed on 13 March 2021).

27. Caporaso, J.G.; Lauber, C.L.; Walters, W.A.; Berg-Lyons, D.; Huntley, J.; Fierer, N.; Owens, S.M.; Betley, J.; Fraser, L.; Bauer, M.; et al. Ultra-high-throughput microbial community analysis on the Illumina HiSeq and MiSeq platforms. *ISME J.* **2012**, *6*, 1621–1624. [CrossRef] [PubMed]

28. Caporaso, J.G.; Kuczynski, J.; Stombaugh, J.; Bittinger, K.; Bushman, F.D.; Costello, E.K.; Fierer, N.; Peña, A.G.; Goodrich, J.K.; Gordon, J.I.; et al. QIIME allows analysis of high- throughput community sequencing data Intensity normalization improves color calling in SOLiD sequencing. *Nat. Publ. Gr.* **2010**, *7*, 335–336.

29. Lozupone, C.; Knight, R. UniFrac: A new phylogenetic method for comparing microbial communities. *Appl. Environ. Microbiol.* **2005**, *71*, 8228–8235. [CrossRef] [PubMed]

30. Middelbos, I.S.; Godoy, M.R.; Fastinger, N.D.; Fahey, G.C. A dose-response evaluation of spray-dried yeast cell wall supplementation of diets fed to adult dogs: Effects on nutrient digestibility, immune indices, and fecal microbial populations. *J. Anim. Sci.* **2007**, *85*, 3022–3032. [CrossRef]

31. Stercova, E.; Kumprechtova, D.; Auclair, E.; Novakova, J. Effects of live yeast dietary supplementation on nutrient digestibility and fecal microflora in beagle dogs. *J. Anim. Sci.* **2016**, *94*, 2909–2918. [CrossRef] [PubMed]

32. Deng, P.; Swanson, K.S. Gut microbiota of humans, dogs and cats: Current knowledge and future opportunities and challenges. *Br. J. Nutr.* **2015**, *113*, 6–17. [CrossRef]

33. El Khoury, D.; Cuda, C.; Luhovyy, B.L.; Anderson, G.H. Beta glucan: Health benefits in obesity and metabolic syndrome. *J. Nutr. Metab.* **2012**, *2012*, 851362. [CrossRef]

34. Cecchi, H.M. *Fundamentos Teóricos e Práticos em Análise de Alimentos*, 2nd ed.; Universidade Estadual de Campinas (UNICAMP), Ed.; Universidade Estadual de Campinas: Campinas, Brazil, 2003.

35. de Oliveira, L.D.; Takakura, F.S.; Kienzle, E.; Brunetto, M.A.; Teshima, E.; Pereira, G.T.; Carciofi, A.C. Fibre analysis and fibre digestibility in pet foods—A comparison of total dietary fibre, neutral and acid detergent fibre and crude fibre. *J. Anim. Physiol. Anim. Nutr.* **2012**, *96*, 895–906. [CrossRef] [PubMed]

36. Hammes, W.P.; Vogel, R.F. The *Genus lactobacillus*. In *The Genera of Lactic Acid Bacteria*; Springer: Boston, MA, USA, 1995; pp. 19–54.

37. Vickers, R.J.; Sunvold, G.D.; Kelley, R.L.; Reinhart, G.A. Comparison of fermentation of selected fructooligosaccharides and other fiber substrates by canine colonic microflora. *Am. J. Vet. Res.* **2001**, *62*, 609–615. [CrossRef]

38. Garner, C.E.; Smith, S.; Lacy, C.B.; White, P.; Spencer, R.; Probert, C.S.J.; Ratcliffem, N.M. Volatile organic compounds from feces and their potential for diagnosis of gastrointestinal disease. *FASEB J.* **2007**, *21*, 1675–1688. [CrossRef]

39. Bastos, T.S.; de Lima, D.C.; Souza, C.M.M.; Maiorka, A.; de Oliveira, S.G.; Bittencourt, L.C.; Félix, A.P. *Bacillus subtilis* and *Bacillus licheniformis* reduce faecal protein catabolites concentration and odour in dogs. *BMC Vet. Res.* **2020**, *16*, 116. [CrossRef] [PubMed]

40. Hesta, M.; Janssens, G.P.J.; Debraekeleer, J.; De Wilde, R. The effect of oligofructose and inulin on faecal characteristics and nutrient digestibility in healthy cats. *J. Anim. Physiol. Anim. Nutr.* **2001**, *85*, 135–141. [CrossRef] [PubMed]

41. Walker, W.A.; Duffy, L.C. Diet and bacterial colonization: Role of probiotics and prebiotics. *J. Nutr. Biochem.* **1998**, *9*, 668–675. [CrossRef]

42. Wollowski, I.; Rechkemmer, G.; Pool-Zobel, B.L. Protective role of probiotics and prebiotics in colon cancer. *Am. J. Clin. Nutr.* **2001**, *73*, 451s–455s. [CrossRef] [PubMed]

43. Cummings, J.H.; Hill, M.J.; Bone, E.S.; Branch, W.J.; Jenkins, D.J. The effect of meat protein and dietary fiber on colonic function and metabolism. II. Bacterial metabolites in feces and urine. *Am. J. Clin. Nutr.* **1979**, *32*, 2094–2101. [CrossRef] [PubMed]

44. Slavin, J. Fiber and prebiotics: Mechanisms and health benefits. *Nutrients* **2013**, *5*, 1417–1435. [CrossRef]

45. Perini, M.P.; Rentas, M.F.; Pedreira, R.; Amaral, A.R.; Zafalon, R.V.A.; Rodrigues, R.B.A.; Henríquez, L.B.F.; Zanini, L.; Vendramini, T.H.A.; Balieiro, J.C.C.; et al. Duration of prebiotic intake is a key-factor for diet-induced modulation of immunity and fecal fermentation products in dogs. *Microorganisms* **2020**, *8*, 1916. [CrossRef]

46. Beloshapka, A.N.; Dowd, S.E.; Suchodolski, J.S.; Steiner, J.M.; Duclos, L.; Swanson, K.S. Fecal microbial communities of healthy adult dogs fed raw meat-based diets with or without inulin or yeast cell wall extracts as assessed by 454 pyrosequencing. *FEMS Microbiol. Ecol.* **2013**, *84*, 532–541. [CrossRef]

47. Possemiers, S.; Pinheiro, I.; Verhelst, A.; Van Den Abbeele, P.; Maignien, L.; Laukens, D.; Reeves, S.G.; Robinson, L.E.; Raas, T.; Schneider, Y.-J.; et al. A dried yeast fermentate selectively modulates both the luminal and mucosal gut microbiota and protects against inflammation, as studied in an integrated in vitro approach. *J. Agric. Food Chem.* **2013**, *61*, 9380–9392. [CrossRef]

48. Middelbos, I.S.; Boler, B.M.V.; Qu, A.; White, B.A.; Swanson, K.S.; Fahey, G.C. Phylogenetic Characterization of Fecal Microbial Communities of Dogs Fed Diets with or without Supplemental Dietary Fiber Using 454 Pyrosequencing. *PLoS ONE* **2010**, *5*, e9768. [CrossRef] [PubMed]

49. Hidaka, H.; Eida, T.; Takizawa, T.; Tokunaga, T.; Tashiro, Y. Effects of Fructooligosaccharides on Intestinal Flora and Human Health. *Bifidobact. Microflora* **1986**, *5*, 37–50. [CrossRef]

50. Barry, K.A.; Middelbos, I.S.; Vester Boler, B.M.; Dowd, S.E.; Suchodolski, J.S.; Henrissat, B.; Coutinho, P.M.; White, B.A.; Fahey, J.G.C.; Swanson, K.S. Effects of dietary fiber on the feline gastrointestinal metagenome. *J. Proteome Res.* **2012**, *11*, 5924–5933. [CrossRef]

51. Kelder, T.; Stroeve, J.H.M.; Bijlsma, S.; Radonjic, M.; Roeselers, G. Correlation network analysis reveals relationships between diet-induced changes in human gut microbiota and metabolic health. *Nutr. Diabetes* **2014**, *4*, 122–127. [CrossRef] [PubMed]

52. Van Rijssel, M.; Hansen, T.A. Clostridium thermosaccharolyticum strain. *FEMS Microbiol. Lett.* **1989**, *61*, 41–46. [CrossRef]

53. Ravussin, Y.; Koren, O.; Spor, A.; Leduc, C.; Gutman, R.; Stombaugh, J.; Knight, R.; Ley, R.E.; Leibel, R.L. Responses of gut microbiota to diet composition and weight loss in lean and obese mice. *Obesity* **2012**, *20*, 738–747. [CrossRef]

54. Warren, Y.A.; Tyrrell, K.L.; Citron, D.M.; Goldstein, E.J.C. Clostridium aldenense sp. nov. and *Clostridium citroniae* sp. nov. isolated from human clinical infections. *J. Clin. Microbiol.* **2006**, *44*, 2416–2422. [CrossRef]

55. Swidsinski, A.; Dörffel, Y.; Loening-Baucke, V.; Theissig, F.; Rückert, J.C.; Ismail, M.; Rau, W.A.; Gaschler, D.; Weizenegger, M.; Kühn, S.; et al. Acute appendicitis is characterised by local invasion with *Fusobacterium nucleatum/necrophorum*. *Gut* **2011**, *60*, 34–40. [CrossRef]

56. Tremaroli, V.; Bäckhed, F. Functional interactions between the gut microbiota and host metabolism. *Nature* **2012**, *489*, 242–249. [CrossRef] [PubMed]

57. Xu, Z.; Knight, R. Dietary effects on human gut microbiome diversity. *Br. J. Nutr.* **2015**, *113*, 1–5. [CrossRef]

58. Rentas, M.F.; Pedreira, R.S.; Perini, M.P.; Risolia, L.W.; Zafalon, R.V.A.; Alvarenga, I.C.; Vendramini, T.H.A.; Balieiro, J.C.D.C.; Pontieri, C.F.F.; Brunetto, M.A. Galactoligosaccharide and a prebiotic blend improve colonic health and immunity of adult dogs. *PLoS ONE* **2020**, *15*, e0238006. [CrossRef]

59. Lin, C.Y.; Carroll, M.Q.; Miller, M.J.; Rabot, R.; Swanson, K.S. Supplementation of Yeast Cell Wall Fraction Tends to Improve Intestinal Health in Adult Dogs Undergoing an Abrupt Diet Transition. *Front. Vet. Sci.* **2020**, *7*, 905. [CrossRef] [PubMed]

60. Kubala, L.; Ruzickova, J.; Nickova, K.; Sandula, J.; Ciz, M.; Lojek, A. The effect of (1→3)-β-D-glucans, carboxymethylglucan and schizophyllan on human leukocytes in vitro. *Carbohydr. Res.* **2003**, *338*, 2835–2840. [CrossRef]

61. Magnani, M.; Castro-Gómez, R.J.H. β-glucana from Saccharomyces cerevisiae: Constitution, bioactivity and obtaining [β-glucana de *Saccharomyces cerevisiae*: Constituição, bioatividade e obtenção]. *Semin. Agrar.* **2008**, *29*, 631–650. [CrossRef]

 *fermentation*

Review

Converting Sugars into Cannabinoids—The State-of-the-Art of Heterologous Production in Microorganisms

Gabriel Rodrigues Favero, Gilberto Vinícius de Melo Pereira *, Júlio Cesar de Carvalho, Dão Pedro de Carvalho Neto and Carlos Ricardo Soccol

Department of Bioprocess Engineering and Biotechnology, Federal University of Paraná (UFPR), Curitiba 80060-000, Brazil; gabrielrfavero@gmail.com (G.R.F.); jccarvalho@ufpr.br (J.C.d.C.); daopcn@gmail.com (D.P.d.C.N.); soccol@ufpr.br (C.R.S.)
* Correspondence: gilbertovinicius@gmail.com or gilbertovinicius@ufpr.br

Abstract: The legal cannabis market worldwide is facing new challenges regarding innovation in the production of cannabinoid-based drugs. The usual cannabinoid production involves growing *Cannabis sativa* L. outdoor or in dedicated indoor growing facilities, followed by isolation and purification steps. This process is limited by the growth cycles of the plant, where the cannabinoid content can deeply vary from each harvest. A game change approach that does not involve growing a single plant has gained the attention of the industry: cannabinoids fermentation. From recombinant yeasts and bacteria, researchers are able to reproduce the biosynthetic pathway to generate cannabinoids, such as (-)-Δ^9-tetrahydrocannabinol (Δ^9-THC), cannabidiol (CBD), and (-)-Δ^9-tetrahydrocannabivarin (Δ^9-THCV). This approach avoids pesticides, and natural resources such as water, land, and energy are reduced. Compared to growing cannabis, fermentation is a much faster process, although its limitation regarding the phytochemical broad range of molecules naturally present in cannabis. So far, there is not a consolidated process for this brand-new approach, being an emerging and promising concept for countries in which cultivation of *Cannabis sativa* L. is illegal. This survey discusses the techniques and microorganisms already established to accomplish the task and those yet in seeing for the future, exploring upsides and limitations about metabolic pathways, toxicity, and downstream recovery of cannabinoids throughout heterologous production. Therapeutic potential applications of cannabinoids and in silico methodology toward optimization of metabolic pathways are also explored. Moreover, conceptual downstream analysis is proposed to illustrate the recovery and purification of cannabinoids through the fermentation process, and a patent landscape is presented to provide the state-of-the-art of the transfer of knowledge from the scientific sphere to the industrial application.

Keywords: cannabinoids biosynthesis; *Cannabis sativa*; cannabidiol; fermentation; heterologous expression; metabolic engineering; tetrahydrocannabinol; tetrahydrocannabivarin

Citation: Favero, G.R.; de Melo Pereira, G.V.; de Carvalho, J.C.; de Carvalho Neto, D.P.; Soccol, C.R. Converting Sugars into Cannabinoids—The State-of-the-Art of Heterologous Production in Microorganisms. *Fermentation* **2022**, *8*, 84. https://doi.org/10.3390/fermentation8020084

Academic Editor: Ronnie G. Willaert

Received: 25 January 2022
Accepted: 14 February 2022
Published: 17 February 2022

Publisher's Note: MDPI stays neutral with regard to jurisdictional claims in published maps and institutional affiliations.

Copyright: © 2022 by the authors. Licensee MDPI, Basel, Switzerland. This article is an open access article distributed under the terms and conditions of the Creative Commons Attribution (CC BY) license (https://creativecommons.org/licenses/by/4.0/).

1. Introduction

The global cannabis and cannabinoids market has undergone a great increase in recent years with legalization for medical and recreational purposes in different U.S. states and countries. In 1996, California (CA) was the first U.S. state to legalize medical cannabis use [1]. Five years later, Canada was the first country in the modern era to legalize medical cannabis nationwide, establishing public policies that became a reference in this subject [2]. The recreational use of cannabis was not accepted in the USA until 2012 when Washington (WA) [3] and Colorado (CO) [4] passed a ballot initiative for this purpose. In a global scenario, Uruguay was the first country to legalize the recreational use of cannabis nationwide in 2013 [5], followed by Canada in 2018 [6].

Although the global cannabidiol (CBD) market has been valued at US$2.8 billion in 2020 and has a compound annual growth rate (CAGR) of 21.2% projected to 2028 [7], its commercialization is still restrictive. Furthermore, the usual cannabinoid production is attached with the agricultural process of growing *Cannabis sativa* L., either in outdoor fields or in dedicated indoor growing facilities. The flowers are harvested and the active compounds are isolated through chemical (e.g., extraction with ethanol, ethyl acetate, butane, and CO_2) or physical (such as heated press) processes to take cannabinoids out of the vegetal biomass [8].

The agriculture-based process requires a significant amount of energy, especially light, and chemical fertilizers. As with any agricultural commodity, it is limited by the slow growth cycles of the plant, where the cannabinoid content can vary from one cycle to another, and are susceptible to pests, weather, and environmental specificities [9]. As a matter of fact, environmental conditions play an important role in mineral nutrient availability, affecting secondary metabolites' final concentration in plants. The work of Shiponi and Bernstein [10] evaluated the hypothesis that phosphorous (P) uptake, distribution, and availability in the plant affects cannabinoids' biosynthesis. By analyzing two genotypes of medical "drug-type" cannabis grown under five P concentrations (5, 15, 30, 60, and 90 mg/L), it was noted that the values lower than 15 mg/L were insufficient to support optimal plant function, with reduced physiological responses, whereas values between 30 and 90 mg/L were within the optimal range for plant development, increasing total cannabinoids content per plant. With that, the regime of mineral nutrients must be adjusted to account for production goals and the genetic specificities of the strain. Moreover, the indoor production of cannabis is responsible for greenhouse gas (GHG) emissions that range between 2 and 5 tons of CO_2-equivalent per kg of dried flower—attributed to electricity and natural gas consumption from indoor environmental controls, high-intensity grow lights, and supply of CO_2 to accelerate plant growth [11].

With the advance of metabolic engineering and synthetic biology, the tailor-made design of cell factories became a reality, providing a remarkable opportunity for the biosynthesis of cannabinoids and analogs, especially for those found in small quantities in cannabis. As matter of fact, the expression of tetrahydrocannabinol synthase (THCAS) was already achieved using *P. pastoris* as host [12]. With cannabigerolic acid (CBGA) being added into the media, Δ^9-tetrahydrocannabinolic acid (Δ^9-THCA) was synthesized. Luo et al. [13] were able to produce several cannabinoids and analogs from the genetic recombination of *Saccharomyces cerevisiae*, generating an yeast that can synthesize cannabinoids from galactose. With specific genetic modifications, cannabinoids that were previously generated in small quantities can now be scaled up. Furthermore, there is no need for pesticides, and the natural resources required (land, water, and energy) and CO_2 footprint are reduced as well.

However, a drawback of fermentation is its limitation to achieve the phytochemical broad range of molecules naturally present in *Cannabis sativa* L., turning full-spectrum extracts (i.e., those with phytocannabinoids and secondary metabolites) unfeasible to be obtained other than by the plant. The term *entourage effect* [14] is often used to refer to potential synergies between chemical compounds present in cannabis, such as cannabinoids-cannabinoids interactions [15–17] and the presence of other secondary metabolites such as terpenes/terpenoids [18]. The list of terpenes/terpenoids found in cannabis is vast due to differences between strains, chemotypes, and environmental conditions, but in general, the most common terpenes/terpenoids found are β-myrcene, limonene, linalool, β-caryophyllene, α-pinene, β-ocimene, terpinolene, and geraniol [18]. They are mainly responsible for the odor and taste present in cannabis flowers and are used in perfume fragrances and cleaning products worldwide. Besides these organoleptic characteristics, terpenes/terpenoids have been studied for their therapeutic potential, with works analyzing analgesic [19–21], anti-inflammatory [22–26], gastroprotective [27–29], anxiolytic/antidepressant [30–35], apoptotic/antimetastatic [36,37] antinociceptive [38–40], neuroprotective [41–44], sedative/motor relaxant [45–47], and antifungal [48,49] properties. This broad range of metabolites in different concentrations provides unique therapeutic applications for full spectrum extracts.

This review describes the techniques and microorganisms already established to accomplish the task and those yet in seeing for the future, exploring upsides and limitations regarding metabolic pathways, toxicity, and downstream recovery of cannabinoids throughout heterologous production. Moreover, therapeutic potential applications of cannabinoids, in silico methodology toward optimization of metabolic pathways, and a patent landscape are explored.

2. Biosynthesis of Phytocannabinoids

Cannabinoids are active lipophilic compounds that interact with specific protein receptors in the human body, constituting a system of physiological regulations—the endocannabinoid system. Two receptors for this system are well-known: CB_1, located in the central nervous system (CNS) and peripheral nervous system (PNS), with high density in the basal ganglia, cerebellum, hippocampus, and cortex; and CB_2, restricted to immune tissues and immune cells. Some cannabinoids are produced endogenously in various vertebrates and are known as endocannabinoids, such as anandamide (AEA) and 2-arachidonylglycerol (2-AG) [50]. Other cannabinoids are produced only by plants of the genus *Cannabis* (mainly by *sativa* and *indica* species), including (-)-Δ^9-tetrahydrocannabinol (Δ^9-THC) and cannabidiol (CBD), and are known as phytocannabinoids [18]. Over 500 chemical compounds were identified in *C. sativa* L., including 102 phytocannabinoids, being Δ^9-THC, CBD, cannabigerol (CBG), and cannabichromene (CBC) their main representatives [51]. In the plant, they are usually found in their carboxylated state, including tetrahydrocannabinolic acid (Δ^9-THCA) and cannabidiolic acid (CBDA). Although cannabinol (CBN) is one of the major cannabinoids found in cannabis, it is not directly produced by the plant, being a product of Δ^9-THC oxidation [52]. Phytocannabinoids are separated into families based on their structures such as cannabigerol (CBG)-family, cannabichromene (CBC)-family, cannabidiol (CBD)-family, tetrahydrocannabinol (THC)-family, cannabinol (CBN)-family [53] (Figure 1). They are all composed of a phenolic (resorcinol) moiety and a monoterpene moiety, later described in this survey.

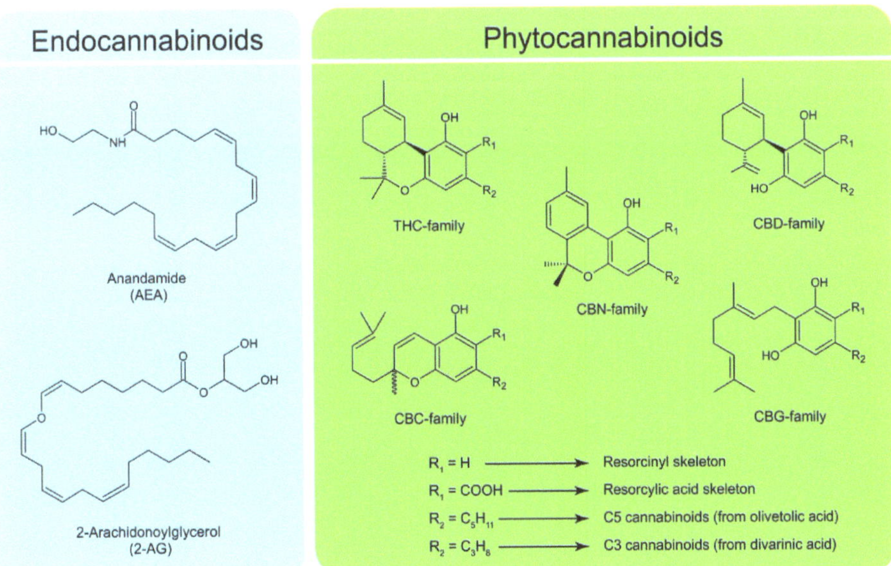

Figure 1. Structures of endocannabinoids and major phytocannabinoids present in *C. sativa* L. THC: tetrahydrocannabinol, CBN: cannabinol, CBD: cannabidiol, CBC: cannabichromene, CBG: cannabigerol [53].

The effects of cannabinoids were studied only from the 20th century, where several analyses resulted in the development of dronabinol (Marinol®; Unimed Pharmaceuticals,

Inc., Marietta, GA, USA). This drug is based on Δ^9-THC, which in 1964—and after decades of attempts to isolate and determine its chemical structure—was identified as the main psychoactive component of cannabis. Together with Cesamet® (Valeant Pharmaceuticals North America, Aliso Viejo, CA, USA), they were the first cannabinoid-based drugs to be prescribed in the United States, presenting antiemetic and appetite-stimulating action for patients with cancer and AIDS [50]. Several studies are being carried out for possible pharmacological applications involving cannabinoids, especially with CBD due to the absence of psychoactive effects. Conditions such as Alzheimer's disease, anxiety, cancer, chronicle pain, depression, epilepsy, inflammatory diseases, multiple sclerosis, and Parkinson's disease are being investigated with promising results [54].

Phytocannabinoids are synthesized and stored within glandular trichomes that are present on cannabis flowers with some extension to other structures, such as leaves and stems, but almost absent in seeds and roots [55]. To produce these compounds in a heterologous host, the genes, metabolic pathways, bottlenecks, and specificities involved during phytocannabinoids biosynthesis in *Cannabis sativa* L. must be comprehended and availed, in order to be further optimized according to the host's characteristics and limitations.

The biosynthesis of cannabinoids begins with metabolic pathways to produce geranyl pyrophosphate (GPP) and olivetolic acid (OA) as shown in Figure 2 [56]. Geranyl pyrophosphate (GPP) is mainly biosynthesized via the 2C-methyl-D-erythritol-4-phosphate (MEP) pathway, also known as non-mevalonate or 1-deoxy-D-xylulose-5-phosphate (DOXP) pathway, and in a small extension through mevalonate (MVA) pathway [56,57]. The final products, isopentenyl pyrophosphate (IPP) and dimethylallyl pyrophosphate (DMAPP), are catalyzed to GPP by the action of geranyl pyrophosphate synthase (GPPS), providing the monoterpene moiety of phytocannabinoids [58]. In parallel, the polyketide pathway toward OA starts with hexanoic acid produced either by an early termination of fatty acid biosynthesis or by the breakdown of C18 unsaturated fatty acids via the lipoxygenase pathway [59]. The hexanoic acid is converted to hexanoyl-CoA by the action of an acyl-activating enzyme type 1 (AAE1) found in *Cannabis sativa* (*Cs*AAE). Then, a type III tetraketide synthase (*Cs*TKS), also known as olivetol synthase (OLS), promotes the aldol condensation of hexanoyl-CoA with three molecules of malonyl-CoA, producing olivetol, followed by the C2–C7 aldol cyclization to OA carried by a polyketide cyclase (*Cs*OAC) [60]. With an olivetolic acid pool, the phenolic (resorcinol) moiety is available to be further converted into cannabinoids. More details regarding MEP/DOXP pathway, MVA pathway, fatty acid biosynthesis, and lipoxygenase pathway are summarized in several reviews [61–64] with higher plants metabolism focus.

With the availability of the precursors, an aromatic prenyltransferase named geranylpyrophosphate:olivetolate geranyltransferase (GOT) is responsible to convert GPP and OA into cannabigerolic acid (CBGA) [13], the central precursor for phytocannabinoids biosynthesis. This enzyme was detected in 1998 and is assumed to be an integral membrane protein, although some activity was found in soluble fractions [65,66].

With an appropriated CBGA pool, enzymes such as THCA synthase, CBDA synthase, and CBCA synthase promote an oxidative cyclization of the monoterpene moiety of the substrate, generating Δ^9-THCA, CBDA, and CBCA, respectively. In the plant, the phytocannabinoids are stored as carboxylic acid; they can be decarboxylated to their corresponding neutral form through drying, heating, or combustion [67].

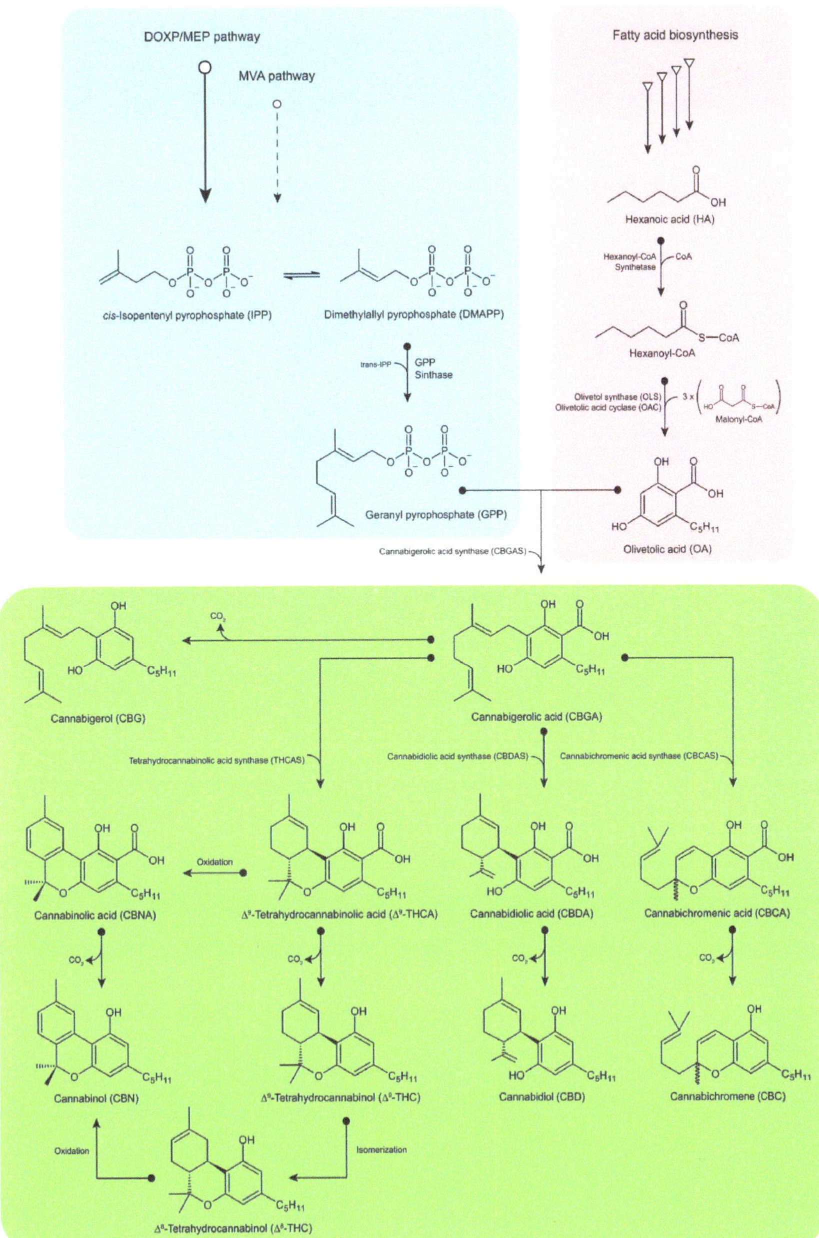

Figure 2. Phytocannabinoids biosynthesis in *Cannabis sativa* L. The monoterpene moiety is provided majoritarian through the MEP/DOXP pathway, and in small extension through the MVA pathway, in which geranyl pyrophosphate (GPP) is synthesized. Parallel to that, fatty acids metabolism uses hexanoic acid as a substrate to fulfill the phenolic (resorcinol) moiety of cannabinoids, generating olivetolic acid (OA). Through the action of cannabigerolic acid synthase (CBGAS), GPP and OA are converted into cannabigerolic acid (CBGA), the central precursor for many other cannabinoids, such as Δ^9-tetrahydrocannabinolic acid (Δ^9-THCA), cannabidiolic acid (CBDA), and cannabichromenic acid (CBCA) [13,55,56,67].

Phytocannabinoids such as CBGA, Δ^9-THCA, CBDA, and CBCA are known as C5 phytocannabinoids since they have an n-pentyl side chain in the phenolic moiety. However, there are also C3 phytocannabinoids, or propyl cannabinoids, derived not from olivetolic acid (OA) but from divarinic acid (DA) as illustrated in Figure 3. The prenylation of DA with GPP results in cannabigerovarinic acid (CBGVA), the central precursor for C3 phytocannabinoids biosynthesis. The cannabinoid synthase enzymes are not alkyl length selective and can convert CBGVA into the propyl homologous of THCA, CBDA, and CBCA, known as tetrahydrocannabivarinic acid (Δ^9-THCVA), cannabidivarinic acid (CBDVA), and cannabichromevarinic acid (CBCVA), respectively [68]. Since these compounds are not commonly produced by cannabis strains due to dissimilar enzyme specificities at the level of CBGA or CBGA-analogs formation [69], the analysis and studies of its therapeutic value are impaired. Nevertheless, the agricultural-based method has the genetic restrictions imposed by the plant, with selective breeding as the main resource to achieve better yields of a specific compound, despite its limited randomness expressed in the next offspring. With that, chemotype inheritance and genetic engineering are the objects of study to manipulate secondary metabolites' final concentration and can be conferred in recent works [68,70].

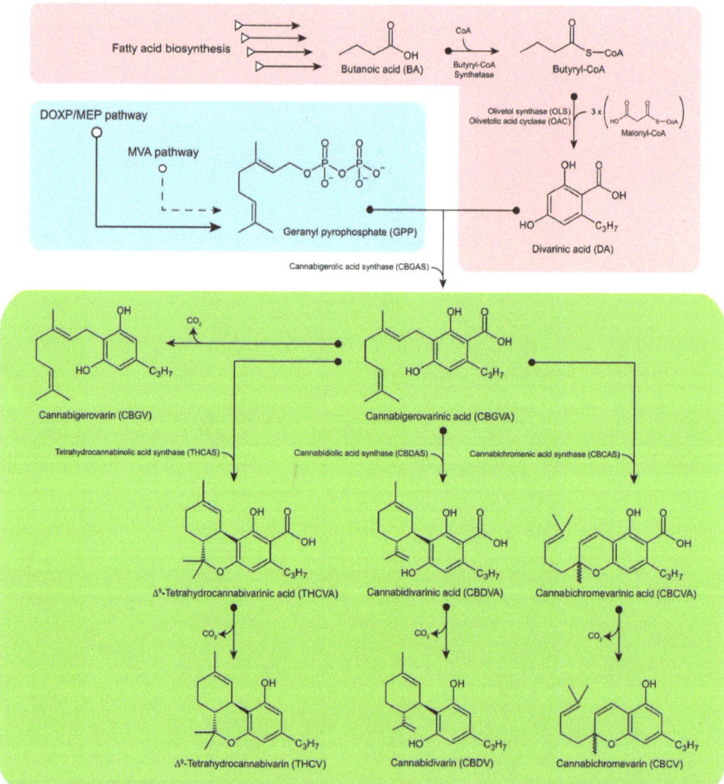

Figure 3. Propyl phytocannabinoids (C3) biosynthesis in *Cannabis sativa* L. Monoterpene moiety is provided majoritarian through the MEP/DOXP pathway, and in small extension through the MVA pathway, in which geranyl pyrophosphate (GPP) is synthesized. The fatty acids metabolism uses butanoic acid as a substrate to fulfill the phenolic (resorcinol) moiety of cannabinoids, generating divarinic acid (DA). Through the action of cannabigerolic acid synthase (CBGAS), GPP and DA are converted into cannabigerovarinic acid (CBGVA), the central precursor for many other C3 cannabinoids, such as Δ^9-tetrahydrocannabivarinic acid (Δ^9-THCVA), cannabidivarinic acid (CBDA) and cannabichromevarinic acid (CBCVA) [13,55,56,67].

3. Metabolic Engineering towards Phytocannabinoids Biosynthesis in Microorganisms

3.1. Design of a Suitable Host

A better approach to target the production of non-common cannabinoids can be achieved through the aid of metabolic engineering and synthetic biology. Usually, a safe and well-described cell is chosen as a "cell factory", a chassis for the production of the desired chemical compound. The chosen cell can express the pathways needed to achieve the product, but, typically, the flux toward the product is naturally low. Using classic strain improvement or directed genetic modifications (i.e., metabolic engineering), it is possible to increase the flux toward the product. If the cell does not naturally produce the compound of interest, the insertion of a synthetic pathway is necessary. Normally, the product will be generated in small amounts, but the pathway can be optimized to ensure a high flux toward the target, using concepts from both metabolic engineering and synthetic biology. Finally, a complete synthetic cell can be constructed in a manner that its pathways are tailored for the desired product, achieving great yields and low concentration of by-products [71].

Since fermentation of cannabinoids is a relatively new approach, there is no consensus on the best microorganism yet. The first step is to determine which microorganisms can be tailored for heterologous biosynthesis of these compounds. A review published by Carvalho et al. [72] covers some of the main host characteristics, such as genetic tools available for the microorganism, plant protein expression capacity, possibility of posttranslational modifications, and specific biosynthetic pathways. The microorganisms analyzed in this survey were *Escherichia coli*, *Saccharomyces cerevisiae*, *Komagataella phaffii* (*Pichia pastoris*), and *Kluyveromyces marxianus*, with qualitative indicators regarding hosts characteristics aforementioned. It was noticed that *E. coli* has significant genetic tools reported, and an arsenal of strains, promoters, and vectors, but its limited posttranslational modifications make it unlikeable to be a suitable host. All the other microorganisms are yeasts, with *S. cerevisiae* and *K. phaffii* (*P. pastoris*) being the most widely reported in the literature. The yeast *K. marxianus* has been reported to present an efficient hexanoic acid pathway [73], which could solve the low-availability pool of this metabolite during heterologous biosynthesis of cannabinoids.

3.2. From Sugar to Cannabinoids

The main intermediates and genes during phytocannabinoids biosynthesis in *S. cerevisiae* were recently reported by Luo et al. [13]. The chosen substrate for the microorganism was galactose. The GPP was produced with the introduction of the Ef*mvaE* and Ef*mvaS* genes of *Enterococcus faecalis* (an acetyl-CoA acetyltransferase/HMG-CoA reductase and an hydroxymethylglutaryl-CoA synthase [74]), and by overexpressing the genes of the mevalonate pathway (*ERG12*, *ERG8*, *ERG19*, and *IDI1*) [75] and a mutated *ERG20*$^{F96W/N127W}$ gene (erg20∗) that preferentially produces GPP over FPP [76]. Hexanoyl-CoA was produced heterologously using genes from *Ralstonia eutropha* (Re*bktB*, a β-keto thiolase from *Ralstonia eutropha* H16 that catalyzes condensation reactions between acetyl-CoA with acyl-CoA molecules [77]), *Cupriavidus necator* (Cn*paaH1*, an NADH-dependent 3-hydroxyacyl-CoA dehydrogenase [78]), *Clostridium acetobutylicum* (Ca*crt*, a crotonase that catalyzes the dehydration of 3-hydroxybutyryl-CoA to crotonyl-CoA in the n-butanol biosynthetic pathway [79]) and *Treponema denticola* (Td*ter*, a trans-enoyl-CoA reductase [80]), or feeding hexanoic acid as a substrate for AAE (encoded by Cs*AAE1* from *Cannabis*). Expression of the genes encoding Cs*TKS* and Cs*OAC* produced olivetolic acid, which was prenylated by Cs*PT4-T*, a geranylpyrophosphate:olivetolate geranyltransferase activity. The resulting CBGA was transformed into Δ^9-THCA and CBDA using THCAS and CBDAS. After exposure to heat, Δ^9-THCA and CBDA were decarboxylated to Δ^9-THC and CBD, respectively. As both Δ^9-THC and CBD come from CBGA, the insertion of gene copies that encode THCAS or CBDAS will determine which final product is going to be synthesized. The final concentration obtained of Δ^9-THCA and CBDA was 8.0 mg/L and 4.3 µg/L, respectively. In addition to cannabinoids derived from olivetolic acid, Luo et al. [13] also produced propyl cannabinoids (from divarinic acid). The hexanoic acid was replaced by butanoic acid, providing an appropriate butanoyl-CoA pool for the synthesis of C3 cannabinoids. Thus, Δ^9-THCVA and CBDVA were produced with concentrations of 4.8 mg/L and 6.0 µg/L, respectively [13].

A list of the enzymes involved during heterologous biosynthesis of phytocannabinoids by *S. cerevisiae* with their respective accession numbers on GenBank is available in Table 1.

Table 1. List of enzymes and corresponding GenBank accession numbers involved in heterologous expression of phytocannabinoids in *S. cerevisiae*.

Enzyme	Abbreviation	Accession No.	EC No.	References
Acyl activating enzyme 1	AAE	AFD33345.1	6.2.1.1	[81]
Olivetol synthase (tetraketide synthase 3)	OLS (TKS)	AB164375	2.3.1.206	[82]
Olivetolic cyclase	OAC	AFN42527.1	4.4.1.26	[60]
Geranylpyrophosphate:olivetolate geranyltransferase	GOT (CsPT4-T)	US10975379B2 [a]	2.5.1.102	[13]
Tetrahydrocannabinolic acid synthase	THCAS	AB057805	1.21.3.7	[83]
Cannabidiolic acid synthase	CBDAS	AB292682	1.21.3.8	[84]
Cannabichromenic acid synthase	CBCAS	WO2015196275A1 [b]	1.3.3-	[85,86]

[a] Patent number, [b] Application number.

Zirpel et al. [12] tested the production of Δ^9-THCA by heterologous hosts such as *E. coli*, *S. cerevisiae*, and *K. phaffii* [12], in which *S. cerevisiae* and *K. phaffii* showed THCA synthase activity after addition of 1 mM CBGA, leading to a Δ^9-THCA production of 0.36 g/L in *K. phaffii*. No functional expression of THCA synthase could be found in *E. coli*, hence it was concluded by the authors that functional expression of THCAS might require eukaryotic chaperones to facilitate covalent binding of FAD to the THCAS or glycosylation of the protein.

Renew Biopharma chose the green alga *Chlamydomonas reinhardtii* as a host, stating that it is capable of compartmentalizing the biosynthesis of cannabinoids in its chloroplasts, which protects the rest of the cellular structures [9]. This approach resulted in a more expensive downstream since microalgae are known to have a complex cellular wall. For instance, *Chlamydomonas reinhardtii* has a multilayered extracellular matrix, which requires physical and chemical agents to rupture it and access the cannabinoids [87].

3.3. Patent Prospection

A survey on the free access Patent Inspiration database was conducted using the term (*cannabi**) as a keyword for search on Title or Abstract, while the terms *microorganism* AND *yeast* AND *production* have been searched on Abstract and Description. The initial results revealed a total of 58 documents filled over the past 20 years proposing the protection of new technologies associated with the biotechnological production of cannabinoids or their derivates. However, after a thorough analysis, only 16 patens actually protected processes and methods related to the prospected theme (Table 2).

Table 2. Data obtained from the patent survey on Patent Inspiration database on cannabinoid biosynthesis via microbial source.

Patent Number	Title	Resume	Country	Applicants	Granted	Citations	Year (Publication)
US9546362B2	Genes and proteins for alkanoyl-coa synthesis	Proposition of genetic engineering of plant, yeast, or bacterial cells with a cassette comprising 13 where homologous, isolated, and/or purified sequences of *Cannabis sativa* for the production of cannabinoids using carboxylic acids as substrate	Canada	University of Saskatchewan and Natural Resources Council	Yes	18	2014
EP3067058A1	Biological composition based on engineered *Lactobacillus paracasei* subsp. *paracasei* f19 for the biosynthesis of cannabinoids	Discloses the use of the strain *Lactobacillus paracasei* subsp. *paracasei* f19 as a suitable host for *Cannabis sativa* gene incorporation	Italy	Farmagens Health Care SRL	No	8	2016
US10801049B2	Production of cannabinoids in microorganisms from a carbon sugar precursor	Claims the application of the insertion of the *pgi*, *zwf*, and *gltA* genes and the mutation of the *fadD* gene to the synthesis of the hexanoyl-CoA precursor from simple sugar sources	USA	Syntiva Therapeutics Inc.	Yes	0	2019
US10392635B2	Production of Tetrahydrocannabinolic Acid in Yeast	Insertion of a mutant aromatic prenyltransferase in yeast models, resulting in a higher yield of geranyl pyrophosphate, an important precursor of the cannabinoids	USA	Librede Inc.	Yes	1	2019
US10837031B2	Recombinant production systems for prenylated polyketides of the cannabinoid family	Proposes the recombinant production of cannabinoids in yeasts and filamentous fungi through the production of cannabinoid precursors when grown in the presence of exogenous prenol and isoprenol	USA	Baymedica Inc.	Yes	3	2019
US2020370073A1	Biosynthetic cannabinoid production methods	Proposes the commercial-scale production and processing of biosynthetic cannabinoids produced by growing genetically modified microalgae in a photo-bioreactor and the posterior recovery of the cannabinoid via extraction and distillation	USA	Insectergy LLC	No	0	2020
US2020340026A1	Neurotransmitters and Methods of Making the Same	Discloses the modification of microalgae for the expression of *Cannabis sativa*-encoding genes	USA	Purissima Inc.	No	0	2020

Table 2. *Cont.*

Patent Number	Title	Resume	Country	Applicants	Granted	Citations	Year (Publication)
US2020325508A1	Genes and proteins for aromatic polyketide synthesis	Expression or over-expression of the enzyme that catalyzes the synthesis of aromatic polyketides (e.g., olivetolic acid) which may result in increased production of cannabinoid compounds	Canada	University of Saskatchewan and Natural Resources Council	No	0	2020
US2020291434A1	Metabolic engineering of *E. coli* for the biosynthesis of cannabinoid products	Insertion of an overexpressed, bifunctional enzyme ispDF responsible for the synthesis of isoprene, terpenoids, and cannabinoids	Canada	Inmed Pharmaceuticals Inc.	No	0	2020
US2020224231A1	Production of cannabinoids in yeast using a fatty acid feedstock	Modification of the peroxisomal β-oxidation in yeasts to provide an affordable and sustainable production of cannabinoids using vegetable oil or animal fat	USA	Levadura Biotechnology Inc.	No	0	2020
US10975379B2	Microorganisms and methods for producing cannabinoids, and cannabinoid derivatives	Proposes the recombinant expression of a geranyl pyrophosphate: olivetolic acid geranyltransferase (GOT) to produce cannabinoids molecules, precursors, or its derivatives	USA	University of California	Yes	0	2020
US2020165644A1	Production of cannabinoids in yeast	Heterologous synthesis of cannabinoids using 5% of fatty acids in genetically modified yeasts containing one or more genes responsible for the production of GPP producing; two or more olivetolic acid-producing genes; one or more cannabinoid precursor or cannabinoid producing genes; and one or more Hexanoyl-CoA producing genes	USA	Biomedican Inc.	No	0	2020
US2020165641A1	Bidirectional multi-enzymatic scaffolds for biosynthesizing of cannabinoids	Metabolic engineering of yeasts and bacteria using a complex system of 15 enzyme-encoding sequences for the production of a wide range of cannabinoids using glucose as carbon source via hexanoyl-CoA, malonyl-CoA, or mevalonate pathways	USA	Khona Pharms LLC	No	0	2020

Table 2. *Cont.*

Patent Number	Title	Resume	Country	Applicants	Granted	Citations	Year (Publication)
US2020080115A1	Cannabinoid Production by Synthetic In Vivo Means	Transformation of yeast cells with three or more vectors comprising for the enhanced GPP production, production of OTA and GOT activity	USA	Biotic Sciences LLC	No	0	2020
US2020071732A1	Production of Cannabinoids in Yeast	Genetic engineering of yeast cells with the inclusion of the GPP pathway genes, allowing a superior yield of cannabinoids and use of glucose as carbon source	USA	Librede Inc.	No	0	2020
US10954534B2	Production of Cannabigerolic Acid in Yeast	Claims the heterologous expression of cannabigerolic acid in yeasts and bacteria through the insertion of *Cannabis sativa* acyl-activating enzyme, mutant prenyltransferase, olivetolic synthase, olivetolic acid cyclase, and aromatic prenyltransferase	USA	Librede Inc.	Yes	0	2020

The analysis of the International Patent Classification (IPC) revealed that the introduction of foreign genes synthesizing transferases (C12N9/10) or lyases (C12N9/88) in vectors or expression systems specially adapted for *E. coli* (C12N15/70) are the main areas investigated (data not shown). Although not being able to perform post-translational modification as yeasts and higher eukaryotic cells, due to *E. coli* superior growth rate, low nutritional requirement, higher yield, and its extensive genetic information turned into a preferable host for tailoring new metabolic pathways for the industrial production of cannabinoids [72,88,89].

The technology of the cannabinoids biosynthesis was first protected by the University of Saskatchewan in association with the Natural Resources Council of Canada, where homologous, isolated, and/or purified sequences of *Cannabis sativa* alkanoyl-CoA synthetases, type III polyketide synthase, polyketide cyclase, aromatic prenyltransferase, and a cannabinoid-forming oxidocylase were used as target genes for the cannabinoid production in *E. coli* and *Saccharomyces cerevisiae* systems (US9546362B2). According to the granted patent, carboxylic acids (C5–C20) and coenzyme-A are required as substrates, which directly impacts the cost of the final product. Similar plasmid vector configurations were later proposed with the addition of inducible galactose operons (US10392635B2), substitution of alkanoyl-CoA synthetases for prenol or isoprenol kinases (US10837031B2), or proposition of new host cells, such as *Lactobacillus paracasei* subsp. *paracasei* (EP3067058A1).

A recently granted patent by the American company Syntiva Therapeutics Inc. (US10801049B2) discloses the incorporation of phosphoglucose isomerase (*pgi*), glucose 6-phosphate dehydrogenase (*zwf*), and citrate synthase (*gltA*) genes in yeasts, which allows the heterologous production of hexanoate synthesis during the stationary phase using simple sugars. In addition, the overexpression of the long-chain fatty acid-CoA (*fadD*) ligase gene, responsible for the conversion of hexanoate into hexanoyl-CoA, also resulted in the silencing of the *fadE* gene associated with the degradation of this precursor. Such genetic modifications achieved costs inferior to US$1.000 per kilogram of purified cannabinoid and significantly increased the yield of the process [90]. This disruptive technology led to a significant leap in the number of filed patents, from 4 documents between 2014 and 2019 to 11 only in the last year (Table 2). The incremental changes proposed by these recent patents include the modification of the peroxisomal β-oxidation in yeasts to allow the use of fatty acids and affordable sources of vegetable and animal fat (US2020224231A1) and the inclusion of different pathways that allows the conversion of glucose into cannabinoids via acetoacetyl-CoA, malonyl-CoA, or mevalonate (US2020071732A1; US2020165641A1).

Our survey revealed that only three countries detain the technology for the heterologous production of cannabinoids, being the United States the major contributor with 12 filled documents, followed by Canada and Italy with three and one documents, respectively. The presence of Canada in this selective group is supported by the Cannabis Act [6], a jurisdictional regulation that establishes production guidelines, licenses, and requirements for cannabis-derived products, providing regulatory approval for both plant cultivation and the heterologous expression. USA and Italy, on the other hand, only have parameters defined by law regarding the cultivation and usage of the source material (i.e., cannabis plants with Δ^9-THC content of 0.3% or 0.2–0.6%, respectively), leaving the microbial production under an unregulated ground [91,92]. However, the allowance of cannabinoids prescription from a licensed healthcare provider [92,93] and the approval of the first CBD-containing drug (Epidiolex®; GW Pharmaceuticals, Cambridge, UK) by the FDA in 2018 for treating severe seizures in patients above one year old [94] creates a prone environment for the development of biosynthetic cannabinoid industry in these countries. This statement is supported by the nature of the applicants in the prospected patents, which are majorly represented by private companies.

3.4. Culture Medium, Production System, and Broth Composition

The production of phytocannabinoids by heterologous expression in yeasts has been accomplished through fed-batch liquid cultures [12,13]. This production system is indicated for fermentations in which substances are periodically added to the medium to fulfill the chemical demand of the target microorganism. The interval between applications avoids excess toxic substances in the medium, preventing possible detours during biosynthesis or even cell death. As shown by Coral et al. [95] the medium composition plays an important role to determine the optimal point between biomass and product concentration.

Luo et al. [13] worked with recombinant *S. cerevisiae* in liquid culture medium. Strains were pre-grown in yeast peptone dextrose extract (YPD) medium overnight and then back-diluted to OD_{600} = 0.2 into yeast peptone galactose extract (YPG), a non-selective culture medium for *Candida*, *Pichia*, *Saccharomyces*, and *Zygosaccharomyces* containing 20 g/L of peptone, 10 g/L of yeast extract, and 20 g/L of galactose. The medium was supplemented with 1 mM olivetolic acid or corresponding fatty acid (such as hexanoic, pentanoic, and butanoic acid). Strains were incubated for 24 h, 48 h, or 96 h in 24-deep-well plates (800 r.p.m.) at 30 °C while supplementing with 2% (*w/v*) galactose every 24 h.

Zirpel et al. [12] worked with recombinant *E. coli*, *P. pastoris*, and *S. cerevisiae*. Recombinant *E. coli* cells were grown in 1 l flasks, containing 100 mL LB-medium (50 µg kanamycin mL^{-1}, 33 µg chloramphenicol mL^{-1}, 100 µg spectinomycin mL^{-1}) at 37 °C and 200 rpm to an OD_{600} of 0.6. THCAS expression was started by the addition of 1 mM IPTG and cells grown for 16 h at 20 °C. Recombinant *S. cerevisiae* cells were grown in minimal medium without leucine at 30 °C and 200 rpm for 24 h. Cells were used to inoculate 100 mL of 2 × YPAD medium at an OD_{600} of 0.5 and incubated with 0.5 % (*w/v*) galactose at 20 °C and 200 rpm for 144 h. Recombinant *P. pastoris* cells were grown in BMGY at 30 °C and 200 rpm for 24 h. Afterward, cells were harvested by centrifugation at 5000× *g* for 5 min and resuspended in modified BMMY (mBMMY) [96] to an OD_{600} of 20. *Pichia* cells were cultivated at 15 °C and 200 rpm until no increase in THCAS activity could be observed and supplemented with 0.5% (*v/v*) methanol every 24 h for protein expression.

3.5. Metabolic Engineering In Silico

Despite the remarkable work accomplished by Luo et al. [13] the titers of Δ^9-THCA (8.0 mg/L^{-1}) and CBDA (4.4 µg/L^{-1}) obtained were low, making the process economically unfeasible to be scaled up into industrial levels. Improvement and redesign of metabolic pathways toward the product is the main strategy to enhance higher concentrations of cannabinoids. In fact, metabolic bottlenecks for the biosynthesis of Δ^9-THCA have been recently analyzed in silico and reported [66] for an engineered *S. cerevisiae* strain. The kinetics of reactions toward cannabinoids were modeled using MATLAB® (version 9.4, The Mathworks, Inc., Natick, MA, USA) with the SimBiology extension [97], in which Δ^9-THCA was produced from glucose instead of galactose—a much-appreciated upgrade since galactose is up to 100-fold more expensive than glucose. Nevertheless, a high glucose concentration at the beginning leads to respiratory inhibition known as the Crabtree effect [98], in which ethanol is produced and the growth rates are slowed.

The first challenge lies in acetyl-CoA, the committed precursor for mevalonate and olivetolic acid pathways, responsible for the GPP and OA pool, respectively. Thomas et al. [66] suggested the replacement of acetaldehyde dehydrogenase (ADH) as well as acetyl-CoA synthetase with aldehyde dehydrogenase acylating (ADA) from *Dickeya zeae*, an optimization that grants higher specific activity, demands less energy, and prevents acetate formation. Moreover, the ethanol generated by aerobic cultivation on glucose can be converted back into acetaldehyde with the addition of ADH2 under specific promoter control. In parallel, non-essential pathways can be muted to enhance the carbon flux toward cannabinoids. The peroxisomal citrate synthase and cytosolic malate synthase consume cytosolic acetyl-CoA, being the genes *CIT2* and *MLS1* excellent targets to be muted to improve acetyl-CoA pool.

The hexanoic acid production is another metabolic bottleneck referring to the limited pool of acetyl-CoA and down related to OA. The low specificity of OAC turns only 5% of all the hexanoic acid into OA and the remaining 95% into olivetol. The feeding of hexanoic acid is advantageous but limited to up to 1 mM due to cell toxicity and slower growth rates. OA feeding is also not recommended due to its high cost, low absorbance by the yeast, and chemical instability. Moreover, the CBGA production is a limiting step toward the optimization of the process as shown by Thomas et al. [66] in a sensitivity analysis. The membrane-bound enzyme CBGAS from *Cannabis sativa* L. was replaced by the soluble prenyltransferase NphB present in *Streptomyces* spp., especially due to a CBGA-specific variant recently reported [99].

In conclusion, the low availability of acetyl-CoA and hexanoic acid with the low specificity of OAC are the main limiting factors for higher yields. Nevertheless, the Δ^9-THCA titer predicted in silico after 40 h of fermentation was 299.8 mg/L^{-1}, a 37-fold increase compared to Luo et al. [13]. Although this value is small close to Δ^9-THCA and CBDA present in plants (5–20% in dry weight of extract), it is a great opportunity for the biosynthesis of non-common cannabinoids such as Δ^9-THCVA and CBDVA (<1% in dry weight).

4. Conceptual Downstream Analysis

4.1. Process Flowchart

A process flowchart was proposed to illustrate the downstream procedures involved during cannabinoids purification via heterologous expression in *S. cerevisiae* (Figure 4). It is considered that the engineered yeast produces Δ^9-THCA. As aforementioned, the downstream unit operations' choices rely on microorganism specificities, and although this is a simplified model, it accounts for the main steps and operations toward the purification of cannabinoids on an industrial scale. With the development of pilot-scale experiments, kinetical and transport parameters can be better estimated for decision-making.

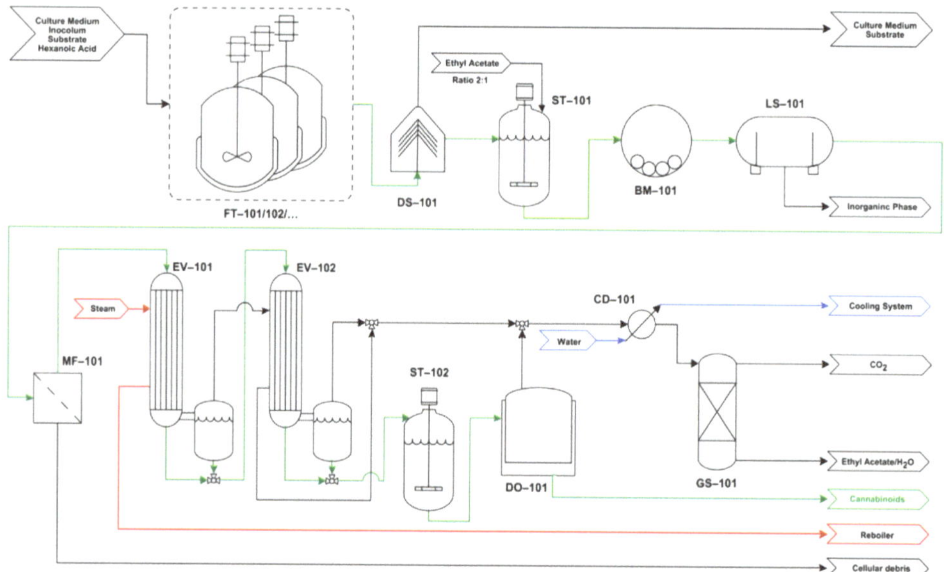

Figure 4. Process flowchart listing the downstream operations required to purify Δ^9-THCA from fermented broth and achieve high-quality Δ^9-THC. FT: fermentation tank; DS: disc-stack centrifuge; ST: settling tank; BM: ball/beads mill; LS: liquid-liquid separator; MF: microfiltration unit; EV: falling film evaporator; DO: decarboxylation oven; GS: gas-liquid separator.

4.2. Process Description

The separation procedures chosen were based on the works of Zirpel et al. [12] and Luo et al. [13], whereas the scaling up of the process were based on the works of Magalhães et al. [100], and Poulos and Farnia [101], although some changes have been proposed to scale up the process. A staggered set of fermentation tanks (FT-101/102/ ...) is considered. After the fermentation time, the fermentation broth is sent to a disk-stack centrifuge (DS-101), responsible for the removal of culture medium and substrate not consumed during fermentation. Centrifugation is a suitable option due to *S. cerevisiae* high density (1.1 g cm^{-3}) and sedimentation radius (2.5 μm) [87]. Another option for this step is microfiltration, although the high-volume flow would require several membrane units to supply it. The cells can be dried in a low-temperature oven to remove the remaining water.

The cells are then sent to a settling tank (ST-101) in which ethyl acetate is used with a 2:1 ratio to resuspend the cells and subsequently promote liquid-liquid extraction. Ethyl acetate was chosen as the solvent due to its high capability to solubilize cannabinoids [102], and also because it is only partially soluble with water (8.3 g/L at $20\ ^\circ$C), which allows the use of liquid–liquid separators at the downstream. As previously mentioned, ethanol is also suitable for cannabinoids extraction, but its high water solubility impairs the subsequent steps. Moreover, ethyl acetate is FDA approved for use in food as a flavor/fragrance enhancer and solvent [102].

The suspension is sent to a ball/bead mill (BM-01) to promote cell lysis. Since *S. cerevisiae* is disproved of a complex polysaccharide cell wall, the physical disruption should be enough, although chemical methods (e.g., detergents, enzymes, chelating agents, and/or solvents) can complement this process. Alternatively, high-pressure homogenizers can be used. In this stage, the cells are broken and the cannabinoids are dispersed in the medium. It is a relatively quick process on a laboratory scale (30 s^{-1} over 3 min) [13]. The biphasic mixture passes through a liquid–liquid separator (LS-101), wherein the upper (organic) phase contains cannabinoids, ethyl acetate, and the lower phase is composed of water, ethyl acetate, and nutrients/culture medium. The lower phase is sent to the solvent recovery area.

The organic phase is forwarded to a microfiltration unit (MF-101) to remove cellular debris. For this operation, the filter membrane needs to have a pore size between 0.2 and 0.45 μm [13] used polyvinylidene difluoride (PVDF) membranes during its polishing steps prior to HPLC analysis.

The filtrate is then sent to a set of multiple effects falling film evaporators (EV-01/02) to remove part of the solvent and prepare the product for the decarboxylation step. Vacuum is used to boil the mixture in low temperatures, avoiding Δ^9-THCA oxidation into CBNA and other secondary reactions [103]. It is known that CBNA/CBN is formed during the long-time storage of cannabis [104], although its rate is reduced in the absence of oxygen and light [52]. The vapor from the first effect is used as a heat duty stream to the second effect. Due to the high boiling points of cannabinoids, losses involved during evaporation are minimal. The vapor and condensate from the second effect are sent to a condenser (CD-101).

The concentrate is forwarded to a settling tank (ST-102) avoiding process discontinuity by upstream delays. The last step is to remove the residual solvent in the product and promote the decarboxylation of Δ^9-THCA into Δ^9-THC. For this step, a decarboxylation vacuum oven (DO-101) is proposed, in which the mixture is dispersed into trays with temperature close to $120\ ^\circ$C for up to one hour [105]. As shown by Wang et al. [52], it is possible to obtain pure Δ^9-THC from Δ^9-THCA by heating the extract to $110\ ^\circ$C for 40 min, under vacuum and absence of light. Even though no significant amount of CBN was detected, a relative loss in total molar concentration of 7.94% was noted, indicating that part of the reactant or product is being consumed by a secondary mechanism (e.g., a side reaction with an unstable intermediate and/or product).

After the decarboxylation step, the Δ^9-THC extract is almost completely pure. The final product consists of Δ^9-THC with residual ethyl acetate. As decarboxylation involves the loss of a carboxyl group, the molar mass of Δ^9-THCA goes from 358.48 g/mol to 314.47 g/mol, causing a reduction in the mass of the final product by 12.3%.

As a complementary procedure, the concentrate can be sent to a fine separation involving chromatography, such as high-performance liquid chromatography (HPLC), counter-current chromatography (CCC), and centrifugal partition chromatography (CPC).

These techniques show high separation capacity and the possibility of scaling. CPC was chosen because of its advantages over CCC, such as a higher flow for the same volume. On a laboratory scale, 250 mL centrifugal partition chromatography has an ideal flow rate of 5–15 mL/min, while 250 mL counter flow chromatography has an ideal flow rate of 1 to 3 mL/min. On an industrial scale, 25 L counter-current chromatography has an ideal flow rate of 100 to 300 mL/min, whereas 25 L centrifugal partition chromatography has an ideal flow rate of 1000 to 3000 mL/min. This ensures greater productivity (due to higher flow and faster separation time), allowing the process to be scalable to up to tons per month [106]. RotaChrom Technologies LLC (Budapest, Hungary) developed an industrial scale CPC, the iCPC®, which can deliver a flow rate of up to 2.5 L/min, achieving 50–500 kg of purified product per month [107]. The final product is resuspended in anhydrous ethanol or formulated in capsules/pills as desired.

5. Further Analysis and Improvements

The production of cannabinoids through heterologous expression in *S. cerevisiae* is feasible, although its low yields and metabolic bottlenecks adds complexity to scale up the process. Although fermentation can supply several cannabinoids, full-spectrum extracts (i.e., those with phytocannabinoids and secondary metabolites) are unlikely to be achieved, especially due to metabolic network complexity and microorganism expression limitations. In the future, it is important to analyze the limiting factors of cannabinoid production in the recombinant microorganism, and even reassess whether *S. cerevisiae* is the best candidate for this task.

To optimize the fermentative production of cannabinoids in recombinant microorganisms, different parameters need to be considered at genetic, metabolic, and technological levels. The first one refers to the expression of genes and pathways for the conversion of glucose into cannabinoids. The metabolic level is responsible for the better understanding of pathway interactions, allowing the characterization of metabolic bottlenecks to be further engineered and optimized. As noticed, the low acetyl-CoA and hexanoic acid availability for subsequent pathways are the main bottlenecks for the biosynthesis of Δ^9-THCA in *S. cerevisiae*. The technological level refers to the downstream procedures needed to achieve high-purity cannabinoids on an industrial scale, avoiding unnecessary losses and providing a final product with accessible cost.

Nevertheless, cannabinoids fermentation is an exciting and brand-new niche arriving that can substantially change the availability of those compounds, providing a high-quality drug at a reasonable price, especially for non-common cannabinoids, such as C3 cannabinoids, novel cannabinoids, and analogs.

Author Contributions: Conceptualization, G.R.F. and G.V.d.M.P.; investigation, G.R.F., G.V.d.M.P. and D.P.d.C.N.; resources, C.R.S.; writing—original draft preparation, G.R.F.; writing—review and editing, G.R.F., G.V.d.M.P., J.C.d.C., D.P.d.C.N. and C.R.S.; visualization, G.V.d.M.P. and J.C.d.C.; supervision, G.V.d.M.P. and J.C.d.C.; project administration, C.R.S.; funding acquisition, C.R.S. All authors have read and agreed to the published version of the manuscript.

Funding: This work was funded by Coordenação de Aperfeiçoamento de Pessoal de Nível Superior (CAPES) and Conselho Nacional de Desenvolvimento Científico e Tecnológico do Brasil (CNPq).

Institutional Review Board Statement: Not applicable.

Informed Consent Statement: Not applicable.

Data Availability Statement: Not applicable.

Conflicts of Interest: The authors declare no conflict of interest.

References

1. California Department of Public Health. Compassionate Use Act of 1996. Available online: https://leginfo.legislature.ca.gov/faces/codes_displaySection.xhtml?sectionNum=11362.5.&lawCode=HSC (accessed on 15 November 2021).
2. Government of Canada. Marihuana Medical Access Regulations (SOR/2001-227). Available online: https://laws-lois.justice.gc.ca/eng/regulations/SOR-2001-227/index.html (accessed on 15 November 2021).
3. State of Washington. Washington Initiative 502 (I502). Available online: https://www.sos.wa.gov/_assets/elections/initiatives/i502.pdf (accessed on 21 November 2021).
4. State of Colorado. Amendment 64. Available online: https://www.colorado.gov/pacific/sites/default/files/13%20Amendment%2064%20LEGIS.pdf (accessed on 1 December 2021).
5. Parliament of Uruguay. Ley 19.172. Available online: https://www.impo.com.uy/bases/leyes/19172-2013 (accessed on 1 December 2021).
6. Government of Canada. Cannabis Act (S.C. 2018, c. 16). Available online: https://laws-lois.justice.gc.ca/eng/acts/c-24.5/ (accessed on 1 December 2021).
7. Grand View Research. Cannabidiol Market Size, Share & Trends Analysis Report by Source Type (Hemp, Marijuana), by Distribution Channel (B2B, B2C), by End Use, by Region, and Segment Forecasts, 2019–2025. *Gd. View Res.* **2019**, *2028*, 2021–2028.
8. Rosenthal, E.; Downs, D. *Beyond Buds: Marijuana Extracts Hash, Vaping, Dabbing, Edibles and Medicines*; Quick Trading Company: Medley, FL, USA, 2014.
9. Dolgin, E. The Bioengineering of Cannabis. *Nature* **2019**, *572*, S5–S7. [CrossRef]
10. Shiponi, S.; Bernstein, N. The Highs and Lows of P Supply in Medical Cannabis: Effects on Cannabinoids, the Ionome, and Morpho-Physiology. *Front. Plant Sci.* **2021**, *12*, 657323. [CrossRef]
11. Summers, H.M.; Sproul, E.; Quinn, J.C. The Greenhouse Gas Emissions of Indoor Cannabis Production in the United States. *Nat. Sustain.* **2021**, *4*, 644–650. [CrossRef]
12. Zirpel, B.; Stehle, F.; Kayser, O. Production of Δ9-Tetrahydrocannabinolic Acid from Cannabigerolic Acid by Whole Cells of *Pichia* (*Komagataella*) *Pastoris* Expressing Δ9-Tetrahydrocannabinolic Acid Synthase from *Cannabis sativa* L. *Biotechnol. Lett.* **2015**, *37*, 1869–1875. [CrossRef]
13. Luo, X.; Reiter, M.A.; d'Espaux, L.; Wong, J.; Denby, C.M.; Lechner, A.; Zhang, Y.; Grzybowski, A.T.; Harth, S.; Lin, W.; et al. Complete Biosynthesis of Cannabinoids and Their Unnatural Analogues in Yeast. *Nature* **2019**, *567*, 123–126. [CrossRef]
14. Ben-Shabat, S.; Fride, E.; Sheskin, T.; Tamiri, T.; Rhee, M.-H.; Vogel, Z.; Bisogno, T.; De Petrocellis, L.; Di Marzo, V.; Mechoulam, R. An Entourage Effect: Inactive Endogenous Fatty Acid Glycerol Esters Enhance 2-Arachidonoyl-Glycerol Cannabinoid Activity. *Eur. J. Pharmacol.* **1998**, *353*, 23–31. [CrossRef]
15. Gallily, R.; Yekhtin, Z.; Hanuš, L.O. Overcoming the Bell-Shaped Dose-Response of Cannabidiol by Using *Cannabis* Extract Enriched in Cannabidiol. *Pharmacol. Pharm.* **2015**, *6*, 75–85. [CrossRef]
16. Johnson, J.R.; Burnell-Nugent, M.; Lossignol, D.; Ganae-Motan, E.D.; Potts, R.; Fallon, M.T. Multicenter, Double-Blind, Random-ized, Placebo-Controlled, Parallel-Group Study of the Efficacy, Safety, and Tolerability of THC:CBD Extract and THC Extract in Patients with Intractable Cancer-Related Pain. *J. Pain Symptom Manag.* **2010**, *39*, 167–179. [CrossRef]
17. Blasco-Benito, S.; Seijo-Vila, M.; Caro-Villalobos, M.; Tundidor, I.; Andradas, C.; García-Taboada, E.; Wade, J.; Smith, S.; Guzmán, M.; Pérez-Gómez, E.; et al. Appraising the "Entourage Effect": Antitumor Action of a Pure Cannabinoid versus a Botanical Drug Preparation in Preclinical Models of Breast Cancer. *Biochem. Pharmacol.* **2018**, *157*, 285–293. [CrossRef]
18. Russo, E.B. Taming THC: Potential Cannabis Synergy and Phytocannabinoid-Terpenoid Entourage Effects. *Br. J. Pharmacol.* **2011**, *163*, 1344–1364. [CrossRef] [PubMed]
19. Quintão, N.L.; da Silva, G.F.; Antonialli, C.S.; Rocha, L.W.; Cechinel Filho, V.; Cicció, J.F. Chemical Composition and Evaluation of the Anti-Hypernociceptive Effect of the Essential Oil Extracted from the Leaves of Ugni Myricoides on Inflammatory and Neuropathic Models of Pain in Mice. *Planta Med.* **2010**, *76*, 1411–1418. [CrossRef] [PubMed]
20. Fidyt, K.; Fiedorowicz, A.; Strząbała, L.; Szumny, A. B-Caryophyllene and B-Caryophyllene Oxide—Natural Compounds of Anticancer and Analgesic Properties. *Cancer Med.* **2016**, *5*, 3007–3017. [CrossRef] [PubMed]
21. Nascimento, S.S.; Camargo, E.A.; Desantana, J.M.; Araújo, A.A.S.; Menezes, P.P.; Lucca-Júnior, W.; Albuquerque-Júnior, R.L.C.; Bonjardim, L.R.; Quintans-Júnior, L.J. Linalool and Linalool Complexed in β-Cyclodextrin Produce Anti-Hyperalgesic Activity and Increase Fos Protein Expression in Animal Model for Fibromyalgia. *Naunyn. Schmiedebergs. Arch. Pharmacol.* **2014**, *387*, 935–942. [CrossRef]
22. Rufino, A.T.; Ribeiro, M.; Judas, F.; Salgueiro, L.; Lopes, M.C.; Cavaleiro, C.; Mendes, A.F. Anti-Inflammatory and Chondroprotec-tive Activity of (+)-α-Pinene: Structural and Enantiomeric Selectivity. *J. Nat. Prod.* **2014**, *77*, 264–269. [CrossRef]
23. Souza, M.C.; Siani, A.C.; Ramos, M.F.S.; Menezes-de-Lima, O.J.; Henriques, M.G.M.O. Evaluation of Anti-Inflammatory Activity of Essential Oils from Two Asteraceae Species. *Pharmazie* **2003**, *58*, 582–586.
24. Lei, Y.; Fu, P.; Jun, X.; Cheng, P. Pharmacological Properties of Geraniol—A Review. *Planta Med.* **2019**, *85*, 48–55. [CrossRef]
25. Mukhtar, Y.M.; Adu-Frimpong, M.; Xu, X.; Yu, J. Biochemical Significance of Limonene and Its Metabolites: Future Prospects for Designing and Developing Highly Potent Anticancer Drugs. *Biosci. Rep.* **2018**, *38*, BSR20181253. [CrossRef]
26. Valente, J.; Zuzarte, M.; Gonçalves, M.J.; Lopes, M.C.; Cavaleiro, C.; Salgueiro, L.; Cruz, M.T. Antifungal, Antioxidant and Anti-Inflammatory Activities of *Oenanthe crocata* L. Essential Oil. *Food Chem. Toxicol.* **2013**, *62*, 349–354. [CrossRef]

27. Takayama, C.; De-Faria, F.M.; de Almeida, A.C.A.; Valim-Araújo, D.D.A.E.O.; Rehen, C.S.; Dunder, R.J.; Socca, E.A.R.; Manzo, L.P.; Rozza, A.L.; Salvador, M.J.; et al. Gastroprotective and Ulcer Healing Effects of Essential Oil from *Hyptis Spicigera* Lam. (*Lamiaceae*). *J. Ethnopharmacol.* **2011**, *135*, 147–155. [CrossRef]

28. Tambe, Y.; Tsujiuchi, H.; Honda, G.; Ikeshiro, Y.; Tanaka, S. Gastric Cytoprotection of the Non-Steroidal Anti-Inflammatory Sesquiterpene, Beta-Caryophyllene. *Planta Med.* **1996**, *62*, 469–470. [CrossRef] [PubMed]

29. De Souza, M.C.; Vieira, A.J.; Beserra, F.P.; Pellizzon, C.H.; Nóbrega, R.H.; Rozza, A.L. Gastroprotective Effect of Limonene in Rats: Influence on Oxidative Stress, Inflammation and Gene Expression. *Phytomedicine* **2019**, *53*, 37–42. [CrossRef] [PubMed]

30. Satou, T.; Kasuya, H.; Maeda, K.; Koike, K. Daily Inhalation of α-Pinene in Mice: Effects on Behavior and Organ Accumulation. *Phytother. Res.* **2014**, *28*, 1284–1287. [CrossRef] [PubMed]

31. Bahi, A.; Al Mansouri, S.; Al Memari, E.; Al Ameri, M.; Nurulain, S.M.; Ojha, S. β-Caryophyllene, a CB2 Receptor Agonist Produces Multiple Behavioral Changes Relevant to Anxiety and Depression in Mice. *Physiol. Behav.* **2014**, *135*, 119–124. [CrossRef] [PubMed]

32. Zhang, L.-L.; Yang, Z.-Y.; Fan, G.; Ren, J.-N.; Yin, K.-J.; Pan, S.-Y. Antidepressant-like Effect of *Citrus sinensis* (L.) Osbeck Essential Oil and Its Main Component Limonene on Mice. *J. Agric. Food Chem.* **2019**, *67*, 13817–13828. [CrossRef] [PubMed]

33. Lima, N.G.P.B.; De Sousa, D.P.; Pimenta, F.C.F.; Alves, M.F.; De Souza, F.S.; Macedo, R.O.; Cardoso, R.B.; de Morais, L.C.S.L.; Melo Diniz, M.D.F.F.; de Almeida, R.N. Anxiolytic-like Activity and GC-MS Analysis of (R)-(+)-Limonene Fragrance, a Natural Compound Found in Foods and Plants. *Pharmacol. Biochem. Behav.* **2013**, *103*, 450–454. [CrossRef] [PubMed]

34. Souto-Maior, F.N.; De Carvalho, F.L.D.; De Morais, L.C.S.L.; Netto, S.M.; De Sousa, D.P.; De Almeida, R.N. Anxiolytic-like Effects of Inhaled Linalool Oxide in Experimental Mouse Anxiety Models. *Pharmacol. Biochem. Behav.* **2011**, *100*, 259–263. [CrossRef] [PubMed]

35. Guzmán-Gutiérrez, S.L.; Bonilla-Jaime, H.; Gómez-Cansino, R.; Reyes-Chilpa, R. Linalool and β-Pinene Exert Their Antidepressant-like Activity through the Monoaminergic Pathway. *Life Sci.* **2015**, *128*, 24–29. [CrossRef]

36. Matsuo, A.L.; Figueiredo, C.R.; Arruda, D.C.; Pereira, F.V.; Scutti, J.A.B.; Massaoka, M.H.; Travassos, L.R.; Sartorelli, P.; Lago, J.H.G. α-Pinene Isolated from *Schinus terebinthifolius* Raddi (Anacardiaceae) Induces Apoptosis and Confers Antimetastatic Protection in a Melanoma Model. *Biochem. Biophys. Res. Commun.* **2011**, *411*, 449–454. [CrossRef]

37. Cho, M.; So, I.; Chun, J.N.; Jeon, J.-H. The Antitumor Effects of Geraniol: Modulation of Cancer Hallmark Pathways (Review). *Int. J. Oncol.* **2016**, *48*, 1772–1782. [CrossRef]

38. Katsuyama, S.; Mizoguchi, H.; Kuwahata, H.; Komatsu, T.; Nagaoka, K.; Nakamura, H.; Bagetta, G.; Sakurada, T.; Sakurada, S. Involvement of Peripheral Cannabinoid and Opioid Receptors in β-Caryophyllene-Induced Antinociception. *Eur. J. Pain* **2013**, *17*, 664–675. [CrossRef] [PubMed]

39. Rao, V.S.; Menezes, A.M.; Viana, G.S. Effect of Myrcene on Nociception in Mice. *J. Pharm. Pharmacol.* **1990**, *42*, 877–878. [CrossRef] [PubMed]

40. Souto-Maior, F.N.; Da Fonsêca, D.V.; Salgado, P.R.R.; Monte, L.D.O.; De Sousa, D.P.; De Almeida, R.N. Antinociceptive and Anticonvulsant Effects of the Monoterpene Linalool Oxide. *Pharm. Biol.* **2017**, *55*, 63–67. [CrossRef] [PubMed]

41. Machado, K.D.C.; Islam, M.T.; Ali, E.S.; Rouf, R.; Uddin, S.J.; Dev, S.; Shilpi, J.A.; Shill, M.C.; Reza, H.M.; Das, A.K.; et al. A Systematic Review on the Neuroprotective Perspectives of Beta-Caryophyllene. *Phytother. Res.* **2018**, *32*, 2376–2388. [CrossRef] [PubMed]

42. Ciftci, O.; Oztanir, M.N.; Cetin, A. Neuroprotective Effects of β-Myrcene Following Global Cerebral Ischemia/Reperfusion-Mediated Oxidative and Neuronal Damage in a C57BL/J6 Mouse. *Neurochem. Res.* **2014**, *39*, 1717–1723. [CrossRef] [PubMed]

43. Shin, M.; Liu, Q.F.; Choi, B.; Shin, C.; Lee, B.; Yuan, C.; Song, Y.J.; Yun, H.S.; Lee, I.-S.; Koo, B.-S.; et al. Neuroprotective Effects of Limonene (+) against Aβ42-Induced Neurotoxicity in a Drosophila Model of Alzheimer's Disease. *Biol. Pharm. Bull.* **2020**, *43*, 409–417. [CrossRef]

44. Xu, P.; Wang, K.; Lu, C.; Dong, L.; Gao, L.; Yan, M.; Aibai, S.; Yang, Y.; Liu, X. The Protective Effect of Lavender Essential Oil and Its Main Component Linalool against the Cognitive Deficits Induced by D-Galactose and Aluminum Trichloride in Mice. *Evid.-Based Complement. Altern. Med.* **2017**, *2017*, 7426538. [CrossRef]

45. Do Vale, T.G.; Furtado, E.C.; Santos, J.G.; Viana, G.S.B. Central Effects of Citral, Myrcene and Limonene, Constituents of Essential Oil Chemotypes from *Lippia alba* (Mill.) N.E. Brown. *Phytomedicine* **2002**, *9*, 709–714. [CrossRef]

46. Gastón, M.S.; Cid, M.P.; Vázquez, A.M.; Decarlini, M.F.; Demmel, G.I.; Rossi, L.I.; Aimar, M.L.; Salvatierra, N.A. Sedative Effect of Central Administration of *Coriandrum Sativum* Essential Oil and Its Major Component Linalool in Neonatal Chicks. *Pharm. Biol.* **2016**, *54*, 1954–1961. [CrossRef]

47. Ito, K.; Ito, M. The Sedative Effect of Inhaled Terpinolene in Mice and Its Structure-Activity Relationships. *J. Nat. Med.* **2013**, *67*, 833–837. [CrossRef]

48. Cavaleiro, C.; Salgueiro, L.; Gonçalves, M.-J.; Hrimpeng, K.; Pinto, J.; Pinto, E. Antifungal Activity of the Essential Oil of *Angelica major* against *Candida*, *Cryptococcus*, *Aspergillus* and *Dermatophyte* Species. *J. Nat. Med.* **2015**, *69*, 241–248. [CrossRef] [PubMed]

49. Pinto, Â.V.; de Oliveira, J.C.; Costa de Medeiros, C.A.; Silva, S.L.; Pereira, F.O. Potentiation of Antifungal Activity of Terbinafine by Dihydrojasmone and Terpinolene against Dermatophytes. *Lett. Appl. Microbiol.* **2020**, *72*, 292–298. [CrossRef] [PubMed]

50. Di Marzo, V.; Bifulco, M.; De Petrocellis, L. The Endocannabinoid System and Its Therapeutic Exploitation. *Nat. Rev. Drug Discov.* **2004**, *3*, 771–784. [CrossRef] [PubMed]

51. ElSohly, M.; Gul, W. Constiuents of *Cannabis Sativa*. In *Handbook of Cannabis*; Oxford University Press: Oxford, UK, 2015; pp. 115–136. [CrossRef]
52. Wang, M.; Wang, Y.-H.; Avula, B.; Radwan, M.M.; Wanas, A.S.; van Antwerp, J.; Parcher, J.F.; ElSohly, M.A.; Khan, I.A. Decarboxylation Study of Acidic Cannabinoids: A Novel Approach Using Ultra-High-Performance Supercritical Fluid Chromatography/Photodiode Array-Mass Spectrometry. *Cannabis Cannabinoid Res.* **2016**, *1*, 262–271. [CrossRef]
53. Hanuš, L.O.; Meyer, S.M.; Muñoz, E.; Taglialatela-Scafati, O.; Appendino, G. Phytocannabinoids: A Unified Critical Inventory. *Nat. Prod. Rep.* **2016**, *33*, 1357–1392. [CrossRef]
54. Pisanti, S.; Malfitano, A.M.; Ciaglia, E.; Lamberti, A.; Ranieri, R.; Cuomo, G.; Abate, M.; Faggiana, G.; Proto, M.C.; Fiore, D.; et al. Cannabidiol: State of the Art and New Challenges for Therapeutic Applications. *Pharmacol. Ther.* **2017**, *175*, 133–150. [CrossRef]
55. Degenhardt, F.; Stehle, F.; Kayser, O. The Biosynthesis of Cannabinoids. In *Handbook of Cannabis and Related Pathologies: Biology, Pharmacology, Diagnosis, and Treatment*; Elsevier: Amsterdam, The Netherlands, 2017; pp. 13–23. [CrossRef]
56. Fellermeier, M.; Eisenreich, W.; Bacher, A.; Zenk, M.H. Biosynthesis of Cannabinoids. *Eur. J. Biochem.* **2001**, *268*, 1596–1604. [CrossRef]
57. Shoyama, Y.; Yagi, M.; Nishioka, I.; Yamauchi, T. Biosynthesis of Cannabinoid Acids. *Phytochemistry* **1975**, *14*, 2189–2192. [CrossRef]
58. Burke, C.C.; Wildung, M.R.; Croteau, R. Geranyl Diphosphate Synthase: Cloning, Expression, and Characterization of This Prenyltransferase as a Heterodimer. *Proc. Natl. Acad. Sci. USA* **1999**, *96*, 13062–13067. [CrossRef]
59. Marks, M.D.; Tian, L.; Wenger, J.P.; Omburo, S.N.; Soto-Fuentes, W.; He, J.; Gang, D.R.; Weiblen, G.D.; Dixon, R.A. Identification of Candidate Genes Affecting Δ9-Tetrahydrocannabinol Biosynthesis in *Cannabis sativa*. *J. Exp. Bot.* **2009**, *60*, 3715–3726. [CrossRef]
60. Gagne, S.J.; Stout, J.M.; Liu, E.; Boubakir, Z.; Clark, S.M.; Page, J.E. Identification of Olivetolic Acid Cyclase from *Cannabis Sativa* Reveals a Unique Catalytic Route to Plant Polyketides. *Proc. Natl. Acad. Sci. USA* **2012**, *109*, 12811–12816. [CrossRef] [PubMed]
61. Dubey, V.S.; Bhalla, R.; Luthra, R. An Overview of the Non-Mevalonate Pathway for Terpenoid Biosynthesis in Plants. *J. Biosci.* **2003**, *28*, 637–646. [CrossRef]
62. Kempinski, C.; Jiang, Z.; Bell, S.; Chappell, J. *Metabolic Engineering of Higher Plants and Algae for Isoprenoid Production*; Springer: Berlin/Heidelberg, Germany, 2015; pp. 161–199. [CrossRef]
63. Moses, T.; Pollier, J.; Thevelein, J.M.; Goossens, A. Bioengineering of Plant (Tri)Terpenoids: From Metabolic Engineering of Plants to Synthetic Biology in Vivo and In Vitro. *New Phytol.* **2013**, *200*, 27–43. [CrossRef]
64. Gülck, T.; Møller, B.L. Phytocannabinoids: Origins and Biosynthesis. *Trends Plant Sci.* **2020**, *25*, 985–1004. [CrossRef] [PubMed]
65. Fellermeier, M.; Zenk, M.H. Prenylation of Olivetolate by a Hemp Transferase Yields Cannabigerolic Acid, the Precursor of Tetrahydrocannabinol. *FEBS Lett.* **1998**, *427*, 283–285. [CrossRef]
66. Thomas, F.; Schmidt, C.; Kayser, O. Bioengineering Studies and Pathway Modeling of the Heterologous Biosynthesis of Tetrahydrocannabinolic Acid in Yeast. *Appl. Microbiol. Biotechnol.* **2020**, *104*, 9551–9563. [CrossRef] [PubMed]
67. Thomas, B.F.; ElSohly, M.A. Biosynthesis and Pharmacology of Phytocannabinoids and Related Chemical Constituents. In *The Analytical Chemistry of Cannabis*; Elsevier: Amsterdam, The Netherlands, 2016; pp. 27–41. [CrossRef]
68. De Meijer, E.P.M.; Hammond, K.M.; Micheler, M. The Inheritance of Chemical Phenotype in *Cannabis Sativa* L. (III): Variation in Cannabichromene Proportion. *Euphytica* **2009**, *165*, 293–311. [CrossRef]
69. Shoyama, Y.; Hirano, H.; Nishioka, I. Biosynthesis of Propyl Cannabinoid Acid and Its Biosynthetic Relationship with Pentyl and Methyl Cannabinoid Acids. *Phytochemistry* **1984**, *23*, 1909–1912. [CrossRef]
70. De Meijer, E.P.M.; Hammond, K.M.; Sutton, A. The Inheritance of Chemical Phenotype in *Cannabis Sativa* L. (IV): Cannabinoid-Free Plants. *Euphytica* **2009**, *168*, 95–112. [CrossRef]
71. Nielsen, J.; Keasling, J.D. Synergies between Synthetic Biology and Metabolic Engineering. *Nat. Biotechnol.* **2011**, *29*, 693–695. [CrossRef]
72. Carvalho, Â.; Hansen, E.H.; Kayser, O.; Carlsen, S.; Stehle, F. Designing Microorganisms for Heterologous Biosynthesis of Cannabinoids. *FEMS Yeast Res.* **2017**, *17*, fox037. [CrossRef]
73. Cheon, Y.; Kim, J.-S.; Park, J.-B.; Heo, P.; Lim, J.H.; Jung, G.Y.; Seo, J.-H.; Park, J.H.; Koo, H.M.; Cho, K.M.; et al. A Biosynthetic Pathway for Hexanoic Acid Production in Kluyveromyces Marxianus. *J. Biotechnol.* **2014**, *182–183*, 30–36. [CrossRef] [PubMed]
74. Sutherlin, A.; Hedl, M.; Sanchez-Neri, B.; Burgner, J.W.; Stauffacher, C.V.; Rodwell, V.W. *Enterococcus Faecalis* 3-Hydroxy-3-Methylglutaryl Coenzyme A Synthase, an Enzyme of Isopentenyl Diphosphate Biosynthesis. *J. Bacteriol.* **2002**, *184*, 4065–4070. [CrossRef] [PubMed]
75. Reider Apel, A.; D'Espaux, L.; Wehrs, M.; Sachs, D.; Li, R.A.; Tong, G.J.; Garber, M.; Nnadi, O.; Zhuang, W.; Hillson, N.J.; et al. A Cas9-Based Toolkit to Program Gene Expression in *Saccharomyces cerevisiae*. *Nucleic Acids Res.* **2017**, *45*, 496–508. [CrossRef] [PubMed]
76. Ignea, C.; Pontini, M.; Maffei, M.E.; Makris, A.M.; Kampranis, S.C. Engineering Monoterpene Production in Yeast Using a Synthetic Dominant Negative Geranyl Diphosphate Synthase. *ACS Synth. Biol.* **2014**, *3*, 298–306. [CrossRef] [PubMed]
77. Kim, E.-J.; Son, H.F.; Kim, S.; Ahn, J.-W.; Kim, K.-J. Crystal Structure and Biochemical Characterization of Beta-Keto Thiolase B from Polyhydroxyalkanoate-Producing Bacterium *Ralstonia eutropha* H16. *Biochem. Biophys. Res. Commun.* **2014**, *444*, 365–369. [CrossRef]
78. Segawa, M.; Wen, C.; Orita, I.; Nakamura, S.; Fukui, T. Two NADH-Dependent (S)-3-Hydroxyacyl-CoA Dehydrogenases from Polyhydroxyalkanoate-Producing *Ralstonia eutropha*. *J. Biosci. Bioeng.* **2019**, *127*, 294–300. [CrossRef]

79. Kim, E.-J.; Kim, Y.-J.; Kim, K.-J. Structural Insights into Substrate Specificity of Crotonase from the N-Butanol Producing Bacterium *Clostridium acetobutylicum*. *Biochem. Biophys. Res. Commun.* **2014**, *451*, 431–435. [CrossRef]

80. Bond-Watts, B.B.; Weeks, A.M.; Chang, M.C.Y. Biochemical and Structural Characterization of the *Trans* -Enoyl-CoA Reductase from *Treponema denticola*. *Biochemistry* **2012**, *51*, 6827–6837. [CrossRef]

81. Stout, J.M.; Boubakir, Z.; Ambrose, S.J.; Purves, R.W.; Page, J.E. The Hexanoyl-CoA Precursor for Cannabinoid Biosynthesis Is Formed by an Acyl-Activating Enzyme in *Cannabis sativa* Trichomes. *Plant J.* **2012**, *71*, 353–365. [CrossRef]

82. Taura, F.; Tanaka, S.; Taguchi, C.; Fukamizu, T.; Tanaka, H.; Shoyama, Y.; Morimoto, S. Characterization of Olivetol Synthase, a Polyketide Synthase Putatively Involved in Cannabinoid Biosynthetic Pathway. *FEBS Lett.* **2009**, *583*, 2061–2066. [CrossRef] [PubMed]

83. Sirikantaramas, S.; Morimoto, S.; Shoyama, Y.; Ishikawa, Y.; Wada, Y.; Shoyama, Y.; Taura, F. The Gene Controlling Marijuana Psychoactivity. *J. Biol. Chem.* **2004**, *279*, 39767–39774. [CrossRef] [PubMed]

84. Taura, F.; Sirikantaramas, S.; Shoyama, Y.; Yoshikai, K.; Shoyama, Y.; Morimoto, S. Cannabidiolic-Acid Synthase, the Chemotype-Determining Enzyme in the Fiber-Type *Cannabis sativa*. *FEBS Lett.* **2007**, *581*, 2929–2934. [CrossRef] [PubMed]

85. Morimoto, S.; Komatsu, K.; Taura, F.; Shoyama, Y. Purification and Characterization of Cannabichromenic Acid Synthase from *Cannabis sativa*. *Phytochemistry* **1998**, *49*, 1525–1529. [CrossRef]

86. Page, J.E.; Stout, J.M. Cannabichromenic Acid Synthase from Cannabis Sativa. Br. Patent WO2015196275A1, 30 December 2015.

87. Harrison, R.G.; Todd, P.W.; Rudge, S.R.; Petrides, D.P. *Bioseparations Science and Engineering*, 2nd ed.; Oxford University Press: Oxford, UK, 2015.

88. Zirpel, B. Recombinant Expression and Functional Characterization of Cannabinoid Producing Enzymes in Komagataella Phaffii. Ph.D. Thesis, Technical University of Dortmund, Dortmund, Germany, 2018. [CrossRef]

89. Schwarzhans, J.P.; Luttermann, T.; Geier, M.; Kalinowski, J.; Friehs, K. Towards Systems Metabolic Engineering in *Pichia pastoris*. In *Biotechnology Advances*; Elsevier: Amsterdam, The Netherlands, 2017; pp. 681–710. [CrossRef]

90. Kideckel, D.M.; Pallotta, M.; Hoang, K. *Synthetically-Derived Cannabinoids: The Next Generation of Cannabinoid Production*. Available online: https://www.newcannabisventures.com/wp-content/uploads/2019/02/Synthetically-Derived_Cannabinoids__The_Next_Generation_of_Cannabinoid_Production_February_20_2019.pdf (accessed on 19 December 2021).

91. Government of Italy. Legge n. 242 Del 2 Dicembre 2016. Disposizioni per la Promozione della Coltivazione e della Filiera Agroindustriale della Canapa. Available online: https://www.normattiva.it/uri-res/N2Ls?urn:nir:stato:legge:2016;242 (accessed on 17 December 2021).

92. U.S. Food and Drug Administration (FDA). Regulation of Cannabis and Cannabis-Derived Products, Including Cannabidiol (CBD). Available online: https://www.fda.gov/news-events/public-health-focus/fda-regulation-cannabis-and-cannabis-derived-products-including-cannabidiol-cbd (accessed on 21 December 2021).

93. Zaami, S.; Di Luca, A.; Di Luca, N.M.; Montanari Vergallo, G. Medical Use of Cannabis: Italian and European Legislation. *Eur. Rev. Med. Pharmacol. Sci.* **2018**, *22*, 1161–1167. [CrossRef]

94. U.S. Food and Drug Administration (FDA). FDA Approves First Drug Comprised of an Active Ingredient Derived from Marijuana to Treat Rare, Severe Forms of Epilepsy. Available online: https://www.fda.gov/news-events/press-announcements/fda-approves-first-drug-comprised-active-ingredient-derived-marijuana-treat-rare-severe-forms (accessed on 21 December 2021).

95. Coral, J.; Karp, S.G.; Porto de Souza Vandenberghe, L.; Parada, J.L.; Pandey, A.; Soccol, C.R. Batch Fermentation Model of Propionic Acid Production by *Propionibacterium acidipropionici* in Different Carbon Sources. *Appl. Biochem. Biotechnol.* **2008**, *151*, 333–341. [CrossRef]

96. Taura, F.; Dono, E.; Sirikantaramas, S.; Yoshimura, K.; Shoyama, Y.; Morimoto, S. Production of Δ1-Tetrahydrocannabinolic Acid by the Biosynthetic Enzyme Secreted from Transgenic *Pichia pastoris*. *Biochem. Biophys. Res. Commun.* **2007**, *361*, 675–680. [CrossRef]

97. The MathWorks, Inc. *SimBiology Toolbox*; The MathWorks, Inc.: Natick, MA, USA. Available online: https://www.mathworks.com/products/simbiology.html (accessed on 11 December 2021).

98. De Deken, R.H. The Crabtree Effect: A Regulatory System in Yeast. *J. Gen. Microbiol.* **1966**, *44*, 149–156. [CrossRef]

99. Valliere, M.A.; Korman, T.P.; Woodall, N.B.; Khitrov, G.A.; Taylor, R.E.; Baker, D.; Bowie, J.U. A Cell-Free Platform for the Prenylation of Natural Products and Application to Cannabinoid Production. *Nat. Commun.* **2019**, *10*, 565. [CrossRef]

100. Magalhães, A.I.; de Carvalho, J.C.; Medina, J.D.C.; Soccol, C.R. Downstream Process Development in Biotechnological Itaconic Acid Manufacturing. *Appl. Microbiol. Biotechnol.* **2017**, *101*, 3227–3235. [CrossRef]

101. Poulos, J.L.; Farnia, A.N. Production of Tetrahydrocannabinolic Acid in Yeast. U.S. Patent US10392635B2, 9 July 2015.

102. Towle, T.R. Raising the Bar for Cannabis Extraction Methods: Introducing a Novel, Safe, Efficient, and Environmentally Friendly Approach to Extracting High Quality Cannabis Resins. In *Fall 2019 Symposia, Proceedings of the Cannabis Chemistry Subdivision (CANN-ACS)*; 2019. Available online: https://cdn.technologynetworks.com/ac/resources/pdf/fall-2019-symposia-proceedings-of-the-cannabis-chemistry-subdivision-312347.pdf (accessed on 28 December 2021).

103. Turner, C.E.; Elsohly, M.A. Constituents of *Cannabis Sativa* L. XVI. A Possible Decomposition Pathway of Δ 9-Tetrahydrocannabinol to Cannabinol. *J. Heterocycl. Chem.* **1979**, *16*, 1667–1668. [CrossRef]

104. Repka, M.A.; Munjal, M.; ElSohly, M.A.; Ross, S.A. Temperature Stability and Bioadhesive Properties of Δ 9-Tetrahydrocannabinol Incorporated Hydroxypropylcellulose Polymer Matrix Systems. *Drug Dev. Ind. Pharm.* **2006**, *32*, 21–32. [CrossRef] [PubMed]

105. Chambers, A. Delta Separations. Decarboxylation's Importance in the Cannabis Extraction Process. Available online: https://deltaseparations.com/understanding-decarboxylations-importance-in-your-ethanol-extraction-process (accessed on 25 November 2021).
106. Anderson, D. Progress in Cannabinoid Purification and Testing. Available online: https://www.labx.com/resources/progress-in-cannabinoid-purification-and-testing/80 (accessed on 25 November 2021).
107. RotaChrom. Adaptability and Scale Up. Available online: https://rotachrom.com/products/adaptability-and-scale-up (accessed on 25 November 2021).

 *fermentation*

MDPI

Review

Yeast Hybrids in Brewing

Matthew J. Winans [1,2]

1   Imperial Yeast, Research & Development Laboratory, 19677 NE San Rafael Street, Portland, OR 97230, USA; matt.w@imperialyeast.com or mwinans@mix.wvu.edu; Tel.: +1-971-808-6688 or +1-304-483-1786

2   Biology Department, Eberly College of Arts and Sciences, West Virginia University, 53 Campus Drive, Morgantown, WV 26506, USA

Abstract: Microbiology has long been a keystone in fermentation, and innovative yeast molecular biotechnology continues to represent a fruitful frontier in brewing science. Consequently, modern understanding of brewer's yeast has undergone significant refinement over the last few decades. This publication presents a condensed summation of *Saccharomyces* species dynamics with an emphasis on the relationship between; traditional *Saccharomyces cerevisiae* ale yeast, *S. pastorianus* interspecific hybrids used in lager production, and novel hybrid yeast progress. Moreover, introgression from other *Saccharomyces* species is briefly addressed. The unique history of *Saccharomyces cerevisiae* and *Saccharomyces* hybrids is exemplified by recent genomic sequencing studies aimed at categorizing brewing strains through phylogeny and redefining *Saccharomyces* species boundaries. Phylogenetic investigations highlight the genomic diversity of *Saccharomyces cerevisiae* ale strains long known to brewers for their fermentation characteristics and phenotypes. The discovery of genomic contributions from interspecific *Saccharomyces* species into the genome of *S. cerevisiae* strains is ever more apparent with increasing research investigating the hybrid nature of modern industrial and historical fermentation yeast.

Keywords: hybrid; lager; yeast; introgression; interspecific; domestication; phylogeny; brewing; molecular; genomics

Citation: Winans, M.J. Yeast Hybrids in Brewing. *Fermentation* **2022**, *8*, 87. https://doi.org/10.3390/fermentation8020087

Academic Editor: Ronnie G. Willaert

Received: 4 January 2022
Accepted: 14 February 2022
Published: 18 February 2022

Publisher's Note: MDPI stays neutral with regard to jurisdictional claims in published maps and institutional affiliations.

Copyright: © 2022 by the author. Licensee MDPI, Basel, Switzerland. This article is an open access article distributed under the terms and conditions of the Creative Commons Attribution (CC BY) license (https://creativecommons.org/licenses/by/4.0/).

1. Species of *Saccharomyces*

Saccharomyces cerevisiae may be one of the oldest domesticated organisms known to humans. Domestication events imposed on brewing strains of the budding yeast species *S. cerevisiae* resulted in unique strains similar to the divergence seen in animal lineages of *Canis familiaris* breeds or the plant lineage of *Brassica oleracea* foods. It has been suggested that *S. cerevisiae* behaved as a synanthropic species, following human settlements as a commensal organism residing in gardens and vineyards, although the time period and location of the yeast's origins has been the subject of much debate throughout history. Domesticated *Saccharomyces* brewing strains feature flocculation capabilities, fast fermentation rates, malt sugar utilization, pleasant aromas, and are largely negative for production of phenolic off flavors (POF) [1,2].

Nearly two centuries have passed since the first accessible description was produced regarding brewer's yeast and its recognition in fermentation [3,4]. Recently, phylogenic research utilizing genomics and modern molecular biology techniques has shed some light on the historically convoluted nomenclature surrounding this budding yeast. Genomic analysis of the *Saccharomyces* genus has consolidated many variations into eight individual species: *S. cerevisiae, S. paradoxus (syn. S. cariocanus, S. cerevisiae var. tetraspora, S. cerevisiae var. terrestris, S. douglasii), S. uvarum (syn. S. bayanus var. uvarum), S. mikatae, S. kudriavzevii, S. arboricola (syn. S. arboricolus), S. eubayanus, and S. jurei* [3,5–19] (Table 1). Moreover, two natural hybrids are recognized in the *Saccharomyces* clade: *S. pastorianus* (syn. *S. carlsbergensis, S. monacensis*) and *S. bayanus* [20–22]. Most modern lager fermentations utilize *S. patorianus* yeasts.

Table 1. Current *Saccharomyces* Yeast Species History.

Saccharomyces	Described	Substrate	Location	Reference
cerevisiae	1838	Beer	Germany	[4]
uvarum	1898	*Ribes rubrum*, redcurrant juice	South Holland, The Netherlands	[16,17]
paradoxus	1914	Tree sap	Russia	[15]
kudriavzevii	1991	Decayed leaf	Japan	[18]
mikatae	1993	Decayed leaf	Japan	[19]
arboricola	2008	*Fagaceae* spp.	West China	[6]
eubayanus	2011	*Nothofagus* spp. & parasitic fungi *Cyttaria* spp.	Andean, Patagonia	[11]
jurei	2017	*Quercus robur*	Saint Auban, France	[14]

2. Hybrid Nature of Yeast

Interspecific hybrids are not unique to lager brewing. For example, the livestock and agricultural industries commonly employ selective breeding to alter species' properties or increase yields [23–25]. A time-honored showpiece of hybrid vigor is the mule, a great pack animal known for its hardiness and longevity. For over 4000 years, the mule has been bred as the hybrid progeny of a male donkey and a female horse. Since the early 1900s, maize has been hybridized to increase yields and introduce biodiversity [26]. Similarly, hybrid yeasts have been isolated from fermentation processes on numerous occasions [27] (Figure 1). A hybrid between *S. cerevisiae* and *S. kudriavzevii* was isolated from Belgian Trappist beers [28]. Popular in wine production, strain VIN7 is a hybrid of *S. cerevisiae* and *S. kudriavzevii* [29]. Other interspecific *S. cerevisiae* and *S. uvarum* hybrids are also regularly used for production of wines [29,30]. Spontaneous fermentations have yielded *Pichia apotheca*, a hybrid of *P. membranifaciens* and an unknown species [31]. Hybrid vigor, or heterosis, confers a competitive advantage by facilitating transgressive phenotypes in changing environments, and is known to be a driver of fungal evolution and adaptation [32]. This is especially important for yeast as the many stages of fermentation and maturation create a microbially competitive environment in which rapid adaptation may be advantageous.

The mule of the brewing industry is the lager yeast *S. pastorianus*, an interspecific hybrid that produces the lion's share, in volume, of the global beer production. Although its use is widespread, the biodiversity is limited to two main lineages, Saaz/group I (*syn. S. carlsbergensis*, (L12: Noble, Imperial Yeast Culture Collection, type strain CBS1513) and Frohberg/group II (L13: Global, Imperial Yeast Culture Collection, type strain-Weihenstephan 34/70). Saaz and Frohberg lineages vary in their genomic composition from each parent species, *S. eubayanus* and *S. cerevisiae*, which influences important fermentation characteristics. Genomic analysis demonstrated a genomic composition of 1:2 *S. cerevisiae* to *S. eubayanus* sub genome in the Saaz lineage and 2:2 *S. cerevisiae* to *S. eubayanus* sub genome in the Frohberg lineage supporting the traditional designations used by brewers [33–35]. Saaz lineage hybrids are very well adapted to cold fermentations and many of these strains lack maltotriose utilization [27]. The Frohberg hybrids contain more *S. cerevisiae* genomic content conferring greater attenuation, higher ethanol production, differing ester profiles, and higher typical viabilities [36]. The composition of genetic material transferred and retained in these hybrids impart important fermentation characteristics and phenotypes such as the POF (phenolic off flavor) trait, efficient fermentation of maltose and maltotriose, reduction of diacetyl, flocculation, and production of unique volatile metabolite profiles that are low in off-aroma/flavors. Investigation into these hybrid lineages supports bolstering the fermentation capacity of *S. cerevisiae* with hybrid vigor from *S. eubayanus* incorporation and conveyance of a positive phenotype.

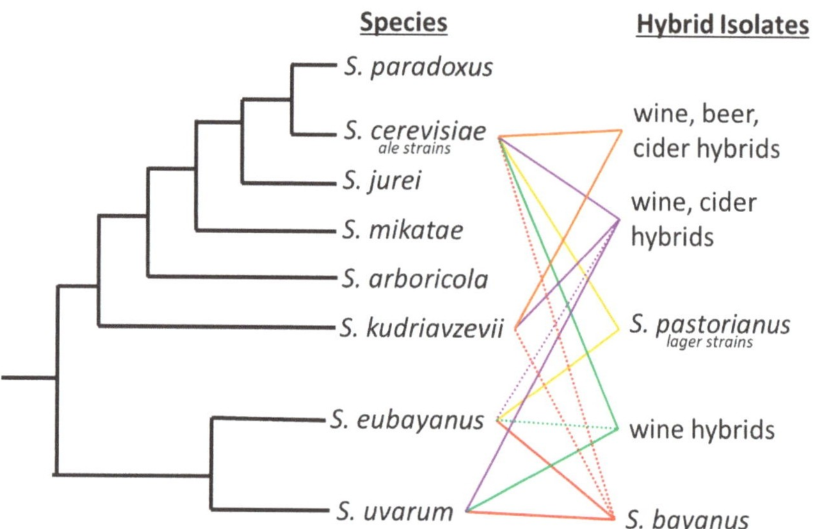

Figure 1. Saccharomyces phylogenetic tree with industrially important hybrids. Industrial hybrids are listed to the right by the fermentation they have predominately been associated with. Solid lines between two species signifies the interspecific hybrids and dashed lines denote introgression from a third or fourth species that may not always be present in each hybrid strain.

3. Novel Hybrid Development

The first yeast breeding experiments aimed at combining desirable traits of brewing strains were conducted by Ojvind Winge during his tenure at the Carlsberg Laboratory in the 1930s [37]. Hybrid yeast development has been carried out for over half a century since then, aimed predominantly at increasing attenuation and fermentation rates via intraspecific crosses with ale and lab strains [38–40]. Modern fermentations benefit from many innate and acquired hybrids that have been isolated or developed [11,27,28,41–52]. Early efforts in brewing science established the fundamentals necessary to explore the phylogeny, genomics, and strain development for *Saccharomyces* fermentation. During typical rich nutrient propagations of yeast in a brewing environment, mother cells reproduce asexually to bud off small daughter clones (Figure 2). Under poor nitrogen conditions, such as proline, yeast growth changes to a pseudohyphal form [53,54]. The complete absence of a nitrogen source and the presence of a non-fermentable carbon source, such as acetate, will sporulate yeast cells [55]. Sporulation transforms the cell wall into the ascus, or sack, that holds four spores termed a tetrad. Analogous to the human egg and sperm, these spores divide equally into mating types as either a or α [56]. When conditions improve for yeast growth, new haploid (*1n*) yeast can conjugate with the opposite mating type yeast as they form a shmoo. Depending on the genomic make up of each parental strain or species, there is some genomic instability or rewiring that occurs during the following mitotic budding growth.

Interspecific hybridization is seen as a valuable tool for yeast strain development, enabling the combination and enhancement of characteristics from both parental strains or species [57]. The development of hybrids is executed via three primary methodologies: spore–spore mating, rare mating, and protoplast fusion (Figure 2). Spore to spore mating is most similar to what would be considered natural mating, as outlined in Figure 2b. This approach bears a high success rate, high genomic stability, and can avoid the aid of selection markers such as drug resistance or autotrophies. Rare mating utilizes a described spontaneous loss of heterozygosity at the mating type locus. Normal diploid cells carry two sets of chromosomes with both the MATa and MATα genetic alleles and do not respond to sex pheromones for mating purposes. The spontaneous loss of either sex allele tolerates yeast mating to a yeast cell of complimentary sex. This results in yeast with high chromosome counts, influencing gene dosage during cellular processes and partially explains the outperformance over a diploid yeast of the same background [57]. Rare mating, as the name implies, is uncommon and selection markers are needed to perform this technique. The frequency of rare mating is estimated to occur in 1 out of 10 million cells [58]. This procedure is beneficial in overcoming poor sporulation but produces hybrids prone to high genomic instability. Lastly, protoplast fusion is performed by removing the cell wall and fusing the protoplasts of two cells together before the cell wall is repaired. This technique generates cells with a high chromosome copy number and higher genomic instability but overcomes low sporulation. This technique can be used in the laboratory to combine yeast from different genera such as the brewer's yeast *S. cerevisiae* and other yeast outside of the *Saccharomyces* genera which are otherwise incompatible [59]. Protoplast fusion is considered genetic engineering in many parts of the world. Recent select investigations into *Saccharomyces* hybrid application in beverage fermentations are listed in table format (Table 2).

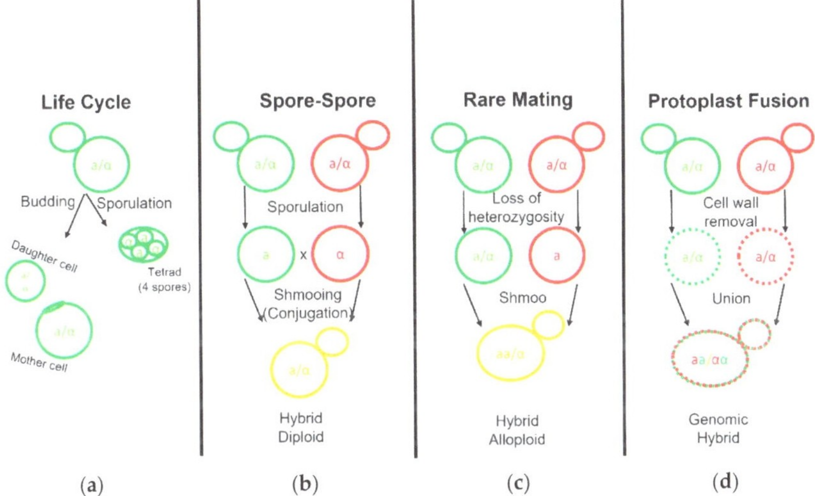

Figure 2. Life and Mating of Saccharomyces Yeast. Diagram pertaining to the clonal growth typical of yeast fermentation cultures and the various known techniques employed to generate yeast hybrids. (**a**) Diploid yeast cells may bud and grow clonally to form a mother and daughter cell or undergo sporulation to form a tetrad. (**b**) Yeast hybridization may form by direct spore to spore mating. (**c**) Yeast hybridization may form by rare mating events in which one or both diploid parent cells gain competency by becoming hemizygous or MATa/MATa and MATα/MATα diploids. (**d**) Yeast hybridization may also form by fusion of two separate yeast cell protoplasts with their cell wall removed.

Utilizing advanced molecular biology techniques for the modification of yeast strains presents an ongoing endeavor by many academic labs and some commercial yeast laboratories. Several research groups have developed protocols that use plasmids carrying genetic markers for drug resistance or functional enzymes that target the mating system in yeast. Industrial strains are well-known for their poor ability to adhere to laboratory techniques including sporulation and transformation [60]. One strategy developed at the University of Wisconsin generates allotetraploid strains of prototrophic yeast without the need for sporulation or modification to the nuclear genome of parental yeast strains [61]. This method leverages a series of inducible plasmids coined HyPr (Hybrid Production) containing compilatory drug markers and an inducible HO cassette. HO encodes an endonuclease that performs a double stranded break in the DNA that determines the *Saccharomyces* sex type. To repair this damage, the yeast cell will copy the silenced sex type already present in the genome, effectively performing a sex change and allowing the cell to mate with other cells of the opposite mating type. Using drug resistance markers and the HO cassette, hybrids are produced and serial growth in the absence of drug selection purges cells of exogenous DNA [61]. Similarly at the Advanced Industrial Science and Technology Center of the Biomedical Research Institute of Japan, another plasmid utilizing system aimed at exploiting the loss of heterozygosity of yeast cells was developed. A series of three to four plasmids allows for isolating a- and α- type cells from mixed cell populations and subsequent continual cross breeding [62]. CRISPR/Cas9 has also been utilized to force double stranded breaks in the MAT locus in order to increase the diversity of industrial yeast strains and their hybrids [63]. In the coming years, industrial yeasts will be designed by utilizing these protocols and molecular toolkits.

Currently, the options for strain selection in *S. cerevisiae* yeast are plentiful, but the criteria for fermentation performance in the brewing environment remains selective. Strains are employed largely by beer style, equipment availability, and supporting knowledge base. Certain beer styles also contain defining features from specific yeast flavor active molecules [64]. Banana and clove flavors are derived from isoamyl acetate and 4-vinyl guaiacol (4VG) in Weissbier [64–66]. Hybridization of yeast bears several advantages in brewing to include transgressive phenotypes such as increased ethanolic fermentation performance or stress tolerance, shifting fermentation temperatures beyond traditional inhibitory conditions, and creating a mosaic blend of parental fermentation profiles [24,51,57,63,67]. Targets of yeast hybridization may include increased formation of glycerol for an enhanced mouthfeel, alternative carbon source metabolic ability, reduced off-flavor production, increased formation of antioxidants that increase beer flavor stability, or increase production of yeast longevity molecules such as trehalose. The methodologies to create yeast hybrids vary in their specificity to target genetic or phenotypic results, but efforts to harness yeast hybrids in brewing broadly increase the biodiversity of fermentation yeast, add depth to the complexity of fermentation profiles, and advance brewing science knowledge.

Table 2. Interspecific Yeast Hybrids in Fermentation.

History	Parents	Reference
Isolated	*S. cerevisiae* × *S. eubayanus*	[11]
Isolated	*S. cerevisiae* × *S. eubayanus* × *S. uvarum*	[41]
Isolated	*S. cerevisiae* × *S. uvarum*	[68]
Isolated	*S. cerevisiae* × *S. kudriavzevii*	[27,28,42]
Isolated	*S. uvarum* × *S. eubayanus*	[27,42]
Developed	*S. cerevisiae* × *S. eubayanus*	[43,45–47,69]
Developed	*S. cerevisiae* × *S. mikatae*	[48]
Developed	*S. cerevisiae* × *S. kudriavzevii* × *S. paradoxus*	[49]
Developed	*S. cerevisiae* × *S. kudriavzevii*	[52]
Developed	*S. cerevisiae* × *S. arboricola*	[48,51]
Developed	*S. cerevisiae* × *S. jurei*	[70]

4. Natural Considerations and State of the Field

Outside of industrial fermentations and laboratory settings, *Saccharomyces* yeast is readily obtained from tree sap exuded seasonally from oaks and beech tree bark, part of the *Fagaceae* family. The natural fluctuation of available carbon, nutrition, and climate influences the state cells between active replication, a non-dividing state termed quiescence, and sporulation [71]. During industrial use, yeast spend their time in active growth and metabolism.

The fermentation of sugars into ethanol, carbon dioxide, and flavor molecules is utilized to prepare food or drink for human consumption. The brewer's and baker's yeast have secured a niche ecological space in industry via an advantage in high sugar environments. The Crabtree effect is a critical microbial factor for beer production and occurs when fermentation of sugars is the preferred metabolic route instead of aerobic respiration in the presence of oxygen. This metabolic capability is estimated to have evolved during the Cretaceous period, ca. 125 million years ago, when modern fruiting plants also appeared and humans were absent [72,73]. Bacteria present in these environments competed for resources, but the make–accumulate–consume provided a winning strategy for yeast. By producing inhibitory concentrations of alcohol for later consumption via a diauxic shift the yeast continued to occupy these environments over time [71,72,74].

The production of fruity esters may be an advantageous evolutionary trait developed to allow the passage of non-mobile unicellular fungi yeast from one rich environment to the next either via insect or brewer [75–78]. One theory suggests insect vectors' intestines act as a vessel for facilitating natural yeast hybridization events [76,77]. Unfortunately, the microscopic nature of yeast impose a limit for scientists to infer retrospectively on yeast mating modes and frequency by analyzing genomic data sets [79,80].

Interspecific hybridizations facilitate exchange of DNA which intertwine lineages and blur traditional species boundaries [32]. In yeast, the newly developed hybrids can rapidly adapt by filtering their diverse chromosomes and retaining advantageous portions via the loss of heterozygosity [81–83]. Cells experiencing a stressor, such as a drug or environmental condition, appear to prioritize genomic filtering early in growth, demonstrated by *Saccharomyces* interspecific hybrids subjected to various temperatures. *S. cerevisiae* and *S. uvarum* hybrids retained the cryotolerant subgenome of *S. uvarum* during cold adaptation, but retained the *S. cerevisiae* subgenome during warm environmental growth [83,84]. The dynamics between interspecific hybridization, adaptation, and evolution in *Saccharomyces* is not fully understood, but recent evidence suggests that this exchange of DNA components has and will continue to play a pivotal role as new yeast hybrids are described.

5. Conclusions

The quest to gain diverse and novel fermentation characteristics from a pure culture remains an overarching goal for brewing molecular biologists. Flavor attributes and temperature phenotypes were investigated by many research groups; however, gaps in knowledge remain concerning yeast fermentation and the evolution field. Efforts large and small are being exerted worldwide to explore the awesome power of yeast genetics in brewing sciences, yet many questions continue to surround the budding fungus akin to the brewers and bakers of the globe. Are interspecific hybrids naturally abundant under the correct environmental conditions or are they rare success stories? To what extent are yeast present in various natural climates and substrates, what vectors are involved in mobility, what interactions occur with other microbes, what is the typical lifecycle timeline of wild yeast, do *Saccharomyces* yeast originate out of Asia, or will researchers ever obtain adequate and representative environmental samples? Yeast phenotypes that, to the authors knowledge, remain unexplored by targeted hybridization include formation of antioxidants that increase flavor stability, formation of glycerol for an enhanced mouthfeel in low alcohol beverages, and reduced off-flavor production.

Many recent interspecific hybrids developed have largely focused on reinventing the lager yeast, *S. pastorianus*, by crossing *S. cerevisiae* with *S. eubayanus*. As research continues, understudied Saccharomyces species may serve as a reservoir for diverse genomic contributions. Natural isolates of Saccharomyces hybrids suggest that interspecific mating is not as uncommon as previously thought and recent investigations continue to redefine species whilst uncovering genetic exchange events. The ability of yeast to participate in interspecific hybridization and avoid hinderance by large genetic distances between parent species is promising and illustrative of the introgression likely yet to be discovered in many preserved isolates. While there are many yeast strains favored for production of quality fermented foods and beverages, their status as species or hybrids may be adjusted over time owing to the continual refinement of phylogenetics, the advancement in genome sequencing technology, and increased accessibility to genomic sequencing capabilities. By increasing yeast sampling, focusing on metabolic specifics, and facilitating collaboration, the yeast of tomorrow will be driven by scientific innovation in the laboratory today.

Funding: This research received no external funding.

Institutional Review Board Statement: Not applicable.

Informed Consent Statement: Not applicable.

Data Availability Statement: Not applicable.

Acknowledgments: A special acknowledgement is deserved to every team member of Imperial Yeast for supporting fundamental yeast research in and outside of the facility. Additional acknowledgment is needed to recognize the time needed to pursue and refine this educational outreach from Jess Caudil, Owen Lingley, and Jason Stepper. Thank you to Scott Forbes for translation of old German manuscripts.

Conflicts of Interest: The authors declare no conflict of interest. The funders had no role in the design of the study; in the collection, analyses, or interpretation of data; in the writing of the manuscript, or in the decision to publish the results.

Nomenclature

Terminology	Description
a/α	The two mating types of *S. cerevisiae* that enable cellular fusion when the complimentary pheromone is detected. The mating type and sexual state is largely determined by the MAT locus on chromosome III.
Allele	A variant of any particular gene found at the same genomic location. The mating type of yeast is determined by which allele, MATa or MATα, is present at the MAT locus.
Alloploids	A hybrid organism or cell composed of two or more sets of chromosomes obtained from two separate species.
Auxotrophy	The loss of a functional gene needed for growth. This is particularly relevant for amino acid synthesis and metabolism during yeast genetics experiments where controlling growth is needed.
Ascus	The sack like structure enclosing the four spores of *Saccharomyces*, this characteristic is important in classification of the *Ascomycota* (sac fungi) yeast.
Brassica oleracea	Important food crop plant species. It is wild cabbage in the uncultivated form, but cultivation has yielded varieties over time to include broccoli, brussels sprouts, kale, cauliflower, cabbage, and collards.
Crabtree Positive	Microorganism that preferentially metabolizes sugar into ethanol in the presence of oxygen instead of cellular respiration.
Diauxic Shift	The shift from fermentation to utilizing ethanol for cellular respiration and cell growth.
Diploid	An organism with a paired set of chromosomes creating a (2*n*) genome, such as a human with a set of genomes from each parent.
Fagaceae spp.	Family of *Angiosperms* (flowering plants) that include the beech and oak trees.
Frohberg	One of two primary lineages of modern lager yeast known to brewers originally used in Germany.

Terminology	Description
Gene	The basic unit of heredity which is composed of a sequence of DNA. It is characterized by a start sequence, genomic segment that often is translatable into a protein of function, and a stop sequence.
Genomics	The study of all of an organism's complete sequences of genetic materials composed of DNA (deoxyribonucleic acid) including structure, function, evolution, mapping, editing, and environmental interactions.
Haploid	An organism with one set of chromosomes composing their genome (*1n*).
Hemizygous	A condition of diploid cells where only one copy of a gene or other genetic component is present.
Hybrid Vigor (*syn.* Heterosis)	The tendency for the hybrid progeny between two species to outperform their parents in traits such as strength, size, or yield.
Interspecific	Arising or existing between separate species.
Intraspecific	Arising or existing within a species or individuals from the same species.
Locus	A specific fixed position on an individual chromosome where particular genetic content is present.
Make Accumulate Consume	*S. cerevisiae* yeast survival strategy that invokes the ability to make ethanol from saccharides, the ability to survive accumulation of the toxin, and the ability to consume ethanol for energy post fermentation.
Maltotriose	Prominent trisaccharide typical of beer wort and is enzymatically derived from starch. It is composed of three glucose molecules linked together by α-1,4 glycosidic bonds.
Nomenclature	The terminology and language used to categorize and communicate in various disciplines.
Out of Asia/Silk Road Hypothesis	The idea that *Saccharomyces* yeast originate from an Asian geographical region because of the abundance natural biodiversity found in that region.
Phenotype	A characterized trait, best exemplified by measurable qualities such as color or yield.
Phylogeny	The lineage and evolution of relative organisms.
Progeny	The offspring or descendants of an organism.
Quiescence	A state of inactivity, dormancy, or a period of idleness in yeast ecology.
Saaz	One of two primary lineages of modern lager yeast known to brewers originally used in Bohemia region of the Czech Republic where the town Žatec/Saaz is located.
Shmoo	The distinct physical form of two *Saccharomyces* yeast cells mating. This terminology originated from the morphological similarity to an Al Capp cartoon popular in America during the early 1950s.

Terminology	Description
Synanthropic Species	An undomesticated organism that habitually exists with human populations and benefit from non-natural environments. Their life cycles are adapted fully or in part to conditions created by human activity.
Tetrad	Four spores produced via meisis of *Ascomycota* yeast, specifically focused on yeast *S. cerevisiae* in this article.
Transgressive Phenotype	Formation of extreme phenotypes that surpass the ability of parental lineages, often found in hybrids and can be positive or negative for fitness of the individual.

References

1. Gallone, B.; Steensels, J.; Baele, G.; Maere, S.; Verstrepen, K.J.; Prahl, T.; Soriaga, L.; Saels, V.; Herrera-Malaver, B.; Merlevede, A.; et al. Domestication and Divergence of *Saccharomyces cerevisiae* Beer Yeasts. *Cell* **2016**, *166*, 1397–1410.e16. [CrossRef] [PubMed]
2. Preiss, R.; Tyrawa, C.; Krogerus, K.; Garshol, L.M.; Van Der Merwe, G. Traditional Norwegian Kveik are a genetically distinct group of domesticated *Saccharomyces cerevisiae* brewing yeasts. *Front. Microbiol.* **2018**, *9*, 2137. [CrossRef] [PubMed]
3. Pasteur, L. Nouveaux faits concernant l'histoire de la fermentation alcoolique. *Comptes Rendus Chim.* **1858**, *47*, 1011–1013.
4. Meyen, F.J.F. Jahresbericht über die Resultate der Arbeiten im Felde der physiologischen Botanik von dem Jahre 1837. In *Berlin: Nicolai'sche Buchhandlung*; Nicolai: Berlin, Germany, 1838; pp. 1–186.
5. Martini, A.V.; Martini, A. Three newly delimited species of *Saccharomyces sensu stricto*. *Antonie Van Leeuwenhoek* **1987**, *53*, 77–84. [CrossRef] [PubMed]
6. Wang, S.A.; Bai, F.Y. *Saccharomyces arboricolus* sp. nov., a yeast species from tree bark. *Int. J. Syst. Evol. Microbiol.* **2008**, *58*, 510–514. [CrossRef]
7. Naumova, E.S.; Roberts, I.N.; James, S.A.; Naumov, G.I.; Louis, E.J. Three new species in the *Saccharomyces* sensu stricto complex: *Saccharomyces cariocanus, Saccharomyces kudriavzevii* and *Saccharomyces mikatae*. *Int. J. Syst. Evol. Microbiol.* **2015**, *50*, 1931–1942. [CrossRef]
8. Naumov, G.I.; Naumova, E.S.; Hagler, A.N.; Mendonça-Hagler, L.C.; Louis, E.J. A new genetically isolated population of the *Saccharomyces* sensu stricto complex from Brazil. *Antonie Van Leeuwenhoek* **1995**, *67*, 351–355. [CrossRef]
9. Naumov, G.I.; Lee, C.F.; Naumova, E.S. Molecular genetic diversity of the *Saccharomyces* yeasts in Taiwan: *Saccharomyces arboricola, Saccharomyces cerevisiae* and *Saccharomyces kudriavzevii*. *Antonie Van Leeuwenhoek Int. J. Gen. Mol. Microbiol.* **2013**, *103*, 217–228. [CrossRef]
10. Naumov, G.I.; Naumova, E.S.; Louis, E.J. Two New Genetically Isolated Popoulations of the *Saccharomyces* Sensu Stricto Complex from Japan. *J. Gen. Appl. Microbiol.* **1995**, *41*, 499–505. [CrossRef]
11. Libkind, D.; Hittinger, C.T.; Valério, E.; Gonçalves, C.; Dover, J.; Johnston, M.; Gonçalves, P.; Sampaio, J.P. Microbe domestication and the identification of the wild genetic stock of lager-brewing yeast. *Proc. Natl. Acad. Sci. USA* **2011**, *108*, 14539–14544. [CrossRef]
12. Boynton, P.J.; Greig, D. The ecology and evolution of non-domesticated *Saccharomyces* species Primrose. *Yeast* **2014**, *31*, 449–462. [CrossRef] [PubMed]
13. Sipiczki, M. Interspecies hybridisation and genome chimerisation in *Saccharomyces*: Combining of gene pools of species and its biotechnological perspectives. *Front. Microbiol.* **2018**, *9*, 3071. [CrossRef] [PubMed]
14. Naseeb, S.; James, S.A.; Alsammar, H.; Michaels, C.J.; Gini, B.; Nueno-Palop, C.; Bond, C.J.; McGhie, H.; Roberts, I.N.; Delneri, D. *Saccharomyces jurei* sp. Nov., isolation and genetic identification of a novel yeast species from *Quercus robur*. *Int. J. Syst. Evol. Microbiol.* **2017**, *67*, 2046–2052. [CrossRef] [PubMed]
15. Batschinskaya, A. Entwicklungsgeschichte und Kultur des neuen Hefepilzes *Saccharomyces paradoxus*. *J. Microbiol. Epidemiol. Immunobiol.* **1914**, *1*, 231–247.
16. Beijerinck, M. Uber Regeneration der Sporenbildung bei Alkohol Hefen, wo diese Funktion im Verchwinden begriffen ist. *Cent. Bakt* **1898**, *4*, 657–731.
17. Beijerinck, M. Sur la generation de la faculte de produire des spores chez des levures en voie de la perdre. *Extr. Des Arch. Neelandaises Des Sci. Exactes Nat. Ser. II* **1898**, *2*, 1–81.
18. Kaneko, Y.; Banno, I. Re-examination of *Saccharomyces bayanus* strains by DNA-DNA hybridization and electrophoretic karyotyping. *IFO Res. Commun.* **1991**, *15*, 30–41.
19. Yamada, Y.; Mikata, K.; Banno, I. Reidentification of 121 strains of the genus *Saccharomyces*. *Bull JFCC* **1993**, *9*, 95–119.
20. Masneuf, I.; Hansen, J.; Groth, C.; Piskur, J.; Dubourdieu, D. New hybrids between *Saccharomyces sensu* stricto yeast species found among wine and cider production strains. *Appl. Environ. Microbiol.* **1998**, *64*, 3887–3892. [CrossRef]
21. Nguyen, H.V.; Legras, J.L.; Neuvéglise, C.; Gaillardin, C. Deciphering the hybridisation history leading to the lager lineage based on the mosaic genomes of *Saccharomyces bayanus* strains NBRC1948 and CBS380 T. *PLoS One* **2011**, *6*, e25821. [CrossRef]

22. Querol, A.; Bond, U. The complex and dynamic genomes of industrial yeasts: MINIREVIEW. *FEMS Microbiol. Lett.* **2009**, *293*, 1–10. [CrossRef]

23. Chen, Z.J. Genomic and epigenetic insights into the molecular bases of heterosis. *Nat. Rev. Genet.* **2013**, *14*, 471–482. [CrossRef]

24. Fu, D.; Xiao, M.; Hayward, A.; Jiang, G.; Zhu, L.; Zhou, Q.; Li, J.; Zhang, M. What is crop heterosis: New insights into an old topic. *J. Appl. Genet.* **2015**, *56*, 1–13. [CrossRef] [PubMed]

25. Schnable, P.S.; Springer, N.M. Progress toward understanding heterosis in crop plants. *Annu. Rev. Plant Biol.* **2013**, *64*, 71–88. [CrossRef] [PubMed]

26. Crow, J.F. 90 Years Ago: The Beginning of Hybrid Maize. *Genetics* **1998**, *148*, 923–928. [CrossRef]

27. Gallone, B.; Steensels, J.; Mertens, S.; Dzialo, M.C.; Gordon, J.L.; Wauters, R.; Theßeling, F.A.; Bellinazzo, F.; Saels, V.; Herrera-Malaver, B.; et al. Interspecific hybridization facilitates niche adaptation in beer yeast. *Nat. Ecol. Evol.* **2019**, *3*, 1562–1575. [CrossRef] [PubMed]

28. González, S.S.; Barrio, E.; Querol, A. Molecular characterization of new natural hybrids of *Saccharomyces cerevisiae* and *S. kudriavzevii* in brewing. *Appl. Environ. Microbiol.* **2008**, *74*, 2314–2320. [CrossRef]

29. Borneman, A.R.; Desany, B.A.; Riches, D.; Affourtit, J.P.; Forgan, A.H.; Pretorius, I.S.; Egholm, M.; Chambers, P.J. The genome sequence of the wine yeast VIN7 reveals an allotriploid hybrid genome with *Saccharomyces cerevisiae* and *Saccharomyces kudriavzevii* origins. *FEMS Yeast Res.* **2012**, *12*, 88–96. [CrossRef]

30. Le Jeune, C.; Lollier, M.; Demuyter, C.; Erny, C.; Legras, J.-L.; Aigle, M.; Masneuf Pomarède, I. Characterization of natural hybrids of *Saccharomyces cerevisiae* and *Saccharomyces bayanus* var. *uvarum*. *FEMS Yeast Res.* **2007**, *7*, 540–549. [CrossRef]

31. Heil, C.S.; Burton, J.N.; Liachko, I.; Friedrich, A.; Hanson, N.A.; Morrise, C.L.; Schacherer, J.; Shendurer, J.; Thomas, J.H.; Maitreya, J. Dunham Identification of a novel interspecific hybrid yeast from a metagenomic spontaneously inoculated beer sample using Hi-C. *Yeast* **2018**, *35*, 71–84. [CrossRef]

32. Steensels, J.; Gallone, B.; Verstrepen, K.J. Interspecific hybridization as a driver of fungal evolution and adaptation. *Nat. Rev. Microbiol.* **2021**, *19*, 485–500. [CrossRef] [PubMed]

33. Liti, G.; Peruffo, A.; James, S.A.; Roberts, I.N.; Louis, E.J. Inferences of evolutionary relationships from a population survey of LTR-retrotransposons and telomeric-associated sequences in the *Saccharomyces sensu stricto* complex. *Yeast* **2005**, *22*, 177–192. [CrossRef] [PubMed]

34. Dunn, B.; Sherlock, G. Reconstruction of the genome origins and evolution of the hybrid lager yeast *Saccharomyces pastorianus*. *Genome Res.* **2008**, *18*, 1610–1623. [CrossRef] [PubMed]

35. Glendinning, T. Some Practical Aspects of the Fermentable Matter. *J. Fed. Inst. Brew.* **1898**, *5*, 20–34. [CrossRef]

36. Walther, A.; Hesselbart, A.; Wendland, J. Genome Sequence of *Saccharomyces carlsbergensis*, the World's First Pure Culture Lager Yeast. *G3: Genes | Genomes | Genet.* **2014**, *4*, 783–793. [CrossRef] [PubMed]

37. Barnett, J.A. A history of research on yeasts 10: Foundations of yeast genetics. *Yeast* **2007**, *24*, 799–845. [CrossRef]

38. Hammond, B.J.R.M.; Eckersley, K.W. Fermentation Properties of Brewing Yeast with Killer Character. *J. Insta. Brew.* **1984**, *90*, 167–177. [CrossRef]

39. Johnston, J.R. Breeding yeasts for brewing. *J. Inst. Brew.* **1965**, *71*, 135–137. [CrossRef]

40. Spencer, J.F.T.; Spencer, D.M. Hybridization of non-sporulating and weakly sporulating strains of brewer's and distiller's yeasts. *J. Inst. Brew.* **1977**, *83*, 287–289. [CrossRef]

41. Peŕez-Través, L.; Lopes, C.A.; Querol, A.; Barrio, E. On the complexity of the *Saccharomyces bayanus* taxon: Hybridization and potential hybrid speciation. *PLoS ONE* **2014**, *9*, e93729. [CrossRef]

42. Langdon, Q.K.; Peris, D.; Baker, E.C.P.; Opulente, D.A.; Nguyen, H.V.; Bond, U.; Gonçalves, P.; Sampaio, J.P.; Libkind, D.; Hittinger, C.T. Fermentation innovation through complex hybridization of wild and domesticated yeasts. *Nat. Ecol. Evol.* **2019**, *3*, 1576–1586. [CrossRef] [PubMed]

43. Mertens, S.; Steensels, J.; Saels, V.; De Rouck, G.; Aerts, G.; Verstrepen, K.J. A large set of newly created interspecific *Saccharomyces* hybrids increases aromatic diversity in lager beers. *Appl. Environ. Microbiol.* **2015**, *81*, 8202–8214. [CrossRef] [PubMed]

44. Magalhães, F.; Krogerus, K.; Vidgren, V.; Sandell, M.; Gibson, B. Improved cider fermentation performance and quality with newly generated *Saccharomyces cerevisiae* × *Saccharomyces eubayanus* hybrids. *J. Ind. Microbiol. Biotechnol.* **2017**, *44*, 1203–1213. [CrossRef] [PubMed]

45. Krogerus, K.; Magalhães, F.; Vidgren, V.; Gibson, B. New lager yeast strains generated by interspecific hybridization. *J. Ind. Microbiol. Biotechnol.* **2015**, *42*, 769–778. [CrossRef]

46. Krogerus, K.; Arvas, M.; De Chiara, M.; Magalhães, F.; Mattinen, L.; Oja, M.; Vidgren, V.; Yue, J.X.; Liti, G.; Gibson, B. Ploidy influences the functional attributes of de novo lager yeast hybrids. *Appl. Microbiol. Biotechnol.* **2016**, *100*, 7203–7222. [CrossRef]

47. Hebly, M.; Brickwedde, A.; Bolat, I.; Driessen, M.R.M.; de Hulster, E.A.F.; van den Broek, M.; Pronk, J.T.; Geertman, J.M.; Daran, J.M.; Daran-Lapujade, P. *S. cerevisiae* × *S. eubayanus* interspecific hybrid, the best of both worlds and beyond. *FEMS Yeast Res.* **2015**, *15*, 1–14. [CrossRef]

48. Nikulin, J.; Krogerus, K.; Gibson, B. Alternative *Saccharomyces* interspecies hybrid combinations and their potential for low-temperature wort fermentation. *Yeast* **2018**, *35*, 113–127. [CrossRef]

49. Bellon, J.R.; Eglinton, J.M.; Siebert, T.E.; Pollnitz, A.P.; Rose, L.; De Barros Lopes, M.; Chambers, P.J. Newly generated interspecific wine yeast hybrids introduce flavour and aroma diversity to wines. *Appl. Microbiol. Biotechnol.* **2011**, *91*, 603–612. [CrossRef]

50. Bellon, J.R.; Schmid, F.; Capone, D.L.; Dunn, B.L.; Chambers, P.J. Introducing a New Breed of Wine Yeast: Interspecific Hybridisation between a Commercial *Saccharomyces cerevisiae* Wine Yeast and *Saccharomyces mikatae*. *PLoS ONE* **2013**, *8*, e62053. [CrossRef]

51. Winans, M.J.; Yamamoto, Y.; Fujimaru, Y.; Kusaba, Y.; Gallagher, J.E.G.; Kitagaki, H. *Saccharomyces arboricola* and Its Hybrids' Propensity for Sake Production: Interspecific hybrids reveal increased fermentation abilities and a mosaic metabolic profile. *Fermentation* **2020**, *6*, 14. [CrossRef]

52. Bizaj, E.; Cordente, A.G.; Bellon, J.R.; Raspor, P.; Curtin, C.D.; Pretorius, I.S. A breeding strategy to harness flavor diversity of *Saccharomyces* interspecific hybrids and minimize hydrogen sulfide production. *FEMS Yeast Res.* **2012**, *12*, 456–465. [CrossRef] [PubMed]

53. Gancedo, J.M. Control of pseudohyphae formation in *Saccharomyces cerevisiae*. *FEMS Microbiol. Rev.* **2001**, *25*, 107–123. [CrossRef] [PubMed]

54. Jones, M.; Pierce, J. Absorption of Amino Acids from Wort by Yeast. *J. Inst. Brew.* **1964**, *70*, 307–315. [CrossRef]

55. Esposito, R.E.; Klapholz, S. Meiosis and Ascospore Development. In *The Molecular Biology of the Yeast Saccharomyces: Life Cycle and Inheritance*; Strathern, J.N., Jones, E.W., Broach, E.R., Eds.; Cold Springs Harbor Laboratory Press: Cold Springs Harbor, NY, USA, 1981; pp. 211–287.

56. Liti, G. The fascinating and secret wild life of the budding yeast *S. cerevisiae*. *Elife* **2015**, *4*, e05835. [CrossRef]

57. Krogerus, K.; Magalhães, F.; Vidgren, V.; Gibson, B. Novel brewing yeast hybrids: Creation and application. *Appl. Microbiol. Biotechnol.* **2017**, *101*, 65–78. [CrossRef] [PubMed]

58. Gunge, N.; Nakatomi, Y. Genetic mechanisms of rare matings of the yeast saccharomyces cerevisiae heterozygous for mating type. *Genetics* **1972**, *70*, 41–58. [CrossRef]

59. Ferenczy, L.; Maráz, A. Transfer of mitochondria by protoplast fusion in *Saccharomyces cerevisiae*. *Nature* **1977**, *268*, 524–525. [CrossRef]

60. Ogata, T.; Okumura, Y.; Tadenuma, M.; Tamura, G. Improving Transformation Method for Industrial Yeasts: Construction of ADH1-APT2 Gene and Using Electroporation. *J. Gen. Appl. Microbiol.* **1993**, *294*, 285–294. [CrossRef]

61. Alexander, W.G.; Peris, D.; Pfannenstiel, B.T.; Opulente, D.A.; Kuang, M.; Hittinger, C.T. Efficient engineering of marker-free synthetic allotetraploids of *Saccharomyces*. *Fungal Genet. Biol.* **2016**, *89*, 10–17. [CrossRef]

62. Fukuda, N.; Kaishima, M.; Ishii, J.; Kondo, A.; Honda, S. Continuous crossbreeding of sake yeasts using growth selection systems for a-type and α-type cells. *AMB Express* **2016**, *6*, 45. [CrossRef]

63. Krogerus, K.; Fletcher, E.; Rettberg, N.; Gibson, B.; Preiss, R. Efficient breeding of industrial brewing yeast strains using CRISPR/Cas9-aided mating-type switching. *Appl. Microbiol. Biotechnol.* **2021**, *105*, 8359–8376. [CrossRef]

64. Strong, G.; England, K. *Beer Style Guidelines—BJCP*; Brewers Association: Boulder, CO, USA, 2021.

65. Verstrepen, K.J.; Derdelinckx, G.; Dufour, J.P.; Winderickx, J.; Thevelein, J.M.; Pretorius, I.S.; Delvaux, F.R. Flavor-active esters: Adding fruitiness to beer. *J. Biosci. Bioeng.* **2003**, *96*, 110–118. [CrossRef]

66. Mcmurrough, I.; Madigan, D.; Donnelly, D.; Hurley, J.; Doyle, A.; Hennigan, G.; McNulty, N. Control of Ferulic Acid and 4-Vinyl Guaiacol in Brewing. *J. Inst. Brew.* **1996**, *102*, 327–332. [CrossRef]

67. Gibson, B.; Geertman, J.M.A.; Hittinger, C.T.; Krogerus, K.; Libkind, D.; Louis, E.J.; Magalhães, F.; Sampaio, J.P. New yeasts-new brews: Modern approaches to brewing yeast design and development. *FEMS Yeast Res.* **2017**, *17*, 1–13. [CrossRef] [PubMed]

68. Krogerus, K.; Preiss, R.; Gibson, B. A unique saccharomyces cerevisiae saccharomyces uvarum hybrid isolated from norwegian farmhouse beer: Characterization and reconstruction. *Front. Microbiol.* **2018**, *9*, 2253. [CrossRef] [PubMed]

69. Magalhães, F.; Krogerus, K.; Castillo, S.; Ortiz-Julien, A.; Dequin, S.; Gibson, B. Exploring the potential of *Saccharomyces eubayanus* as a parent for new interspecies hybrid strains in winemaking. *FEMS Yeast Res.* **2017**, *17*, 1–10. [CrossRef]

70. Giannakou, K.; Visinoni, F.; Zhang, P.; Nathoo, N.; Jones, P.; Cotterrell, M.; Vrhovsek, U.; Delneri, D. Biotechnological exploitation of Saccharomyces jurei and its hybrids in craft beer fermentation uncovers new aroma combinations. *Food Microbiol.* **2021**, *100*, 103838. [CrossRef]

71. Gray, J.V.; Petsko, G.A.; Johnston, G.C.; Ringe, D.; Singer, R.A.; Werner-washburne, M. "Sleeping Beauty": Quiescence in *Saccharomyces cerevisiae*. *Microbiol. Mol. Biol. Rev.* **2004**, *68*, 187–206. [CrossRef]

72. Dashko, S.; Zhou, N.; Compagno, C.; Piškur, J. Why, when, and how did yeast evolve alcoholic fermentation? *FEMS Yeast Res.* **2014**, *14*, 826–832. [CrossRef]

73. Fleming, T.H.; John Kress, W. A brief history of fruits and frugivores. *Acta Oecologica* **2011**, *37*, 521–530. [CrossRef]

74. Hagman, A.; Sall, T.; Compagno, C.; Piskur, J. Yeast "Make-Accumulate-Consume" Life Strategy Evolved as a Multi-Step Process That Predates the Whole Genome Duplication. *PLoS ONE* **2013**, *8*, e68734. [CrossRef]

75. Christiaens, J.F.; Franco, L.M.; Cools, T.L.; de Meester, L.; Michiels, J.; Wenseleers, T.; Hassan, B.A.; Yaksi, E.; Verstrepen, K.J. The fungal aroma gene ATF1 promotes dispersal of yeast cells through insect vectors. *Cell Rep.* **2014**, *9*, 425–432. [CrossRef] [PubMed]

76. Stefanini, I.; Dapporto, L.; Berń, L.; Polsinelli, M.; Turillazzi, S.; Cavalieri, D. Social wasps are a saccharomyces mating nest. *Proc. Natl. Acad. Sci. USA* **2016**, *113*, 2247–2251. [CrossRef] [PubMed]

77. Stefanini, I.; Dapporto, L.; Legras, J.-L.; Calabretta, A.; Di Paola, M.; De Filippo, C.; Viola, R.; Capretti, P.; Polsinelli, M.; Turillazzi, S.; et al. Role of social wasps in *Saccharomyces cerevisiae* ecology and evolution. *Proc. Natl. Acad. Sci. USA* **2012**, *109*, 13398–13403. [CrossRef] [PubMed]

78. Reuter, M.; Bell, G.; Greig, D. Increased outbreeding in yeast in response to dispersal by an insect vector. *Curr. Biol.* **2007**, *17*, R81–R83. [CrossRef]
79. Tsai, I.J.; Bensasson, D.; Burt, A.; Koufopanou, V. Population genomics of the wild yeast *Saccharomyces paradoxus*: Quantifying the life cycle. *Proc. Natl. Acad. Sci. USA* **2008**, *105*, 4957–4962. [CrossRef]
80. Ruderfer, D.M.; Pratt, S.C.; Seidel, H.S.; Kruglyak, L. Population genomic analysis of outcrossing and recombination in yeast. *Nat. Genet.* **2006**, *38*, 1077–1081. [CrossRef]
81. James, T.Y.; Michelotti, L.A.; Glasco, A.D.; Clemons, R.A.; Powers, R.A.; James, E.S.; Simmons, D.R.; Bai, F.; Ge, S. Adaptation by Loss of Heterozygosity in *Saccharomyces cerevisae* Clones Under Divergent Selection. *Genetics* **2019**, *213*, 665–683. [CrossRef]
82. Lancaster, S.M.; Payen, C.; Heil, C.S.; Dunham, M.J. Fitness benefits of loss of heterozygosity in *Saccharomyces* hybrids. *Genome Res.* **2019**, *29*, 1685–1692. [CrossRef]
83. Smukowski Heil, C.S.; DeSevo, C.G.; Pai, D.A.; Tucker, C.M.; Hoang, M.L.; Dunham, M.J. Loss of Heterozygosity Drives Adaptation in Hybrid Yeast. *Mol. Biol. Evol.* **2017**, *34*, 1596–1612. [CrossRef]
84. Smukowski Heil, C.S.; Large, C.R.L.; Patterson, K.; Hickey, A.S.M.; Yeh, C.L.C.; Dunham, M.J. Temperature preference can bias parental genome retention during hybrid evolution. *PLoS Genet.* **2019**, *15*, e1008383. [CrossRef] [PubMed]

 *fermentation*

Communication

The Dynamics of Single-Cell Nanomotion Behaviour of *Saccharomyces cerevisiae* in a Microfluidic Chip for Rapid Antifungal Susceptibility Testing

Vjera Radonicic [1,2,*], Charlotte Yvanoff [1,2], Maria Ines Villalba [2,3], Sandor Kasas [2,3,4,†] and Ronnie G. Willaert [1,2,†]

[1] Research Group Structural Biology Brussels, Alliance Research Group VUB-UGent NanoMicrobiology (NAMI) Vrije Universiteit Brussel, 1050 Brussels, Belgium; charlotte.grabielle.yvanoff@vub.be (C.Y.); ronnie.willaert@vub.be (R.G.W.)
[2] International Joint Research Group VUB-EPFL NanoBiotechnology & NanoMedicine (NANO), Vrije Universiteit Brussel, 1050 Brussels, Belgium; ines.villalba@epfl.ch (M.I.V.); sandor.kasas@epfl.ch (S.K.)
[3] Laboratory of Biological Electron Microscopy, Ecole Polytechnique Fédérale de Lausanne (EPFL), 1015 Lausanne, Switzerland
[4] Centre Universitaire Romand de Médecine Légale, UFAM, Université de Lausanne, 1015 Lausanne, Switzerland
* Correspondence: vjera.radonicic@vub.be
† These authors contributed equally to this work.

Citation: Radonicic, V.; Yvanoff, C.; Villalba, M.I.; Kasas, S.; Willaert, R.G. The Dynamics of Single-Cell Nanomotion Behaviour of *Saccharomyces cerevisiae* in a Microfluidic Chip for Rapid Antifungal Susceptibility Testing. *Fermentation* **2022**, *8*, 195. https://doi.org/10.3390/fermentation8050195

Academic Editor: Manuel Malfeito Ferreira

Received: 2 March 2022
Accepted: 20 April 2022
Published: 26 April 2022

Publisher's Note: MDPI stays neutral with regard to jurisdictional claims in published maps and institutional affiliations.

Copyright: © 2022 by the authors. Licensee MDPI, Basel, Switzerland. This article is an open access article distributed under the terms and conditions of the Creative Commons Attribution (CC BY) license (https://creativecommons.org/licenses/by/4.0/).

Abstract: The fast emergence of multi-resistant pathogenic yeasts is caused by the extensive—and sometimes unnecessary—use of broad-spectrum antimicrobial drugs. To rationalise the use of broad-spectrum antifungals, it is essential to have a rapid and sensitive system to identify the most appropriate drug. Here, we developed a microfluidic chip to apply the recently developed optical nanomotion detection (ONMD) method as a rapid antifungal susceptibility test. The microfluidic chip contains no-flow yeast imaging chambers in which the growth medium can be replaced by an antifungal solution without disturbing the nanomotion of the cells in the imaging chamber. This allows for recording the cellular nanomotion of the same cells at regular time intervals of a few minutes before and throughout the treatment with an antifungal. Hence, the real-time response of individual cells to a killing compound can be quantified. In this way, this killing rate provides a new measure to rapidly assess the susceptibility of a specific antifungal. It also permits the determination of the ratio of antifungal resistant versus sensitive cells in a population.

Keywords: cellular nanomotion; single cell; optical nanomotion detection; microfluidic chip; no-flow chamber; yeast; *Saccharomyces cerevisiae*; antifungal

1. Introduction

Fungal infections are an important growing health problem due to a high mortality rate, which is currently more than 1.6 million people worldwide per year [1–3]. It is becoming increasingly difficult to cure these infections as the microorganisms become more and more resistant to the existing antimicrobial drugs [4]. These resistant microorganisms induce an increase in hospital stays and medical costs. Options to control their proliferation consist of the development of novel molecules or a more targeted use of antimicrobial drugs. However, the choice of available antifungal drugs to treat invasive fungal infections is limited, since only three structural classes of compounds (i.e., polyenes, azoles, and echinocandins) are available [5]. Furthermore, these current antifungal drug compounds can show significant limitations such as toxicity (e.g., amphotericin B displays considerable toxicity and undesirable side effects [6,7]), issues with pharmacokinetic properties and the activity spectrum, a small number of targets [8,9], and a risk of interacting with other drugs, such as chemotherapy agents and immunosuppressants [10,11]. Due to this urgent problem, there is an increasing interest in developing new means to fight fungal infections, one of which is the development of new antifungal compounds. Today, multiple compounds are in the clinical development stage [12–18].

An alternative way to reduce the spread of pathogenic fungi is the development of rapid antifungal sensitivity tests. Such tests should permit the quick identification of the most appropriate drug to fight a specific fungus (ideally in a timeframe of 1–3 h), and they should drastically reduce the use of large-spectrum antifungals that are documented to induce resistance. Standard antifungal susceptibility testing (AFST) methods rely on measuring fungal growth in the presence of antifungals over a few days [19]. Among them, commercially available tests, such as Sensititre YeastOne, Etest, and the fully automated Vitek 2 yeast susceptibility system, which are all easy-to-use modifications from CLSI/EUCAST reference methods, are widely used for testing the antifungal susceptibility of relevant *Candida* and *Aspergillus* species [8]. New diagnostic approaches, based on technologies such as flow cytometry, MALDI-TOF mass spectroscopy, and isothermal microcalorimetry, have been developed to expand, and potentially improve, the capability of the clinical microbiology laboratory to yield AFST results [19]. However, most of these techniques are expensive, time-consuming (up to 7 days), or deliver limited information on microbial phenotypes. Since microfluidics provides several advantages over existing macro-scale methods, several microfluidic platforms have been developed recently to perform rapid antimicrobial susceptibility tests [20]. These platforms are mainly based on the microscopic transmission or fluorescence observation of cells to quantify the effect of the antimicrobial on cell growth or viability. Unfortunately, most of the newly developed devices are focused on antibiotic susceptibility testing (AST), and much less on AFST.

A couple of years ago, atomic force microscope (AFM)-based sensitivity tests were developed to assess the sensitivity of microorganisms to a given drug in a timeframe of minutes [21,22]. The test consisted in attaching the organism of interest onto an AFM cantilever and monitoring its oscillations (referred to as cellular nanomotion since the displacements are of the nanometre order) as a function of time and the presence of antimicrobial drugs. Numerous studies demonstrated that living organisms attached to AFM cantilevers induce oscillations of the lever that immediately stop when the organism dies [23,24]. Hence, this newly developed technology, called cellular nanomotion detection (NMD), detects living organisms in a chemistry-independent manner and offers a clear advantage in terms of test velocity [25,26], but possesses several drawbacks too. AFMs, unfortunately, are expensive and relatively complex devices that require some expertise to operate. In a typical AFM, the cantilever oscillations are detected by monitoring the deflections of a laser beam that is reflected off the very end of the lever. This requires a precise adjustment of the laser as well as of the detector (four-segment photodiode) before each measurement. Finally, such nanomotion detectors are relatively complex to scale up and can only monitor a single bacterial/fungal species at a time.

Other methods have also been used to detect the nanomotion of living microorganisms that are attached to a surface. These include plasmonic imaging of the z motion of attached bacteria [27], sensing of the attached bacterial vibrations with the phase noise of a resonant crystal [28], tracking the x-y motion of attached uropathogenic *Escherichia coli* [29], and subcellular fluctuation imaging, which is based on total internal reflection microscopy (TIRM) [30], as well as optically tracking bacterial responses on micropillar architectures using intrinsic phase-shift spectroscopy [31]. Recently, we found that a simple optical microscope equipped with a video camera can also detect living cells' nanometric scale oscillations at a subpixel resolution using dedicated software [32]. We designated this method as "optical nanomotion detection" (ONMD). The technique does not require attaching the cells to a surface or a cantilever, nor must the cells be labelled, which are major advantages. The measurement is fast, simple, and inexpensive.

Here, we developed a microfluidic chip for the rapid determination of drug suscepti-bilities in fungal isolates and tested its functionality. The microfluidic chip contains no-flow imaging chambers, enabling us to determine the optical nanomotion of single yeast cells before, during, and after the treatment with a chemical compound. A non-pathogenic yeast, i.e., *Saccharomyces cerevisiae*, was used as a model yeast to evaluate its ONMD in the imaging chambers. The cells are pumped into the chambers, and next, exposed to a medium containing the antimicrobial drug or compound. In this study, the cells were killed with ethanol and the antifungal amphotericin B. The effect of these chemical compounds on individual yeast cells was successfully assessed by monitoring their nanomotion pattern as a function of time.

2. Materials and Methods

2.1. Yeast Cultivation

In these experiments, we used the *S. cerevisiae* BY4742 strain. Yeast cells were cultured by inoculating 10 mL of YPD medium (10 g/L yeast extract, 20 g/L peptone, 20 g/L dextrose) with a colony from a YPD agar (YPD containing 20 g/L agar) plate. The cultures were grown overnight in 100 mL Erlenmeyer flasks at 30 °C and 200 revolutions per minute (rpm) (Innova 4400, New Brunswick, Edison, NJ, USA). The overnight cultures were diluted in a YPD medium to obtain an optical density at 600 nm ($OD_{600\,nm}$) of 0.5 and were then grown in Erlenmeyer flasks for 1 h at 30 °C and 200 rpm. The cultures were further diluted afterward, depending on the cell concentration ($OD_{600\,nm}$ value), to obtain an optimal number of cells and allowing for single-cell visualisation in each imaging chamber.

2.2. Microfluidic Chip Construction

The microfluidic chip was fabricated in polydimethylsiloxane (PDMS) by soft lithography. First, the design of the microfluidic chip was created in DesignSpark Mechanical (www.rs-online.com, accessed on 10 September 2021), and then printed onto a film photomask (0.18 mm polyester with photo emulsion layer) using a laser-photoplotter (Selba S.A., Versoix, Switzerland). The mould of the PDMS chip was then made on a silicon wafer (4-inch Test CZ-Si wafer, Microchemicals GmbH, Ulm, Germany), using multi-layer photolithography. The wafer was cleaned in acetone (Carl Roth, Karlsruhe, Germany) and 2-propanol (Carl Roth, Karlsruhe, Germany) for 5 min each, then rinsed with ultrapure water and, finally, air blow-dried. The wafer was then heated on a hot plate (Torrey Pines Scientific, Carlsbad, CA, USA) at 100 °C to ensure that all the residual solvents were evaporated. The wafer was then exposed to air plasma at 100 W, 50 kHz, for 1 min (Plasma System Cute, Femto Science, Dongtangiheung-Ro, Korea). Two different resists were used to make the imaging chambers and the channels of different heights. First, the imaging chambers were created using the negative photoresist GM1050 (Gersteltec Engineering Solutions, Pully, Switzerland). The resist was spin-coated onto the wafer at 940 rpm for 40 s and soft-baked on a hot plate at 120 °C for 2 min to reach an approximative thickness of 8 μm. The wafer and the chambers' photomask were brought into hard contact in the mask aligner UV-KUB3 (Kloé, Saint-Mathieu-de-Tréviers, France) and illuminated at 365 nm ultraviolet (UV) light (intensity of 35 mW/cm^2) for 3 s. Following UV exposure, the wafer was post-baked for 2 min at 65 °C in an oven, and then for 10 min at 95 °C on a hot plate. For the channels, the second layer of SU-8 3050 (Kayaku Advanced Materials, Westborough, MA, USA) was spin-coated onto the wafer at 3500 rpm for 30 s and soft-baked for 6 min at 95°C on a hot plate, to reach an approximative thickness of 60 μm. The wafer was then aligned and brought into hard contact with the channels' photomask and illuminated with 365 nm UV light (intensity of 35 mW/cm^2) for 9 s. Next, the wafer was post-baked for 5 min at 95 °C on a hot plate. Finally, the wafer was immersed in SU-8 developer (Kayaku Advanced Materials, Westborough, MA, USA) for 6 min and then washed with 2-propanol and blow-dried. The dimensions of the channels and imaging chambers were measured with a 3-D profilometer (Profilm 3 D, Filmetrics, San Diego, CA, USA).

The SU-8 mould was cleaned from (in-)organic residue by rinsing it with acetone, 2-propanol, and ultrapure water, and blow-dried with air. The mould was then silanised in a glass bell with TMS (chlorotrimethylsilane, Sigma Aldrich, Overijse, Belgium) for 15 min to reduce the adhesion of PDMS to the SU-8 mould. PDMS (Sylgard 184 Silicone Elastomer Kit, Dow Inc., Midland, MI, USA) was prepared by mixing the base and the curing agent at a 10:1 ratio. After casting the PDMS onto the mould, it was degassed in a glass bell with an applied vacuum for 30 min to 1 h. The PDMS was then cured for 1 h at 100 °C on a hot plate or overnight at 60 °C in an oven. After curing, the PDMS was peeled off the mould and the holes for inlets and outlets were punctured with a 1 mm rapid core sampling tool (Electron Microscopy Sciences, Hatfield, PA, USA). Finally, the microfluidic chip was assembled by bonding the PDMS to a glass coverslip (170 μm thickness, Carl Roth, Karlsruhe, Germany), after plasma-treating them both at 100 W, 50 kHz for 1 min.

2.3. Computational Simulation of Fluid Flow in the Microfluidic Chip

To evaluate the absence of fluid flow in the designed imaging chambers, fluid flow simulations were performed using COMSOL Multiphysics 4.4 (COMSOL, Stockholm, Sweden) by selecting "fluid flow" (single-phase flow, laminar flow) physics. Water at 20 °C was selected as the working liquid. Laminar fluid flow through the channels was simulated with laminar inflow as a boundary condition at the inlet. The inlet flow rate was set at 0.5 μL/min, which corresponds to the inlet flow rate that was used during the actual experiments. Pressure (atmospheric pressure) was selected as the boundary condition at the outlet and backflow was suppressed. The physics-controlled mesh was selected as the mesh setting with a normal element size.

2.4. Determination of the Diffusion Coefficient in the Imaging Chamber

To determine that the mass transport in the imaging chamber occurs only by diffusion, and to estimate the time it takes to obtain a homogeneous concentration in the chamber, a diffusion experiment using 100 μM fluorescein (Agar Scientific Ltd., Stansted, Essex, UK) as a tracer compound was performed. The fluorescent dye was pumped in the flow cell from inlet A to outlet A using a syringe pump (KDS 260, Kd Scientific Inc, Holliston, MA, USA) at a flow rate of 0.5 μL/min. When the dye reached the chip, the flow was stopped, and every 10 s, images were recorded with a PCO Edge 4.2 sCMOS camera (Excelitas PCO GmbH, Kelheim, Germany). The mean grey values of a small area in the centre of the imaging chamber were calculated (ImageJ Fiji, [33]) and normalised to the maximal mean value.

A 3-D finite element model was set up to simulate the diffusion of fluorescein in the imaging chamber (COMSOL Multiphysics 4.4, Stockholm, Sweden) by selecting "transport of diluted species" (time-dependent study) physics. Water at 20 °C was selected as the working liquid. At the entrances of the chamber, the concentration boundary was set at a normalised concentration of 1. The physics-controlled mesh was selected as the mesh setting with an extremely fine element size. The increase of the concentration in the centre of the channel was calculated as a function of time, and the diffusion coefficient was estimated by fitting the calculated curve on the experimentally determined profile.

2.5. Microfluidic Chip Setup

The inlet and outlet of the imaging chambers (inlet/outlet B, Figure 1a) were connected to a pressure-driven pump (LineUp™ Push-Pull, Fluigent, Le Kremlin-Bicêtre, France), and the inlet and outlet of the channels (inlet/outlet A, Figure 1a) were connected to a syringe pump (KDS 260, Kd Scientific Inc, Holliston, MA, USA) via fluorinated ethylene propylene (FEP) tubing with an internal diameter of 0.51 mm. The microfluidic chip setup was then mounted onto an inverted Nikon Eclipse Ti2 epifluorescence microscope (Nikon, Tokyo, Japan) to perform ONMD with bright-field microscopy. The microfluidic chip and tubing were firstly flushed with 2-propanol and then ultrapure water to eliminate and remove air bubbles. Next, the yeast liquid culture in YPD was filled into the imaging chambers from inlet B (Figure 1a) at a flow rate of 5 μL/min using the pressure-driven pump. Both the inlet and outlet B tubings were clamped, and the first ONMD movie was acquired (see Section 2.6). The solution containing the killing compound 70% (v/v) ethanol or 500 µg/mL amphotericin B (Merck, Darmstadt, Germany) was then filled in from inlet A to outlet A using the syringe pump at a flow rate of 0.5 µL/min. As a control, a second chip was filled with YPD growth medium only, and all the inlets and outlets were closed. The time required for the compounds to be pumped in, to replace the liquid in the inlet channel and the flow cell channels around the imaging chambers, to diffuse into the imaging chamber, and to reach the maximum concentration in the centre was estimated at 10 min. This estimate is based on the time it takes to pump the compound through the tubing to the flow cell at a flow rate of 0.5 µL/min, which is 8 min (volume of the internal volume of the tubing was 4 µL); the time to replace the liquid in the inlet channel and flow cell chamber, which is 20 s (volume of 0.15 µL); and the time to reach a maximum concentration in the centre of the imaging chamber by diffusion, which is 75 s (Figure 2c).

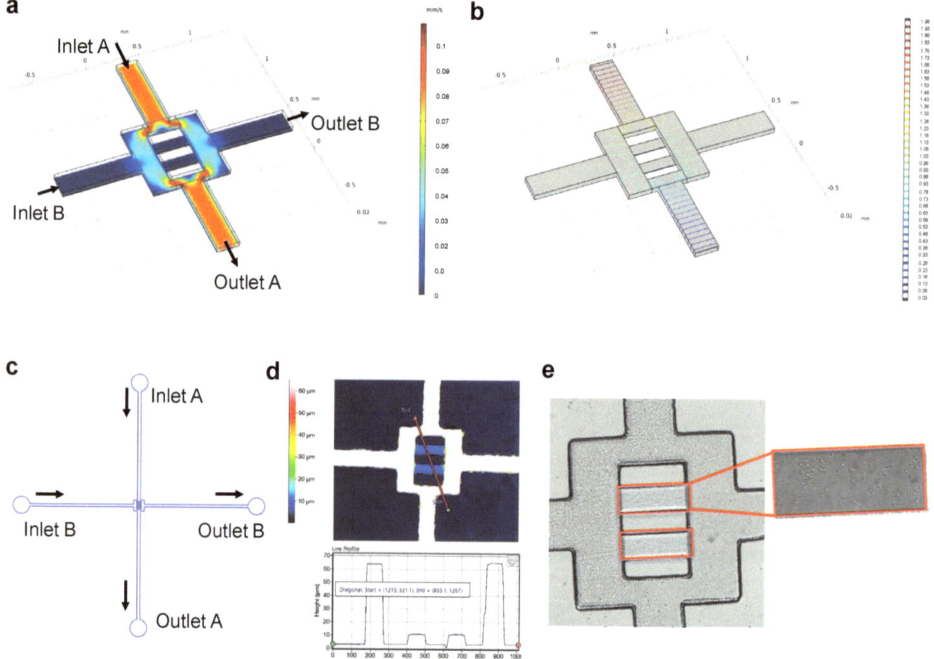

Figure 1. (**a**) Mathematical modelling of the fluid flow velocity at the mid-plane of the imaging chambers (4 µm). (**b**) Pressure distribution over the design. (**c**) Mask design of the flow cell with inlet and outlet channels. (**d**) Dimensions of a cross-section of the 2 imaging chambers and channels (3-D profilometer). (**e**) Constructed flow cell showing the imaging chambers (red rectangles) containing yeast cells.

2.6. Nanomotion Measurement and Analysis

Cellular oscillations were monitored by recording bright-field movies of 500 frames at a framerate of 35 fps with a PCO Edge 4.2 back-illuminated sCMOS camera (Excelitas PCO GmbH, Kelheim, Germany) using a 40× objective. The movies were analysed using a cross-correlation analysis method [34] described previously [32]. The ONMD algorithm (MATLAB, MathWorks) calculates the x-y displacement of individual cells for each frame, and saves the trajectories of the tracked cells, as well as the root mean square of the displacement, to an MS Excel file. The distribution of the displacements per frame was represented as violin and box-and-whisker (10th to 90th percentile) plots, whereas the total displacements for all cells were represented as box-and-whisker plots (Prism8, GraphPad). The software also allows for highlighting pixels that change from frame to frame (differential image) using a colour scale: the pixels that changed the most are indicated in red, whereas the ones that changed the least are shown in blue.

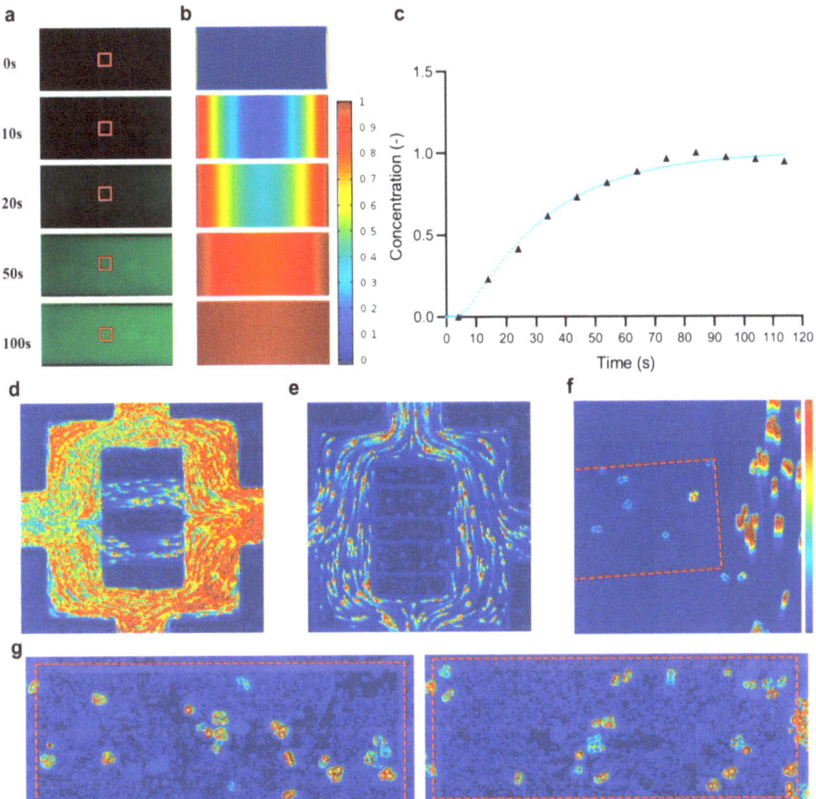

Figure 2. (**a**) Diffusion of fluorescein in the imaging chambers: fluorescent images at different time points; red squares represent the analysed area. (**b**) Simulation of the diffusion in the chamber: evolution of the normalised fluorescein concentration at the same time points as in (**a**). (**c**) Evolution of the experimentally determined fluorescein concentration (black triangles) overlaid with the simulated evolution of the concentration (with a diffusion coefficient of 3.2×10^{-10} m^2/s) (blue dots). (**d**) Differential image of flowing yeast cells in the channels and the chamber during filling the imaging chambers from inlet B. (**e**) Differential image of flowing yeast cells during ethanol flow from inlet A. (**f**) Higher magnification differential image during ethanol flow from inlet A (the imaging chamber edges are indicated with a dashed line). (**g**) Typical differential image of cell positions in the no-flow imaging chamber during cellular nanomotion measurements.

3. Results

3.1. Design and Construction of the Microfluidic Chip

3.1.1. Modelling Fluid Flow and Mass Transport in the Microfluidic Chip

Fluid flow was simulated in the design using finite element modelling. The magnitude of the velocity at a slice positioned at the mid-plane of the imaging chamber (which is at a height of 4 μm) was calculated (Figure 1a). The pressure distribution over the channels and the imaging chambers was also calculated, showing no pressure difference over the chambers (Figure 1b). To experimentally demonstrate that the mass transport in the imaging chambers did occur solely by diffusion, a diffusion experiment was performed using the fluorescent compound fluorescein as a molecular tracer (Figure 2a,c). By comparing the experimentally determined evolution of the concentration in the centre of the imaging chamber to the simulated evolution (Figure 2b,c), a diffusion coefficient of 3.2×10^{-10} m^2/s was estimated. This value corresponds to the range of values, i.e., 4.0×10^{-10} to 6.0×10^{-10} m^2/s, that has been reported using other methods [35–38].

As an additional experimental proof to demonstrate that there is no convective flow in the imaging chambers, the differential image analysis in the ONMD software was used. The software allows for distinguishing cells that are pushed by the convective flow. Figure 2d shows the flow of the cells during the filling process of the imaging chambers and Figure 2e shows the movement of the cells during liquid flow from inlet A. We showed that there is no flow or convection movement of the cells inside the chamber (Figure 2e,f). The cells that are surrounded by a defined highlighted ring are moving in a diffusive environment, whereas cells that have a smear of red colour are affected by fluid flow (such as the cells that are in the channel). Furthermore, these images show that there is no correlation between the cell position and the movement of the cells since the fluid flow through the channels does not seem to affect the movements of cells that are relatively closely located at the channel side (as compared to the more centrally located cells) (Figures 2G and S1).

All these results show that there is no fluid flow in the imaging chambers when a solution containing a killing compound is flown through the channels, and that it is possible to image the same selected cells before and after the treatment.

3.1.2. Microfluidic Chip Construction

The microfluidic chip design consisted of an inlet (B) and an outlet (B) that were used to fill the cells into the imaging chambers (Figures 1a,c and 2d), and an inlet (A) and an outlet (A) that were used to pump in the killing agent (Figures 1a,c and 2e). The fabricated in- and outlet channels had a width of 200 μm, a height of 60 μm, and a length of 7 mm. The vertical channels in the flow cell had a width of 200 μm and a height of 60 μm. The two no-flow imaging chambers had a width of 300 μm, a depth of 100 μm, and a height of 8 μm (Figure 1d,e). The mould for the microfluidic chip was made on a Si wafer in the epoxy-based photoresist SU-8 through a standard photolithography protocol. Using 3-D profilometer imaging, it was shown that the dimensions of the no-flow chamber matched the design specifications (Figure 1d). After casting the PDMS over the mould and curing the PDMS layer, the chip was assembled by pressing it onto a glass coverslip after plasma treatment of both surfaces. The plasma treatment guaranteed that the bonding was leak-free.

3.2. Dynamics of Single-Cell Nanomotion upon Treatment by a Killing Compound

A method was optimised to follow the nanomotion of several cells before and during treatment with a killing compound. After assembling the chip, the in- and outlets were connected by FEP tubing to a pressure-driven pump and a syringe pump. Firstly, the tubing and the chip were rinsed with isopropanol and then water to remove air bubbles.

Secondly, the imaging chambers were then filled with the yeast solution. The filling was ended by stopping the flow and clamping the inlet and outlet tubes. Then, the first movie of the nanomotion of single yeast cells was recorded in the no-flow chambers (time point zero min). The first movie was acquired right after stopping the flow to obtain nanomotion of cells without treatment. Thirdly, the chip was filled with the compound solution from inlet A, using a syringe pump with a low flow rate of 0.5 μL/min for approximately 20 min. Next, the nanomotion of the same individual yeasts was followed as a function of time, with a second movie taken 10 min (±1 min) after the compound reached its maximum concentration in the middle of the chamber. Each subsequent movie was taken at 10 min time intervals.

The first experiment was a control experiment, where the cellular nanomotion of 20 individual yeast cells (Figure S1c) during growth in YPD medium was recorded over a period of 120 min (Figures 3 and S1c). The nanomotion data show that actively growing cells were characterised by an asymmetric distribution of displacement per frame (Figure 3a), which can vary from cell to cell as well as in time. Cells 8 and 13 (Figure 3b,c), and cells 2, 3, 5, and 6 (Figure S2) are characterised by a relatively large variation as a function of time, in contrast to cells 11 and 16 (Figure 3b,c), and cell 12 (Figure S2). Although variations could be seen between individual cells, at the population level (20 cells), no major differences between the distribution of the displacements/frames for all 20 cells and the average total displacements could be observed during the 2 h of growth (Figure 6a,d).

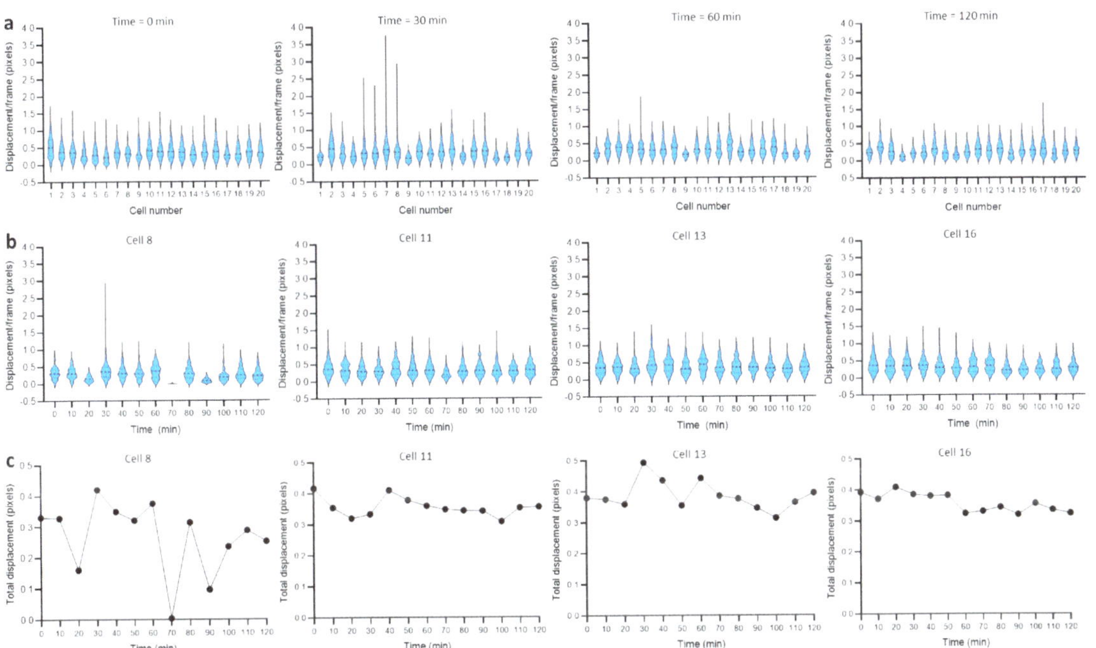

Figure 3. Cellular nanomotion of *S. cerevisiae* cells grown in YPD medium during 2 h. (**a**) Distribution of the displacements per frame for each of the 20 analysed cells at times 0, 30 min, 60 min, and 120 min. (**b**) Distribution of the displacement/frame as a function of time for 4 selected cells: cells 8, 11, 13, and 16. (**c**) Total displacement as a function of time for the selected cells.

In the next experiments, we evaluated the effect of a killing compound on the variation of the cellular nanomotion of 20 individual yeast cells as a function of time. We evaluated 70% (*v/v*) ethanol and the antifungal amphotericin B as killing agents (Figure S1b). Since *S. cerevisiae* cells are quickly killed by ethanol [32], we followed the cellular nanomotion with a time interval of 10 min for only 1 h (Figure 4). The total displacements and the distribution of the displacements/frames showed a reduction already after 10 min of exposure, and stayed low up to the 1 h mark. Most cells died very quickly, as illustrated well by cells 9, 14 (Figure 4b,c), 6, 8, 10, and 12 (Figure S3). Three cells, i.e., cells 11 and 13 (Figure 4b,c), and cell 7 (Figure S3), still showed some nanomotion after 1 h exposure, which indicates that these cells were more ethanol-tolerant than the other cells. The overall susceptibility of the cell population to ethanol is confirmed by the distribution of the displacements/frames graph for all 20 cells, as well the average total displacements graph (Figure 6c,f).

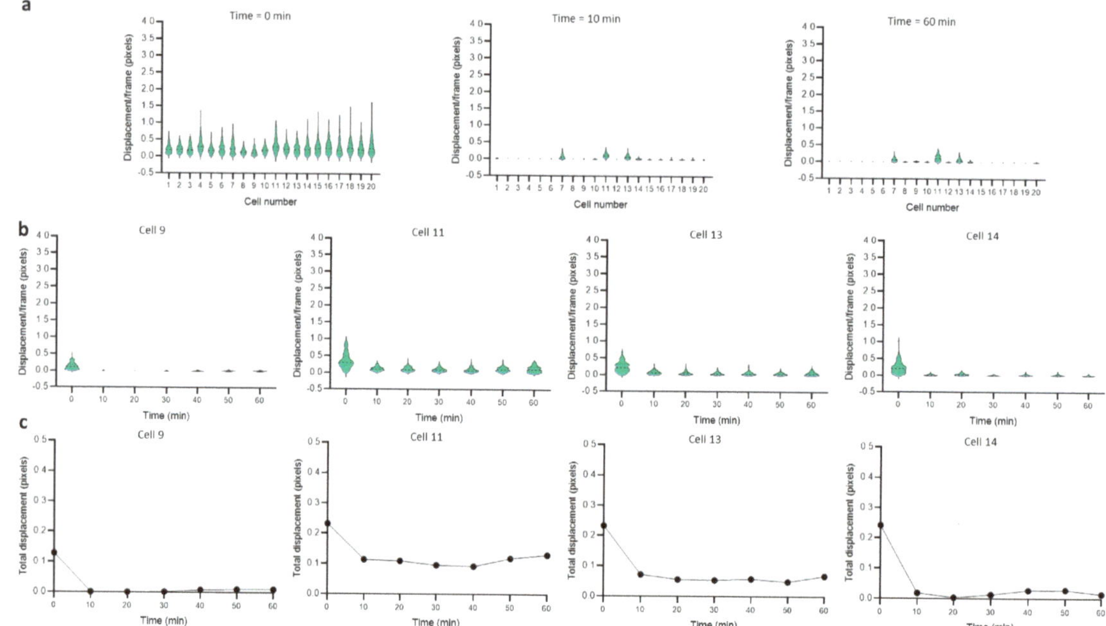

Figure 4. Effect of ethanol (70%) on the cellular nanomotion of *S. cerevisiae*. (**a**) Distribution of the displacements per frame for each of the 20 analysed cells before treatment and after 10 min and 60 min. (**b**) Distribution of the displacements/frames as a function of time for 4 selected cells: cells 9, 11, 13, and 14. (**c**) Total displacements as a function of time for the selected cells.

Next, we evaluated the effect of amphotericin B on *S. cerevisiae* cells. Single-cell displacements were recorded every 10 min for 120 min (Figure S1c). We used a high concentration of the antifungal compound to obtain a fast response [39]. The total displacements and the distribution of the displacements/frames showed a reduction, which was less drastic than in the case of ethanol. Many cells, such as cells 4, 18 (Figure 5b,c), 3, 5, and 19 (Figure S4), were dead after 120 min of treatment. Nevertheless, some cells, i.e., cells 13, 15 (Figure 5b,c), 7, and 16 (Figure S4) still showed some nanomotion after 2 h exposure, which indicates that these cells were less sensitive to the antifungal. At the population level, the dynamics of the killing of the cells by amphotericin B is also demonstrated by the distribution of the displacements/frame graph for all 20 cells, as well the average total displacements graph (Figure 6b,e).

Finally, the sensitivity of the individual yeast cells to ethanol and amphotericin was characterised by analysing the slope of their decrease in nanomotion activity as a function of time. The more sensitive the cells are to the killing compound, the faster they stop moving, and so the more negative the slope of their nanomotion curve is. We calculated the slopes of the total displacement as a function of time for 10, 20, 30, and 40 min during the treatment with ethanol (Figure 7a) and amphotericin (Figure 7b). For the amphotericin treatment, the slopes calculated after 10, 20 and 30 min provide the necessary information to conclude that a cell is sensitive or more resistant (cells 1, 6, 12, 13, and 17 in Figure 7). Since the response to 70% ethanol is very fast, the slope calculated at the 10 min time points already indicates that the cell is sensitive. Based on these results, we can conclude that values of these slopes enable quantifying the effect of a killing compound in a short time frame (10 to 30 min).

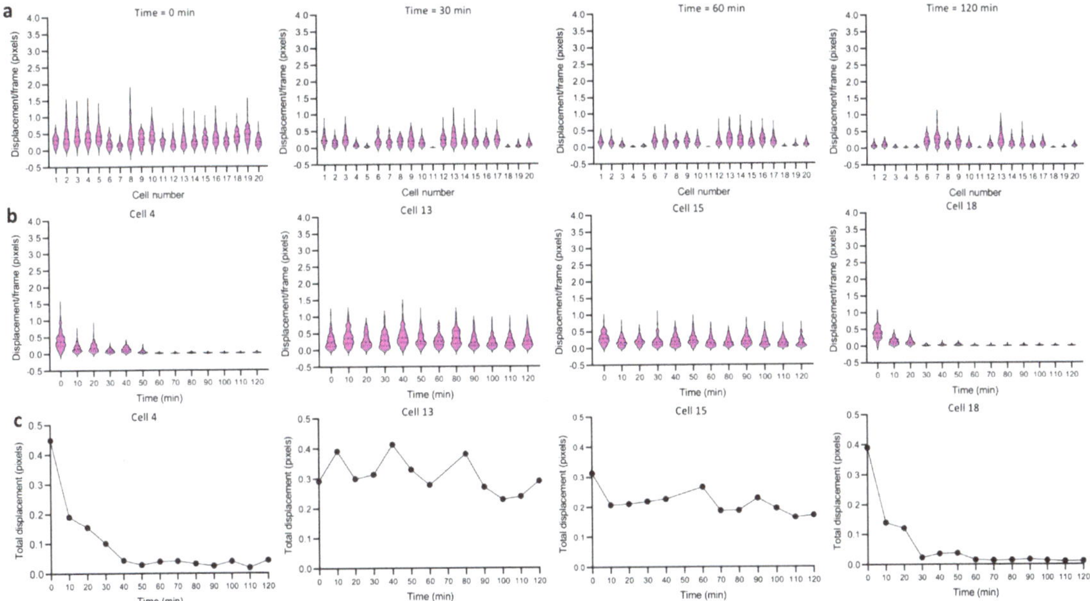

Figure 5. Effect of the antifungal amphotericin B (500 µg/mL) on the cellular nanomotion of *S. cerevisiae*. (**a**) Distribution of the displacements per frame before treatment and after 30 min, 60 min, and 120 min. (**b**) Distribution of the displacements/frames as a function of time for 4 selected cells: cells 4, 13, 15, and 18. (**c**) Total displacements as a function of time for the selected cells.

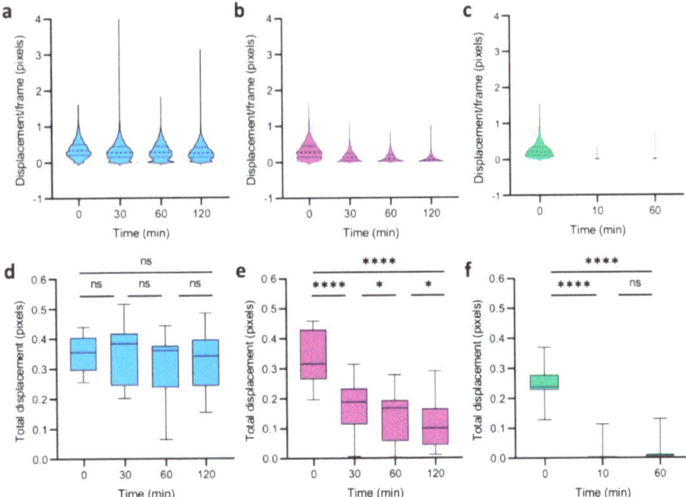

Figure 6. Averaged cellular nanomotion for 20 cells. (**a**) Distribution of the displacements/frames during 2 h growth (control). (**b**) Distribution of the displacements/frames during amphotericin treatment. (**c**) Distribution of the displacements/frames during ethanol treatment. (**d**) The total displacement during 2 h growth (control). (**e**) The total displacement during amphotericin treatment. (**f**) The total displacement during ethanol treatment. Wilcoxon test: **** $p < 0.0001$; * $p < 0.1$; ns: not significant.

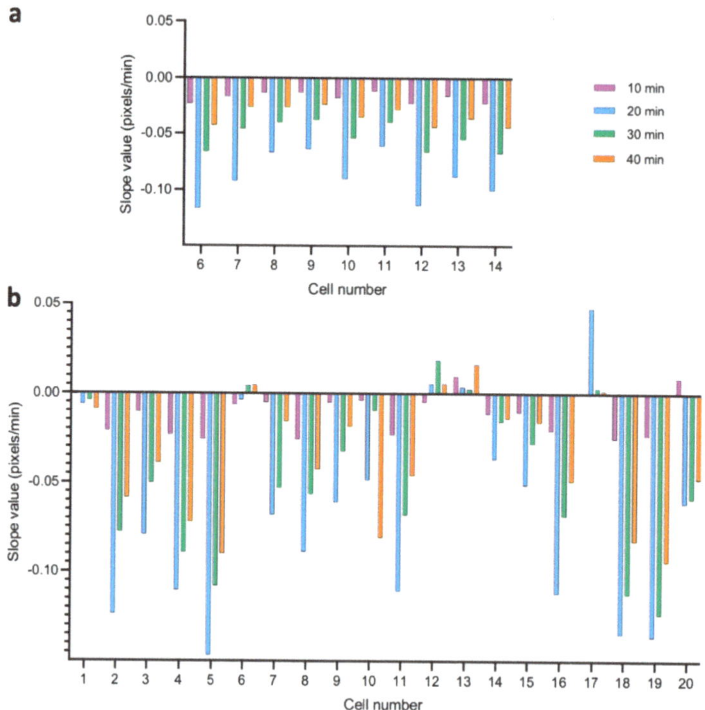

Figure 7. The slopes of the decrease in the total displacement as a function of time are calculated based on the time points 10, 20, 30, and 40 min during (**a**) ethanol and (**b**) amphotericin B trea-ment.

4. Discussion

We developed a microfluidic chip for the optical nanomotion detection (ONMD) method [32]. The microfluidic chip contains no-flow imaging chambers, and the growth medium can be replaced by a treatment solution without disturbing the nanomotion of the cells. The cellular nanomotion of the same cells can be recorded before and during the treatment. Hence, by acquiring ONMD data at regular time intervals of a few minutes, the response of the cells to a chemical compound can be quantified as a function of time.

As a control, we observed the cellular nanomotion for 2 h. The results indicate that the cellular nanomotion reflects the "metabolic state" of the cell and that, in a population of 20 cells, some cells are more metabolically active. Previously, we found that ONMD was sensitive enough to detect differences between the metabolic state of *S. cerevisiae* cells that were grown in a growth medium, compared to cells in a buffer without nutrients [32]. Additionally, the cellular nanomotion of *S. cerevisiae* showed maximal nanomotion at its optimal growth temperature [32]. The variation of the nanomotion as a function of time for each cell could be influenced by the cell cycle. A hypothesis that can explain why the cell displacements are changing over time is that some molecular processes in the cell—such as cytoskeleton rearrangements and organelle transport—are more active at specific phases in the cell cycle. Moreover, major morphogenetic events that could influence the cellular nanomotion occur during the cell cycle, such as the polarisation of the cytoskeleton in late G1, leading to bud emergence, a depolarisation of growth within the bud, leading to uniform bud expansion in early G2, a breakdown of the mother-bud asymmetry in growth in late mitosis, and a refocusing of growth towards the neck upon mitotic exit [40].

We evaluated the sensitivity of the model yeast *S. cerevisiae* to the disinfectant ethanol and the antifungal amphotericin B. The killing of the cells by ethanol was very rapid, as a significant decrease in the cellular nanomotion could be observed within 10 min. A few cells still showed some activity after 60 min. Ethanol tolerance is strain-dependent [41], affects the growth rate, and impairs the cell membrane integrity [42], which results in ionic species permeability and the leakage of metabolites [43], and it freely diffuses inside the cell, where it directly perturbs and denatures intracellular proteins. Additionally, ethanol tolerance is influenced by the cell age [44]. All this could explain the variation that we observed in a population of 20 cells.

Amphotericin B is a polyene antifungal that selectively binds to ergosterol in the cell membrane and causes the formation of pores, which ultimately results in cell death [45]. It is a broad-spectrum fungicidal with activity against many fungi, including *S. cerevisiae* [5]. Most cells were killed within 2 h by treatment with a high concentration of amphotericin B. However, we could detect significant nanomotion of a few cells even after 2 h. These results indicate that there is a significant variation in the sensitivity of the cells in a population. Since ONMD is single-cell-sensitive, the ratio of resistant versus sensitive cells in a population could also be determined. This means that ONMD allows for the detection of one resistant cell in a population. Therefore, this ONMD method is superior to ensemble AFST methods for selecting the optimal antifungal agent.

The slope of the decrease in the nanomotion activity (as measured by the total displacement of the cell) as a function of time provides information on how fast the cell is killed by a specific compound. This "killing rate" is a new measure to rapidly assess AFST.

We selected the model yeast *S. cerevisiae* to evaluate the newly developed microfluidic chip. This chip could also be used to perform AFST on other yeasts such as the pathogenic yeasts *Candida albicans*, *C. glabrata*, and *C. lusitaniae*, since ONMD, as a method to perform rapid AFST, was previously demonstrated on these yeasts [32]. By extension, also AST on bacteria could be performed if the design of the no-flow chamber would be adapted, i.e., by reducing the height of the chamber.

In the future, the number of no-flow chambers could be increased and integrated into a chip with various compartments to assess different concentrations and different antifungals. Alternatively, an antimicrobial concentration gradient over the no-flow chamber could be implemented, which will allow for direct determinations of the minimal inhibitory concentration (MIC). In addition, we will study the variation of the nanomotion of cells during the cell cycle in more detail to find how the cellular nanomotion is influenced by each cell cycle phase.

Supplementary Materials: The following supporting information can be downloaded at: https://www.mdpi.com/article/10.3390/fermentation8050195/s1: Figure S1: Location of the selected cells in the imaging chambers: (**a**) amphotericin B treatment, (**b**) ethanol treatment, and (**c**) control experiment; Figure S2: Cellular nanomotion of *S. cerevisiae* cells grown in YPD medium during 2 h. Distribution of the displacement/frame as a function of time for additional selected cells; Figure S3: Effect of ethanol (70%) on the cellular nanomotion of *S. cerevisiae*. Distribution of the displacement/frame as a function of time for additional selected cells; Figure S4: Effect of the antifungal amphotericin B (500 µg/mL) on the cellular nanomotion of *S. cerevisiae*. Distribution of the displacement/frame as a function of time for additional selected cells.

Author Contributions: Conceptualisation, R.G.W. and S.K.; methodology, R.G.W., V.R., C.Y., M.I.V. and S.K.; software, S.K. and M.I.V.; validation, V.R., C.Y. and R.G.W.; formal analysis, V.R.; resources, R.G.W. and S.K.; data curation, V.R.; writing—original draft preparation, R.G.W., V.R., C.Y., M.I.V. and S.K.; writing—review and editing, R.G.W., V.R., C.Y., M.I.V. and S.K.; visualisation, V.R. and R.G.W.; supervision, R.G.W. and S.K.; project administration, R.G.W. and S.K.; funding acquisition, R.G.W. and S.K. All authors have read and agreed to the published version of the manuscript.

Funding: This research was funded by the Belgian Federal Science Policy Office (Belspo) and the European Space Agency (ESA), grant number PRODEX projects: Yeast Bioreactor and Flumias Nanomotion; FWO, grant numbers AUGE/13/19 and I002620; FWO-SNSF, grant number 310030 L_197946; The FWO-SB-FDW Ph.D. grant of VR. IV and SK were supported by SNSF grants, CRSII5_173863.

Institutional Review Board Statement: Not applicable.

Informed Consent Statement: Not applicable.

Data Availability Statement: All data needed to evaluate the conclusions in the paper are present in the paper and/or the Supplementary Materials. Additional data related to this paper may be requested from the authors.

Acknowledgments: The Research Council of the Vrije Universiteit Brussel (Belgium) and the University of Ghent (Belgium) are acknowledged to support the Alliance Research Group VUB-UGhent NanoMicrobiology (NAMI), and the International Joint Research Group (IJRG) VUBEPFL BioNanotechnology and NanoMedicine (NANO).

Conflicts of Interest: The authors declare no conflict of interest.

References

1. Bongomin, F.; Gago, S.; Oladele, R.O.; Denning, D.W. Global and multi-national prevalence of fungal diseases—Estimate precision. *J. Fungi* **2017**, *3*, 57. [CrossRef] [PubMed]
2. Denning, D.W. Calling upon all public health mycologists: To accompany the country burden papers from 14 countries. *Eur. J. Clin. Microbiol. Infect. Dis.* **2017**, *36*, 923–924. [CrossRef] [PubMed]
3. Geddes-McAlister, J.; Shapiro, R.S. New pathogens, new tricks: Emerging, drug-resistant fungal pathogens and future prospects for antifungal therapeutics. *Ann. N. Y. Acad. Sci.* **2019**, *1435*, 57–78. [CrossRef] [PubMed]
4. Beardsley, J.; Halliday, C.L.; Chen, S.C.-A.; Sorrell, T.C. Responding to the emergence of antifungal drug resistance: Perspectives from the bench and the bedside. *Future Microbiol.* **2018**, *13*, 1175–1191. [CrossRef]
5. Willaert, R. Micro- and Nanoscale Approaches in Antifungal Drug Discovery. *Fermentation* **2018**, *4*, 43. [CrossRef]
6. Lemke, A.; Kiderlen, A.F.; Kayser, O. Amphotericin B. *Appl. Microbiol. Biotechnol.* **2005**, *68*, 151–162. [CrossRef]
7. Bellmann, R. Pharmacodynamics and Pharmacokinetics of Antifungals for Treatment of Invasive Aspergillosis. *Curr. Pharm. Des.* **2013**, *19*, 3629–3647. [CrossRef]
8. Pfaller, M.A. Antifungal Drug Resistance: Mechanisms, Epidemiology, and Consequences for Treatment. *Am. J. Med.* **2012**, *125*, S3–S13. [CrossRef]
9. Pianalto, K.M.; Alspaugh, J.A. New horizons in antifungal therapy. *J. Fungi* **2016**, *2*, 26. [CrossRef]
10. Nivoix, Y.; Ubeaud-Sequier, G.; Engel, P.; Leveque, D.; Herbrecht, R. Drug-Drug Interactions of Triazole Antifungal Agents in Multimorbid Patients and Implications for Patient Care. *Curr. Drug Metab.* **2009**, *10*, 395–409. [CrossRef]
11. Lewis, R.E. Current concepts in antifungal pharmacology. In *Mayo Clinic Proceedings*; Elsevier Ltd.: Amsterdam, The Netherlands, 2011; Volume 86, pp. 805–817.
12. McCarthy, M.W.; Walsh, T.J. Drugs currently under investigation for the treatment of invasive candidiasis. *Expert Opin. Investig. Drugs* **2017**, *26*, 825–831. [CrossRef] [PubMed]
13. Perfect, J.R. The antifungal pipeline: A reality check. *Nat. Rev. Drug Discov.* **2017**, *16*, 603–616. [CrossRef] [PubMed]
14. Fuentefria, A.M.; Pippi, B.; Dalla Lana, D.F.; Donato, K.K.; de Andrade, S.F. Antifungals discovery: An insight into new strategies to combat antifungal resistance. *Lett. Appl. Microbiol.* **2018**, *66*, 2–13. [CrossRef] [PubMed]
15. Silva, L.N.; de Mello, T.P.; de Souza Ramos, L.; Branquinha, M.H.; dos Santos, A.L.S. New and Promising Chemotherapeutics for Emerging Infections Involving Drug-resistant Non-albicans Candida Species. *Curr. Top. Med. Chem.* **2019**, *19*, 2527–2553. [CrossRef] [PubMed]
16. Sousa, F.; Ferreira, D.; Reis, S.; Costa, P. Current insights on antifungal therapy: Novel nanotechnology approaches for drug delivery systems and new drugs from natural sources. *Pharmaceuticals* **2020**, *13*, 248. [CrossRef] [PubMed]
17. Mota Fernandes, C.; Dasilva, D.; Haranahalli, K.; McCarthy, J.B.; Mallamo, J.; Ojima, I.; Del Poeta, M. The Future of Antifungal Drug Therapy: Novel Compounds and Targets. *Antimicrob. Agents Chemother.* **2021**, *65*, 65. [CrossRef] [PubMed]
18. Scorzoni, L.; Fuchs, B.B.; Junqueira, J.C.; Mylonakis, E. Current and promising pharmacotherapeutic options for candidiasis. *Expert Opin. Pharmacother.* **2021**, *22*, 867–887. [CrossRef]
19. Posteraro, B.; Torelli, R.; De Carolis, E.; Posteraro, P.; Sanguinetti, M. Antifungal susceptibility testing: Current role from the clinical laboratory perspective. *Mediterr. J. Hematol. Infect. Dis.* **2014**, *6*, e2014030. [CrossRef]
20. Campbell, J.; McBeth, C.; Kalashnikov, M.; Boardman, A.K.; Sharon, A.; Sauer-Budge, A.F. Microfluidic advances in phenotypic antibiotic susceptibility testing. *Biomed. Microdevices* **2016**, *18*, 103. [CrossRef]
21. Kohler, A.C.; Venturelli, L.; Longo, G.; Dietler, G.; Kasas, S. Nanomotion detection based on atomic force microscopy cantilevers. *Cell Surf.* **2019**, *5*, 100021. [CrossRef]
22. Longo, G.; Alonso-Sarduy, L.; Rio, L.M.; Bizzini, A.; Trampuz, A.; Notz, J.; Dietler, G.; Kasas, S. Rapid detection of bacterial resistance to antibiotics using AFM cantilevers as nanomechanical sensors. *Nat. Nanotechnol.* **2013**, *8*, 522–526. [CrossRef] [PubMed]
23. Kasas, S.; Ruggeri, F.S.; Benadiba, C.; Maillard, C.; Stupar, P.; Tournu, H.; Dietler, G.; Longo, G. Detecting nanoscale vibrations as signature of life. *Proc. Natl. Acad. Sci. USA* **2015**, *112*, 378–381. [CrossRef] [PubMed]

24. Villalba, M.I.; Stupar, P.; Chomicki, W.; Bertacchi, M.; Dietler, G.; Arnal, L.; Vela, M.E.; Yantorno, O.; Kasas, S. Nanomotion Detection Method for Testing Antibiotic Resistance and Susceptibility of Slow-Growing Bacteria. *Small* **2018**, *14*, 1702671. [CrossRef] [PubMed]
25. Venturelli, L.; Kohler, A.C.; Stupar, P.; Villalba, M.I.; Kalauzi, A.; Radotic, K.; Bertacchi, M.; Dinarelli, S.; Girasole, M.; Pešić, M.; et al. A perspective view on the nanomotion detection of living organisms and its features. *J. Mol. Recognit.* **2020**, *33*, e2849. [CrossRef]
26. Kasas, S.; Malovichko, A.; Villalba, M.I.; Vela, M.E.; Yantorno, O.; Willaert, R.G. Nanomotion Detection-Based Rapid Antibiotic Susceptibility Testing. *Antibiotics* **2021**, *10*, 287. [CrossRef]
27. Syal, K.; Iriya, R.; Yang, Y.; Yu, H.; Wang, S.; Haydel, S.E.; Chen, H.-Y.; Tao, N. Antimicrobial Susceptibility Test with Plasmonic Imaging and Tracking of Single Bacterial Motions on Nanometer Scale. *ACS Nano* **2016**, *10*, 845–852. [CrossRef]
28. Johnson, W.L.; France, D.C.; Rentz, N.S.; Cordell, W.T.; Walls, F.L. Sensing bacterial vibrations and early response to antibiotics with phase noise of a resonant crystal. *Sci. Rep.* **2017**, *7*, 12138. [CrossRef]
29. Syal, K.; Shen, S.; Yang, Y.; Wang, S.; Haydel, S.E.; Tao, N. Rapid Antibiotic Susceptibility Testing of Uropathogenic E. coli by Tracking Submicron Scale Motion of Single Bacterial Cells. *ACS Sens.* **2017**, *2*, 1231–1239. [CrossRef]
30. Bermingham, C.R.; Murillo, I.; Payot, A.D.J.; Balram, K.C.; Kloucek, M.B.; Hanna, S.; Redmond, N.M.; Baxter, H.; Oulton, R.; Avison, M.B.; et al. Imaging of sub-cellular fluctuations provides a rapid way to observe bacterial viability and response to antibiotics. *bioRxiv* **2018**, 460139. [CrossRef]
31. Leonard, H.; Halachmi, S.; Ben-Dov, N.; Nativ, O.; Segal, E. Unraveling Antimicrobial Susceptibility of Bacterial Networks on Micropillar Architectures Using Intrinsic Phase-Shift Spectroscopy. *ACS Nano* **2017**, *11*, 6167–6177. [CrossRef]
32. Willaert, R.G.; Vanden Boer, P.; Malovichko, A.; Alioscha-Perez, M.; Radotić, K.; Bartolić, D.; Kalauzi, A.; Villalba, M.I.; Sanglard, D.; Dietler, G.; et al. Single yeast cell nanomotions correlate with cellular activity. *Sci. Adv.* **2020**, *6*, eaba3139. [CrossRef] [PubMed]
33. Schindelin, J.; Arganda-Carreras, I.; Frise, E.; Kaynig, V.; Longair, M.; Pietzsch, T.; Preibisch, S.; Rueden, C.; Saalfeld, S.; Schmid, B.; et al. Fiji: An open-source platform for biological-image analysis. *Nat. Methods* **2012**, *9*, 676–682. [CrossRef] [PubMed]
34. Guizar-Sicairos, M.; Thurman, S.T.; Fienup, J.R. Efficient subpixel image registration algorithms. *Opt. Lett.* **2008**, *33*, 156. [CrossRef] [PubMed]
35. Radomsky, M.L.; Whaley, K.J.; Cone, R.A.; Saltzman, W.M. Macromolecules released from polymers: Diffusion into unstirred fluids. *Biomaterials* **1990**, *11*, 619–624. [CrossRef]
36. Saltzman, W.M.; Radomsky, M.L.; Whaley, K.J.; Cone, R.A. Antibody diffusion in human cervical mucus. *Biophys. J.* **1994**, *66*, 508–515. [CrossRef]
37. Soeller, C.; Jacobs, M.D.; Donaldson, P.J.; Cannell, M.B.; Jones, K.T.; Ellis-Davies, G.C.R. Application of two-photon flash photolysis to reveal intercellular communication and intracellular Ca^{2+} movements. *J. Biomed. Opt.* **2003**, *8*, 418. [CrossRef]
38. Casalini, T.; Salvalaglio, M.; Perale, G.; Masi, M.; Cavallotti, C. Diffusion and aggregation of sodium fluorescein in aqueous solutions. *J. Phys. Chem. B* **2011**, *115*, 12896–12904. [CrossRef]
39. Rosłoń, I.E.; Japaridze, A.; Steeneken, P.G.; Dekker, C.; Alijani, F. Probing nanomotion of single bacteria with graphene drums. *bioRxiv* **2021**, 2021.09.21.461186. [CrossRef]
40. Howell, A.S.; Lew, D.J. Morphogenesis and the Cell Cycle. *Genetics* **2012**, *190*, 51–77. [CrossRef]
41. Pandey, A.K.; Kumar, M.; Kumari, S.; Kumari, P.; Yusuf, F.; Jakeer, S.; Naz, S.; Chandna, P.; Bhatnagar, I.; Gaur, N.A. Evaluation of divergent yeast genera for fermentation-associated stresses and identification of a robust sugarcane distillery waste isolate Saccharomyces cerevisiae NGY10 for lignocellulosic ethanol production in SHF and SSF. *Biotechnol. Biofuels* **2019**, *12*, 40. [CrossRef]
42. Schiavone, M.; Formosa-Dague, C.; Elsztein, C.; Teste, M.-A.; Martin-Yken, H.; De Morais, M.A.; Dague, E.; François, J.M. Evidence for a Role for the Plasma Membrane in the Nanomechanical Properties of the Cell Wall as Revealed by an Atomic Force Microscopy Study of the Response of Saccharomyces cerevisiae to Ethanol Stress. *Appl. Environ. Microbiol.* **2016**, *82*, 4789–4801. [CrossRef] [PubMed]
43. Stanley, D.; Bandara, A.; Fraser, S.; Chambers, P.J.; Stanley, G.A. The ethanol stress response and ethanol tolerance of Saccharomyces cerevisiae. *J. Appl. Microbiol.* **2010**, *109*, 13–24. [CrossRef] [PubMed]
44. Eigenfeld, M.; Kerpes, R.; Becker, T. Understanding the Impact of Industrial Stress Conditions on Replicative Aging in Saccharomyces cerevisiae. *Front. Fungal Biol.* **2021**, *2*, 665490. [CrossRef]
45. Kristanc, L.; Božič, B.; Jokhadar, Š.Z.; Dolenc, M.S.; Gomišček, G. The pore-forming action of polyenes: From model membranes to living organisms. *Biochim. Biophys. Acta-Biomembr.* **2019**, *1861*, 418–430. [CrossRef] [PubMed]

MDPI AG
Grosspeteranlage 5
4052 Basel
Switzerland
Tel.: +41 61 683 77 34

Fermentation Editorial Office
E-mail: fermentation@mdpi.com
www.mdpi.com/journal/fermentation

Disclaimer/Publisher's Note: The statements, opinions and data contained in all publications are solely those of the individual author(s) and contributor(s) and not of MDPI and/or the editor(s). MDPI and/or the editor(s) disclaim responsibility for any injury to people or property resulting from any ideas, methods, instructions or products referred to in the content.

www.ingramcontent.com/pod-product-compliance
Lightning Source LLC
La Vergne TN
LVHW071935160726
843515LV00011B/2614